Assessment and Mitigation of Indoor Radon Problem: A Case Book for Radon Professionals

Authored By

Ashok Kumar

Department of Civil Engineering
The University of Toledo
Toledo, Ohio 43606
USA

&

Akhil Kadiyala

Department of Civil Engineering
The University of Toledo
Toledo, Ohio 43606
USA

CONTENTS

FOREWORD

Do you live or work in a county with indoor radon levels potentially equal to or above the Environmental Protection Agency's (EPA) action level? Do your children go to school or play in areas where radon is a potential problem? Do you know the EPA has a suggested action level for indoor radon (4 pCi/L)? Do you know that EPA and the Surgeon General's office estimate radon is responsible for approximately 21,000 lung cancer deaths each year in the United States?

Not until recently did I become aware of the potential health risks associated with exposure to indoor radon. This awareness occurred after living in metropolitan Atlanta, an area where EPA predicts a high potential for indoor radon, for more than seventeen years. More surprising is during this time I completed my Master of Public Health degree in epidemiology and began my career as an epidemiologist with the Centers for Disease Control and Prevention. Not once did I read, hear, or see an announcement warning me of this potential risk.

When I learned of indoor radon, I began to explore the available epidemiologic literature to understand the adverse health effects associated with such exposure. Similarly, I immersed myself in literature describing the environmental and geological factors influencing potential exposure to radon. This background research led me to initiate a special study, currently underway, to understand our children's potential risk from radon exposure at school. As a part of this process, I became aware of Dr. Kumar's work.

Dr. Kumar's extensive work on the comprehensive homes database maintained by The University of Toledo provides a unique opportunity to learn about the risk associated with residential radon in Ohio. Dr. Kumar, along with his colleagues and students are conducting trailblazing research. This work enhances our understanding of radon in Ohio and is invaluable to our understanding of indoor radon risk across the nation. This work forms the backbone of my radon in schools project.

Assessment and Mitigation of Indoor Radon Problem: A Case Book for Radon Professionals provides the first comprehensive discussion of indoor radon exposure and mitigation systems. The authors do an excellent job describing radon in the environment and explaining how radon becomes a residential problem. Ohio is currently the only state in the country to have this detailed description of radon exposure by county and zip code. Additionally, the evaluation of mitigation systems in Chapter 4 presents compelling evidence in support of specific radon mitigation systems.

According to the EPA's map of radon zones, 34% of all US counties have the potential for indoor radon levels greater than or equal to the action level. The EPA notes that homes with elevated levels of radon have been found in all three zones. The EPA further states that all homes should be tested regardless of geographic location. The work I am currently undertaking and the work of other radon researchers highlight indoor radon issues in buildings located in counties designated as low risk. The widespread occurrence of radon coupled with the World Health Organization's 2009 announcement to reduce the action level for indoor radon to 2.7pCi/L behooves us to increase awareness and promote action to reduce risk. *Assessment and Mitigation of Indoor Radon Problem: A Case Book for Radon Professionals* should inspire continued research in all radon risk zones with the goal of defining indoor radon potential at the zip code level, or by other levels of greater geographic detail, for all states. This will lead the way to developing active surveillance programs for radon detection in private and public buildings as a standard public health practice in our country and globally.

We can do more to ensure greater awareness of the potential risk of radon in our homes and schools. This book will motivate action toward this goal.

Stephanie L. Foster

Lead Spatial Epidemiologist
Geospatial Research, Analysis, and Services Program
Agency for Toxic Substances and Disease Registry-
Centers for Disease Control and Prevention
Atlanta, USA

PREFACE

The study of radon in homes is of utmost importance considering that radon has long been identified as the second major cause of lung cancer incidences after cigarette smoking in conjunction with the fact that people spend nearly 90% of their time indoors. Prior studies on radon have primarily focused on examining the different factors that influence the radon entry rates into a home. Considerable number of studies also modeled the radon entry rates into homes. However, a significant knowledge gap of the lack of a comprehensive study that focused in detail on the aspects of statistical metrics and radon distribution maps for a large area (such as an entire state) that are essential components of a comprehensive assessment of in-home radon problem coupled with the determination of performance of various mitigation systems using statistical tests in reducing the in-home radon levels prompted the authors to compile this book. This book is based on the homes and mitigation databases of the Ohio Radon Information System (ORIS) that comprised of 219,114 data points (1988-2012) and 39,858 data points (2002-2012), respectively. The ORIS is maintained by the Air Pollution Research Group (APRG) of the Civil Engineering Department at The University of Toledo in Toledo, Ohio, USA.

This book is prepared for those who wish to gain a better understanding of the radon problem assessment and mitigation in homes. Our intent is to provide a complete review of the radon problem in homes (Chapter 1), so that the reader acquires a basic understanding of the nature of radon problem. Included in this book are different statistical metrics and Geographic Information System (GIS) maps that summarize the problem of radon in Ohio (Chapters 2-3). Details on the statistics related to performance of different mitigation systems installed in Ohio homes along with a methodology to rank the performance of different mitigation systems are also incorporated in this book (Chapter 4). A summary of the significant findings from the radon assessment coupled with the mitigation systems performance is documented in the end (Chapter 5).

We are indebted to the former and current Civil Engineering graduate students of the APRG, at The University of Toledo, who were involved in this research over the last 30 years and developed the homes database that was used in this study. We appreciate the research grants awarded by the Ohio Department of Health (ODH) and the United States Environmental Protection Agency (U.S. EPA) to The University of Toledo, which made the development of a large radon homes database system possible. We thank Krothapalli Madhusha for her inputs in preparing the GIS maps in this book. We acknowledge ESRI® for providing the ArcGIS software that was used in the preparation of maps, and Microsoft® for providing the Excel spreadsheet that was used in computing the different statistical metrics, associated with radon concentration distributions in Ohio.

The book is likely to serve its intended purpose, even if one reader is able to understand the importance of examining the radon problem in home and get their homes tested and mitigated (in case of any exceedances) for radon. The views expressed in this book are those of the authors alone and do not represent the views of the organizations who, over the years, funded the collection of data.

ACKNOWLEDGEMENTS AND STATEMENT OF CONFLICT OF INTEREST

The authors thank the Ohio Department of Health (ODH) and the United States Environmental Protection Agency (USEPA) for awarding the research grants to The University of Toledo, which made the development of Ohio Radon Information System possible. The contributions of earlier investigators of the grants (Dr. Jim Harrell and Dr. Andrew G. Heydinger, and many graduate students who worked on this project over the years) are all greatly acknowledged. The authors also recognize the contribution of a number of staff members from the ODH. The authors thank Dr. H.R. Olesen for providing the BOOT v2.0 software, ESRI® for providing the ArcGIS software, Minitab® for providing the MINITAB 16 software, Microsoft® for providing the Excel software, and MathWorks® for providing the MATLAB 2010b software that were used in developing this book when performing statistical and geospatial analyses. The authors do not have any financial relationships with the software used in this paper. The views expressed in this book are those of the authors alone.

Ashok Kumar

Department of Civil Engineering
The University of Toledo
Toledo, OH 43606
USA
E-mail: akumar@utnet.utoledo.edu

&

Akhil Kadiyala

Department of Civil Engineering
The University of Toledo
Toledo, OH 43606
USA

CHAPTER 1

The Nature of Radon Problem

Abstract: This chapter provides a comprehensive discussion on the nature of radon problem using relevant literature. This chapter emphasizes details on the aspects of radon formation, units of measurement for radon, pathways for radon entry into homes and the influencing factors, radon monitoring techniques, radon action levels, health effects of radon exposure, and control measures to reduce in-house radon exposure.

Keywords: Radon, ^{222}Rn, radon formation, uranium decay, sources of radon, radon in geological materials, radon in building materials, radon in water, radon in natural gas, radon in outdoor air, radon measurement, emanation coefficients, radon transfer coefficients, radon action limits, USEPA action limit for radon, WHO action limit for radon, radon health effects, radon cancerous risk, radon control measures for new constructions, radon control measures for existing constructions.

1.1. WHAT IS RADON?

Of the many naturally occurring isotopes of radon ($^{\text{atomic weight}}$Rn), ^{222}Rn is the only colorless, odorless, and chemically inert radioactive gas, formed from radium (^{226}Ra), by the breakdown of uranium (^{238}U) in geological materials (rocks, sediments, soils, and water). Low levels of naturally occurring ^{238}U are a common phenomenon in the earth materials. Typically, 'continental surfaces (rocks, sediments, and soils) comprise between 1 and 3 parts per million (ppm) of ^{238}U' (Dyck 1978). Fig. **1** illustrates the radionuclides of ^{238}U natural decay chain. ^{238}U has a half-life of 4.5×10^9 years, *i.e.*, half of a given quantity of ^{238}U breaks down every 4.5×10^9 years. ^{238}U decays to form thorium (^{234}Th) by emitting an alpha particle. ^{234}Th, with a half-life of 24.1 days, further decays to form protactinium (^{234}Pa), by emitting a beta particle. ^{234}Pa, with a half-life of 1.18 minutes, then decays to form a different isotope of Uranium (^{234}U) by emitting a beta particle. ^{234}U, with a half-life of 2.5×10^5 years, further decays to form a different isotope of Thorium (^{230}Th) by emitting an alpha particle; followed by the decay of ^{230}Th (half-life of 8×10^4 years) to form radium (^{226}Ra) by emitting an alpha particle.

Ashok Kumar and Akhil Kadiyala

^{226}Ra, with a half-life of 1.6×10^3 years, decays further to form radon (^{222}Rn) by emitting an alpha particle. ^{222}Rn, with a half-life of 3.8 days, then decays to form polonium (^{218}Po) by emitting an alpha particle. ^{218}Po, with a half-life of 3 minutes, further decays, in succession, to form lead (^{214}Pb) by emitting an alpha particle; followed by the decay of ^{214}Pb (half-life of 27 minutes) to form bismuth (^{214}Bi) by emitting a beta particle and a gamma ray; followed by the decay of ^{214}Bi (half-life of 20 minutes) to form a different isotope of polonium (^{214}Po) by emitting a beta particle and a gamma ray. ^{214}Po, with a half-life of 180 microseconds, then decays to form a different isotope of lead (^{210}Pb) by emitting an alpha particle; followed by the decay of ^{210}Pb (half-life of 22 years) to form a different isotope of bismuth (^{210}Bi) by emitting a beta particle and a gamma ray; followed by the decay of ^{210}Bi (half-life of 5 days) to form a different isotope of polonium (^{210}Po) by emitting a beta particle. ^{210}Po, with a half-life of 138 days, finally decays to form a stable non-radioactive isotope of lead (^{206}Pb) by emitting an alpha particle and a gamma ray. The radionuclides of ^{218}Po, ^{214}Pb, ^{214}Bi, and ^{214}Po are commonly referred as the short-lived radon progeny (formerly designated as the radon daughters), while the radionuclides of ^{210}Pb, ^{210}Bi, ^{210}Po, and ^{206}Pb are generally referred as the long-lived radon progeny. Of the aforementioned radionuclides, ^{222}Rn is the only gas; the other radionuclides being solids.

The concentrations of ^{222}Rn in the air are measured in the units of picocuries per liter (pCi/L) or Becquerels per cubic meter (Bq/m^3). Another, seldom used, indirect measure of ^{222}Rn concentrations in the air is the working level (WL) that represents the concentrations of ^{222}Rn daughters as a quantity of the possible alpha particles energy per liter of air. There is no definitive relationship between WL and pCi/L; the conversion of WL to pCi/L is only approximated at varying equilibrium ratios. Occupational exposure to ^{222}Rn daughters is expressed in the unit of working level month (WLM). Given below are some of the widely used conversions associated with ^{222}Rn measurements in the air.

- 1 Bq = 1 disintegration per second

- 1 Ci = 10^{12} pCi

- 1 Ci = 3.7 × 10^{10} disintegrations per second = 3.7 × 10^{10} Bq

- 1 pCi = 0.037 disintegrations per second = 0.037 Bq

- 1 m^3 = 1000 L

- 1 pCi/L = 37 Bq/m^3

- 1 WL = 200 pCi/L (assumes 50% equilibrium) or 100 pCi/L (assumes 100% equilibrium)

- 1 WLM = 170 hours (21.25 working days/month x 8 hours/day) × 1 WL

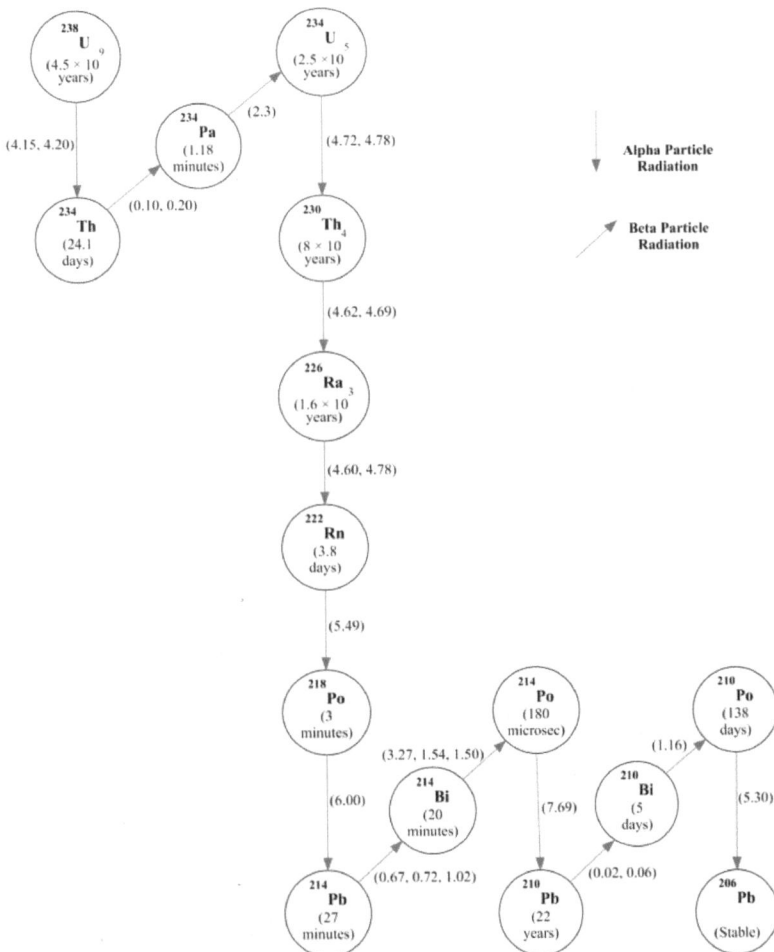

Figure 1: The radionuclides of ^{238}U natural decay chain marked with the corresponding half-life time, alpha energies (in MeV), and beta end-point energies (in MeV).

1.2. HOW DOES ^{222}RN ENTER INTO THE HOMES?

The problem of ^{222}Rn in the homes in the United States of America (USA) was first identified in the December of 1984, when a health physicist at the Limerick Nuclear Generating Station reported to the Pennsylvania Bureau of Radiation Protection about a construction worker of triggering the alarms when passing through the portal radiation monitors of the then incomplete plant. As there were no fission products generated by the incomplete plant, the home of the construction worker was tested for radiation and significantly high ^{222}Rn levels of the order of 13 WL or 2600 pCi/l of ^{222}Rn gas were recorded (Pennsylvania Bureau of Radiation Protection 2012). By the mid 1980's, environmental and health agencies across the world have identified the significance of ^{222}Rn problem inside homes that encouraged the researchers world-wide in determining the possible factors affecting the accumulation of ^{222}Rn concentrations in homes.

Considering that ^{222}Rn is a stable pollutant (owing to its relatively long half-life and chemical activity inertness), the buildup of ^{222}Rn concentrations inside a home is primarily dependent on (i) the entry rate and (ii) the ventilation rate. While the ^{222}Rn entry rates into a building are generally affected by the source type, the availability of a pathway with characteristics that facilitate the movement of ^{22}Rn from the source into the homes, and the availability of entry routes into a building; the ventilation rates are affected by the type of ventilation and the life style of the building occupants. A comprehensive review of the available literature on the ^{222}Rn source types showed that there are a total of five major classifications for ^{222}Rn sources. They are (i) soil, rocks, and sediments, (ii) building materials, (iii) water, (iv) natural gas, and, (v) outdoor air. Additional details on the accumulation of ^{222}Rn concentrations inside a home in context of each of the above mentioned source types are summarized in the subsequent sections.

1.2.1. Rocks, Soil, and Sediments

'^{222}Rn drifts upward through the ground to the surface of the soil, and flows into the outdoor air, or seeps into the buildings through the foundation cracks and other openings' (BEIR VI Report 1999). Direct ingression from the rocks, soil, and sediments is the primary pathway by which ^{222}Rn concentrations gets accumulated inside homes. The amount of ^{222}Rn that reaches the earth's crust is primarily dependent on the geological factors at the site. 'The geological variables that influence the ^{222}Rn concentration and its rate of migration in the subsurface include the composition, the texture, the porosity, the permeability, the moisture content, and the emanation power of the subsurface materials' (Alexander and Devocelle 1997). These geological variables can vary significantly from one place to another place. The amount of ^{222}Rn that enters a home after reaching the earth's surfacial soil is dependent on the type of substructure, the type of ventilation, and the life style of the building occupants. Further details on the factors affecting the movement of ^{222}Rn into homes from the soil, and modeling the ^{222}Rn migration from soils into homes are summarized below.

1.2.1.1. Composition

The rock types of carbonaceous shale, glauconitic sandstone, fluvial sandstone, phosphate, chalk, limestone, dolomite, glacial deposits, bauxite, lignite, coal, granite, gneiss, phyllite, schist, volcanic rocks, and rocks that have been faulted produced higher than the average ^{222}Rn concentrations (Gundersen *et al.* 1992; Tanner 1986). One can anticipate higher averaged ^{222}Rn concentrations in the areas with soils and sediments derived from the above mentioned rock types.

1.2.1.2. Texture, Porosity, and Permeability

The porosity (ratio of volume of all the pores in a material to the volume of the whole) and the permeability (property of a material to allow the passage of liquids or gases through it), which determine the distance that ^{222}Rn can migrate before decaying into other substances is influenced by the texture (the composition and relative distribution of various soil or sediment grain sizes) of the soil. The larger soil or sediment grain sizes have less surface area per unit volume, while the smaller grain sizes have more surface area per unit volume available to contribute ^{222}Rn gas to the pore space. Table **1** summarizes the representative porosity values for the three major divisions of soil classification based on texture, namely, sand, silt, and clay.

The larger the soil or the sediment grain sizes and the greater the uniformity of grain sizes, the greater is the permeability and the greater is the distance that ^{222}Rn can migrate within the soil. On the other hand, a mixture of wide-ranging grain sizes in soil or sediment reduces the pore space, thereby limiting the permeability of the material that restricts the distance to which ^{222}Rn can migrate. As the permeability increases, the convective flow rate also increases, thereby resulting in increased ^{222}Rn movement from the soil to the substructure. Table **2** summarizes the representative permeability values for different soil classifications.

Table 1: Porosity values for different soil classifications

Soil Classification	Porosity
Sand	0.25^a - 0.4^b
Silt	0.4^a - 0.5^b
Clay	0.45^a - 0.6^b

Source: (^{a}Tanner 1993; ^{b}Nazaroff *et al.* 1988a)

Table 2: Permeability values for different soil classifications

Soil Classification	Permeability (m^2)
Clean gravel	10^{-9} - 10^{-7}
Clean sands, clean sand and gravel mixtures, fine sands	10^{-12} - 10^{-9}
Very fine sands, organic and inorganic silts, mixtures of sands, silt and clay, *etc.*	10^{-16} - 10^{-12}
Homogeneous clays	$< 10^{-16}$

Source: (Data extracted from a figure in Nazaroff *et al.* 1988a; Tanner 1993)

1.2.1.3. Moisture Content and Emanation Power

The moisture content (amount of water present in the pore space) in the soil or sediment affects the permeability and the emanating power. 'Excessive moisture in the soil and sediment occupies the pore space and reduces the permeability; thereby limiting the migration of ^{222}Rn. Emanating power is referred to as the fraction of ^{222}Rn formed in the rock or the sediment that escapes from the solid material into the pore space. When ^{226}Ra decays to form ^{222}Rn by emitting an alpha particle, the resulting ^{222}Rn ion "recoils" that causes a significant percentage of the resulting ^{222}Rn ions to "imbed" themselves into adjacent rocks or sediments. In the presence of dry rocks and sediments, only one percent of ^{222}Rn ions escape into pore space' (Alexander and Devocelle 1997). On the other hand, 'the presence of moisture in the pore space of adjacent rocks or sediments "cushions" the collision between the recoiling ^{222}Rn ions and the rock or the sediment. This cushioning effect results in a smaller percentage of the recoiling ^{222}Rn ions getting imbedded into the rock or the sediment, and a greater percentage escaping into the pore space' (Tanner 1993). 'The emanating power of ^{222}Rn soils and sediments were noted to initially increase with an increase in the soil moisture content of up to about 15 to 17 percent by weight and then remains constant' (USEPA Report 1986). Some studies have identified the emanating power to decrease at higher moisture contents (Damkjaer and Korsbech 1985; Lindmark and Rosen 1985).

1.2.1.4. Type of the Substructure

^{222}Rn will enter the lowest level of a building using available pathways. The property of ^{222}Rn being a single atom gas significantly enhances its penetration capability into the building materials like gypsum boards, concrete blocks, mortar, wood panels, and other insulations. The amount of ^{222}Rn entering a home through the substructure is defined by the size and location of the openings in the substructure, and the degree of air movement between the indoor microenvironment and the soil. For structures with basements (floor level is below the surrounding soil surface) or slab-on-grade foundations (floor built to the same level as the surrounding soil surface), the entry points include (i) cracks and pores in floor slabs, walls, and floor-wall joints; and (ii) openings around sump pumps, floor drains, and pipes penetrating floors and walls. Structures with suspended wooden floors, *i.e.*, crawl spaces, between the ground and the lowest floor level are less vulnerable to ^{222}Rn that tends to escape to the outside air when appropriate vents are installed; nonetheless can admit some of the ^{222}Rn through cracks in the flooring, along the foundation sill plate, around service pipes for plumbing and sewerage, and heating and air-conditioning ducts when they pass through the floor.

1.2.1.5. Type of the Ventilation and Life Style of the Building Occupants

Infiltration (uncontrolled entry of the outdoor air into a building through cracks and leakage points of the structure), natural ventilation (flow of outdoor air into a building through opened doors and windows), and mechanical ventilation (the use of blowers or fans to either add or remove the air from a building) are the three fundamental components of ventilation that influence the in-house ^{222}Rn concentrations. Life styles of the occupants of a building also influence the indoor ^{222}Rn concentrations, to some extent.

Infiltration is a result of the air convection into and out of the house, driven by pressure variations across the building shell that result due to the forces exerted by the wind (intensity and direction) and the temperature difference between indoors and outdoors; the mathematical representation of infiltration component of the ventilation rate is denoted by equation 1.1 as documented by Nero (1988).

$$v_i = A_0 \left[(f_w V)^2 + \left(f_s dT^{1/2} \right)^2 \right]^{1/2} \tag{1.1}$$

where,

v_i is the infiltration component of ventilation rate; A_0 is the effective leakage area; f_w is the parameter that accounts for local and terrain shielding effects, the distribution of leakage area, and the height of the building relative to which the wind speed is measured; V is the wind speed; f_s is the stack parameter accounting for the building height and the distribution for leakage area; dT is the temperature difference.

'High winds around and over a structure create a low pressure zone laminar to the walls and the roof that results in the creation of partial vacuum within a home that reduces the interior pressure and aggravates the stack effect' (Hoffmann and May 1997). The stack effect is referred to as the movement of air into and out of the house that is governed by the buoyancy forces, which are directly proportional to the thermal difference and the height of the structure. The stack effect is predominant in the winter months as compared to the summer months, due to the relatively larger differentials between indoor and outdoor temperatures, and reduced ventilation (air exchange rate). The stack effect is further enhanced by the use of exhaust fans in kitchens and bathrooms, air distribution blowers, and clothes dryers.

Barometric pressure changes were noted to have an impact on the ^{222}Rn infiltration rates to some extent, when there is sustained pressure difference between indoor air and pore air of the nearby soil; the influence of barometric pressure changes being directly proportional to the depth of the soil to an impermeable zone (Nazaroff *et al.* 1988a). The ^{222}Rn infiltration rates were additionally noted to be significantly influenced by the amount of precipitation (Nazaroff *et al.* 1988a; Hoffmann and May 1997). While the precipitation events of only a few tenths of an inch will have little influence on the ^{222}Rn infiltration rates, considerable precipitation ($\geq$ 0.75 inch, approximately) can have a pronounced effect resulting in higher ^{222}Rn infiltration rates. Moderate precipitation initially seals the soil with a gas-impermeable and incompressible liquid. As the water interface percolates downward into the higher permeable dry soil beneath the less permeable wet soil that is on the surface, the ^{222}Rn gas in the soil is forced to move through the penetrations in the substructure of the building.

The summer-time conditions encourage a windows-open lifestyle, which enhances the natural ventilation by accommodating for ample air exchange between the indoor and the outdoor air that inhibits the accumulation of ^{222}Rn concentrations inside a house. During the warm months when buildings are well ventilated, the indoor ^{222}Rn levels are largely determined by the geological factors rather than the mechanical factors. However, in the winter-time conditions, there is reduced ventilation as the occupants of the buildings keep the windows closed to keep themselves warm, which makes the ventilation (air exchange) rates to go down, thereby resulting in the accumulation of ^{222}Rn concentrations inside the house. One can expect to have higher averaged ^{222}Rn concentrations in well-weatherized (tight) houses as compared with poorly-weatherized (drafty) houses.

A balanced mechanical ventilation system (with equivalent supply and exhaust air flow) does not have any substantial effect on the pressure distributions along the base of the building walls. On the other hand, a supply-dominant mechanical ventilation system causes an increase in the pressure within the house, which forces the removal of air inside the building through leaks or designed outlet points, thereby preventing the accumulation of ^{222}Rn concentrations inside the home (Nazaroff *et al.* 1988a). American homes generally receive the supply air from the basement itself. In such cases, the mechanical ventilation system helps in drawing more ^{222}Rn from the basement floor, thereby, resulting in the accumulation of ^{222}Rn concentrations within the home.

1.2.1.6. Modeling ^{222}Rn Migration from the Soils into Homes

The mathematical computation of ^{222}Rn migration in soils can be performed by accounting for (i) the diffusive transport that is governed by the Fick's Law (diffusional flux density is proportional to the concentration gradient), and (ii) the convective transport that is governed by the Darcy's Law (the volumetric flow rate is directly proportional to the difference of the fluid heads at the inlet and the outlet, and the cross-sectional area, and is inversely proportional to the length). The general transport equation for radon migration in soils that accounts for both diffusion and convection mechanisms, with negligible and considerable moisture contents are represented by equations 1.2 and 1.3, respectively, as documented by Nazaroff *et al.* (1988a).

$$\frac{\partial I_{Rn}}{\partial t} = \nabla . D_e \nabla I_{Rn} - \nabla . I_{Rn} \frac{v}{\epsilon} + f\rho_s \frac{1-\epsilon}{\epsilon} A_{Ra}\lambda_{Rn} - \lambda_{Rn}I_{Rn} \tag{1.2}$$

$$\frac{1}{\epsilon}\frac{\partial(I_a\epsilon_a + I_w\epsilon_w)}{\partial t} = \nabla . D'_e \nabla I_a - \nabla . I_a \frac{v'}{\epsilon_a} + f'\rho_s \frac{1-\epsilon}{\epsilon} A_{Ra}\lambda_{Rn} - \frac{1}{\epsilon}\lambda_{Rn}(I_a\epsilon_a + I_w\epsilon_w) \tag{1.3}$$

where,

I_{Rn} is the radon activity concentration; ∇ is the three dimensional gradient operator; D_e is the effective diffusion coefficient; v = N/c is the net air velocity; ϵ is the porosity; f is the emanation power; ρ_s is the density of soil grains; A_{Ra} is the radium activity concentration in soil; λ_{Rn} is the decay constant of radon (2.1×10^{-6} per second for ^{222}Rn); I_a and I_w are the radon concentrations in the air volume and the water volume; ϵ_a and ϵ_w are the air and water porosities such that $\epsilon_a + \epsilon_w = \epsilon$; and the primes indicate the new values accounting for moisture content and where there has been a neglection of any moisture migration and migration of radon within the water. Note that in the right hand part of equations 1.2 and 1.3, the first term corresponds to the diffusive transport, the second to the advective transport, the third to the ^{222}Rn generation, and the fourth to the ^{222}Rn radioactive decay.

It is a well-established fact that in majority of the homes with elevated ^{222}Rn levels, the contribution of ^{222}Rn entry through the process of diffusion is negligible when compared with the process of convection (major ^{222}Rn entry mechanism into homes). Consequently, majority of the earlier modeling studies related to ^{222}Rn migration from the soils into homes neglected the diffusive components of the transport equation and solved for the convective components of the transport equation. A complete review of the available analytical models, lumped parameter models, and numerical models that accounted for the convective ^{222}Rn entry into homes by solving the transport equations under more or less approximations to the physical structure of the soil, entry pathways, and boundary conditions was discussed elsewhere (Gadgil 1992). Loureiro *et al.* (1990) is one of the first studies to model the ^{222}Rn entry into homes by accounting for both the diffusion and the convection mechanisms. With the advancement of cost-effective computational methods, numerical approaches are increasingly being used nowadays to examine the ^{222}Rn migration in soils. Refer to the literature (Nazaroff 1988; Garbesi and Sextro 1989; Loureiro *et al.* 1990; Revzan *et al.* 1991; Andersen 2001; Wang and Ward 2000, 2002) for additional information on the modeling of ^{222}Rn migration in soils and entry into the buildings.

1.2.2. Building Materials

Building materials are generally the second main source of ^{222}Rn indoors, even though they were once considered to be the primary source (UNSCEAR Report 1977). The transport of ^{222}Rn in building materials occurs on the basis of (i) flow: ^{222}Rn is transmitted by the liquid water, air, and water vapor within the pore spaces of the building materials, and (ii) diffusion: ^{222}Rn moves relative to the fluid in the internal pores of building materials; the latter being the predominate transport mechanism as majority of the materials that produce ^{222}Rn have very low permeability (Stranden 1988). ^{222}Rn exhalation from building materials depends not only on the radium content, but also on the emanation coefficients (fraction of the ^{222}Rn produced from the material that enters the pores), porosity of the material, density of the material, and radon decay constant; the equations for ^{222}Rn production rate and the diffusive solution are given by equations 1.4 and 1.5, respectively, as documented by Stranden (1988). Stranden (1988) also estimated the contribution of building materials to indoor radon as described by equation 1.6.

$$f = \frac{1}{\epsilon}\lambda_{Rn}I\rho\eta \tag{1.4}$$

$$E_d = \epsilon f L \tanh\left(\frac{d}{2L}\right) \tag{1.5}$$

$$C = \frac{E_d}{V}\frac{S}{\lambda_V} \tag{1.6}$$

where,

f is the ^{222}Rn production rate; ϵ is the porosity of the material; λ_{Rn} is the decay constant of radon (2.1×10^{-6} per second for ^{222}Rn); I is the ^{226}Ra concentration in the material; ρ is the density of the material; η is the emanation coefficient; E_d is the exhalation rate per unit area; L is the diffusive length $= \sqrt{\frac{D_e}{\lambda}}$ with D_e being the effective diffusion coefficient; d is the wall thickness; C is the indoor radon concentration contributed by building materials; S is the exhalation area; V is the volume of the room; and λ_V is the air exchange rate.

Tables **3** through **6** provide a summary of the radium contents, emanation coefficients, porosity, and effective diffusion coefficients for different building materials obtained from the literature.

Table 3: ^{226}Ra contents for different building materials

Building Material	Range (Bq per kg)
Concrete	10-80[a], 300-2500 (alum-shale-based lightweight)[a], 11-26[b], 58-66[c], 86-110 (block)[d], 13-98[d], 58-66[e], 36-87[f]
Brick	20-200 (clay)[a], 67-89 (red)[d], 18-78[d]
Gypsum	5-20[a], 26[d], 610-1160 (phospho)[g]
Cement	10-50[a], 31-41[d]
Fly ash	164[d]
Bottom ash	82[d]
Granite	100-200[a], 149-211 (chip)[d]
Tuff	100-600[a]
Sand	7 (sea)[d], 38-50 (river)[d]
Aggregate	70-202[d]

Source: ([a]Stranden 1988; [b]Roelofs and Scholten 1994; [c]Gadd and Borak 1995; [d]Tso *et al.* 1994; [e]Ward *et al.* 1993; [f]Rogers and Nielson 1992; [g]Rutherford *et al.* 1995)

Table 4: Emanation coefficients for different building materials

Building Material	Range
Concrete	0.1-0.4[a], 0.1-0.4[b], 0.17-0.21[c], 0.07-0.13[d], 0.31 (lightweight)[d], 0.02-0.09[e]
Brick	0.02-0.1[a], 0.02-0.1 (clay)[b],0.05-0.11 (red)[d], 0.09-0.13 (sand)[d]
Gypsum	0.03-0.2[a]
Cement	0.02-0.05[a]
Fly ash	0.002-0.02[a]
Phosphogypsum	0.19-0.2[f]
Bottom ash block	0.07[d]
Granite	0.01-0.03[d]

Source: ([a]Stranden 1988; [b]UNSCEAR Report 1993; [c]Gadd and Borak 1995; [d]Tso *et al.* 1994; [e]Rogers and Nielson 1992; [f]Rutherford *et al.* 1995)

Table 5: Porosity values for different building materials

Building Material	Range
Concrete	0.11-0.15[a], 0.13-0.27 (residential)[b], 0.16-0.24 (aged)[c], 0.17-0.25[d], 0.12-0.2[e], 0.17-0.26[f]
Building materials	0.01-0.7[g]
Red brick	0.24-0.26[d]

Source: ([a]Gadd and Borak 1995; [b]Rogers *et al.* 1994; [c]Rogers *et al.* 1995; [d]Tso *et al.* 1994; [e]Renken and Rosenberg 1995; [f]Rogers and Nielson 1992; [g]UNSCEAR Report 1993)

Table 6: Effective diffusion coefficient values for different building materials

Building Material	Range (square meters per second)
Concrete	$7.6\times10^{-9} - 8.4\times10^{-8}$ [a], $1.13\times10^{-7} - 1.91\times10^{-7}$ [b], $7.2\times10^{-9} - 5.4\times10^{-7}$ [c], $1.1\times10^{-7} - 2.1\times10^{-6}$ [d], $2.1\times10^{-8} - 5.2\times10^{-7}$ (residential)[e], $1.5\times10^{-7} - 5.5\times10^{-7}$ (aged)[f]
Brick	$8.4\times10^{-8} - 3.4\times10^{-7}$ [a]
Gypsum	$1.3\times10^{-6} - 3.6\times10^{-6}$ [a]

Source: ([a]Stranden 1988; [b]Gadd and Borak 1995; [c]Renken and Rosenberg 1995; [d]Rogers and Nielson 1992; [e]Rogers *et al.* 1994; [f]Rogers *et al.* 1995)

1.2.3. Water

Considerable amounts of ^{222}Rn can also enter into the homes through the water supply system. 'The world average ^{222}Rn concentration in all types of water supplies is assumed to be 10 kBq/m³' (UNSCEAR Report 1993). The source of water supply system for a house, *i.e.*, the surface or the groundwater supply system, determines the extent of ^{222}Rn concentration buildup within the homes. Surface water generally contains a low concentration of dissolved ^{222}Rn, as a consequence of the degasification of the ^{222}Rn gas from water into the air. Consequently, there is very little ^{222}Rn in surfacial water (lakes, rivers, reservoirs) and the homes that rely on the surface water supply system usually do not have a ^{222}Rn problem from their water. Public drinking water from surface water sources or wells is normally treated in ways that reduce radon at or near the water source before it is distributed to homes. In big cities, water is processed in large municipal systems that aerate the water, thereby, facilitating the waterborne ^{222}Rn to degasify. The delay in supplying the water to a house from the municipal systems rather than directly from the source also provides additional time that allows for most of the remaining ^{222}Rn to be decayed. In contrast to the surface water supply system, ^{222}Rn is found in substantially higher levels in the groundwater supply system *via* migration of the ^{222}Rn gas in the soil and/or rock. '^{222}Rn in the tap-water from deep wells can range from 100 kBq/m^3 to 100 MBq/m^3' (UNSCEAR Report 1988). Small public water works and private domestic wells often have closed systems and short transit times that do not allow ^{222}Rn to decay to harmless by-products before entering a home. Waterborne ^{222}Rn that gets into the homes through the groundwater supply system directly escapes into the indoor air during activities associated with agitation of water, such as the use of showers, dishwashers, laundry, and toilets, and during bath and cleaning activities. Thus, the ^{222}Rn released from the groundwater supply system can contribute significantly to a total inhalation risk from the indoor air. The areas most prone to have problems with ^{222}Rn in groundwater are those with high levels of radium in the underlying rocks. It takes about 10,000 pCi/l of radon in water to raise the radon in indoor air by 1 pCi/l.

Nazaroff *et al.* (1988b) derived an equation for computing the incremental average increase in indoor ^{222}Rn concentrations from water use on the basis of mass balance modeling approach, as represented by equation 1.7, and noted that the factors affecting the in-house airborne ^{222}Rn concentrations from water supply systems are (i) the residential volumes, (ii) the air exchange rates, (iii) the in-house water use rates, and (iv) the water-to-air transfer coefficients.

$$C_{avg} = \frac{C_w W e}{V \lambda_v} \tag{1.7}$$

where,

C_{avg} is the incremental average increase in indoor radon concentrations from water use; C_w is the radon concentration in water entering the house; W is the time-averaged water use rate per resident; e is the use-weighted average transfer efficiency of radon from water to air; V is the volume per resident of the dwelling; λ_v is the air exchange rate of the residence.

The ratio of the incremental average increase in indoor radon concentrations from water use to the radon concentration in water entering the house is defined as the transfer coefficient. There are limited studies that determined the transfer coefficients for varying household activities. Table **7** presents a summary of the transfer coefficients in relation to varying household activities. It must be noted that the transfer coefficient values obtained by Gesell and Prichard (1980) are overestimated, considering that they assumed all the indoor ^{222}Rn to be a result of the degasification from water use within the building.

Table 7: Transfer coefficients of radon from water to air during various household activities

Household Activity	Transfer Coefficients
Dishwashing	0.98[a], 0.90[b], 0.98[c], 0.95[d]
Laundry	0.95[a], 0.90[b], 0.90 (estimated)[c], 0.92[d]
Shower	0.71[a], 0.63[b], 0.65[c], 0.66[d]
Bath	0.50[a], 0.47[b], 0.30 (estimated)[c], 0.42[d]
Toilet	0.29[a], 0.30[b], 0.30 (estimated)[c], 0.30[d]
Cleaning	0.28[a], 0.45[b], 0.1-0.5 (estimated)[c], 0.34[d]

Source: ([a]Partridge *et al.* 1979; [b]Gessell and Prichard 1980; [c]Hess *et al.* 1982; [d]Nazaroff *et al.* 1988b)

1.2.4. Natural Gas

^{222}Rn can also enter into the homes through the natural gas supplied for domestic use, which is released under combustion. The contribution of natural gas supply to indoor ^{222}Rn concentrations is not significant. The typical value of ^{222}Rn in natural gas is generally around 1000 Bq/m^3, and the ^{222}Rn gas transmitted through natural gas supply system varies from one place to another depending on the location of the natural gas source, *i.e.*, the extraction well (UNSCEAR Report 1993). An average of 37 pCi/l of ^{222}Rn concentrations were noted in the natural gas production wells across the USA (Kraemer 1986); while the production wells in the Europe, Canada, and the USA accounted for ^{222}Rn concentrations of the order of 40-3400 Bq/m^3, 150-30000 Bq/m^3, and 40-54000 Bq/m^3, respectively (Dixon 2001).

1.2.5. Outdoor Air

Outdoor air generally acts as a diluting factor for the indoor ^{222}Rn. However, in certain circumstances, such as the case of high rise buildings with building materials having low ^{226}Ra content, outdoor air can serve as a source. The ^{222}Rn concentrations in the outdoor air are dependent on the atmospheric pressure, the time of the day (in case of non-perturbative weather), and the weather conditions (such as thermal inversion, precipitation). Gessell (1983) reported the average ^{222}Rn gas concentration in the atmosphere in the USA to be of the order of 4-15 Bq/m^3; while other studies (Hopper *et al.* 1991, Robe *et al.* 1992, Deyuan 1993, Grasty 1994, Price *et al.* 1994) reported the average ^{222}Rn gas concentrations to vary by tens of Bq/m^3. ^{222}Rn gas concentrations of higher magnitude were observed in the case of mine tailings (Tyson *et al.* 1993) and in weather conditions of thermal inversion or very low precipitation (Grasty 1994). The ambient air over oceans has very low ^{222}Rn concentrations of the order of 0.1 Bq/m^3. This is a result of the minimum presence of ^{226}Ra in the sea water and the high solubility of ^{222}Rn in water at low temperatures. 'The mean value of outdoor ^{222}Rn concentrations is 10 Bq/m^3 for continental areas and somewhat less in coastal regions' (UNSCEAR Report 1993).

1.3. HOW ARE ^{222}RN CONCENTRATIONS MEASURED INSIDE HOMES?

The measurement of ^{222}Rn and its progeny in homes can be classified into four categories based on the duration of measurement, as long-term (few months to one year), short-term (one to ten days), very short-term (minutes to tens of minutes), and continuous monitoring (in association with the short-term and grab-sampling measurements). The locations of the measuring points are to be decided in accordance with the purpose of conducting the measurements and it is advisable to have a quality assurance program to guarantee the validity of the measurements as recommended by USEPA Report (1992). The ^{222}Rn and its progeny measurement techniques can also be classified as being either active or passive. While the active techniques corresponds to the use of measurement devices that require electric power and/or the use of air pumps to collect activity from the air, the passive techniques are those that do not require electric power when installed at the sampling location. Amongst the available wide-ranging instruments to monitor the ^{222}Rn and its progeny concentrations in homes, the alpha track detectors (ATDs), the activated charcoal detectors (ACDs), the electret ion chambers (EICs), the electronic integrating devices (EIDs), and the continuous radon monitors (CRMs) are the most widely used ^{222}Rn measurement devices (WHO Report 2009). While the ATD, the ACD, and the EIC are categorized as passive devices, the EID and the CRM are categorized as active devices. Refer to WHO Report (2009) for additional details on the working mechanism and capabilities of the above mentioned ^{222}Rn measurement devices. Alternatively, one can also get more detailed information on the available wide-ranging ^{222}Rn measurement devices and measurement protocols by referring to the specific publications (OECD Report 1985; NCRP Report 1988; USEPA Report 1992, 1993a,1997).

1.4. WHAT ARE THE ^{222}RN ACTION LEVELS?

'The World Health Organization (WHO) recommends a radon action level of 2.7 pCi/L' (WHO Report 2009), *i.e.,* all the existing homes with ^{222}Rn concentrations exceeding 2.7 pCi/L are to be mitigated, so that the indoor ^{222}Rn concentrations are reduced to less than 2.7 pCi/L, and new constructions are designed by making sure that the indoor ^{222}Rn concentrations are less than 2.7 pCi/L. Several countries have regulated the ^{222}Rn action levels. Table **8** provides a list of the countries that regulated the ^{222}Rn action levels for both the existing dwellings and the new dwellings under construction.

Table 8: ^{222}Rn action levels for different countries

Country	Action Level	
	Existing Dwellings (Bq/m³)	**New Dwellings (Bq/m³)**
Argentina	400	200
Austria	400	200
Belgium	400	200
Bulgaria	500	200
Canada	200	100
China	400 for houses, 1000 for workplaces	200
Czech Republic	400	200
Denmark	200	200
Finland	400	200
Georgia	200	100
Germany	100	100
Greece	400	200
Ireland	200	200
Kyrgyzstan	< 200	200
Lithuania	400	200
Norway	200	200
Romania	400	200
Russia	400	200
Slovenia	400 for houses, schools, and kindergartens, 1000 for companies	400 for houses, schools, and kindergartens, 1000 for companies
Sweden	200	200
Switzerland	1000	400
UK	200	200
USA	148	148

Source: (WHO Report 2007)

1.5. WHAT ARE THE HEALTH EFFECTS OF EXPOSURE TO ^{222}RN?

Residents of a house exposed to ^{222}Rn will inhale the ^{222}Rn gas along with its solid daughters. The alpha particles generated during the decay process are not energetic enough to pass through a person's outer skin, but when released within the lungs upon inhalation can easily penetrate the unprotected lining and pass through the living cells. During this process, the alpha particles damage the cells in a way that can cause some of them to become cancerous. The other forms of radiation (beta particles and gamma rays) produced by the decay of ^{222}Rn and its daughters, have no significant adverse health effects when released inside the lungs. It has been well-established that ^{222}Rn is the second most important cause of lung cancer after smoking and is the primary cause of lung cancer among people who have never smoked. The majority of ^{222}Rn induced lung cancers are a result of the low and moderate ^{222}Rn concentrations rather than by the high ^{222}Rn concentrations. ^{222}Rn induced lung cancers vary between 3% and 14%, and an average of 10 percent increase in the risk in lung cancer per 100 Bq/m^3 increase in ^{222}Rn concentrations was observed on summarizing the in-house ^{222}Rn studies from Europe, North America, and China (WHO Report 2009). There is no safe level of radon, just different levels of risk. Table **9** presents a summary of the ^{222}Rn induced cancerous risks for smokers and non-smokers. It has also been suggested that drinking water containing ^{222}Rn may cause stomach cancer, but this effect has not been conclusively demonstrated and is, in any case, a very minor risk in comparison to that of ^{222}Rn induced lung cancer.

Table 9: ^{222}Rn induced cancerous risk in smokers and non-smokers

Radon Level (pCi/L)	Cancerous Risk (If 1,000 people were exposed to this level over a lifetime)	
	Smokers	**Non-smokers**
20	About 260 people could get lung cancer	About 36 people could get lung cancer
10	About 150 people could get lung cancer	About 18 people could get lung cancer
8	About 120 people could get lung cancer	About 15 people could get lung cancer
4	About 62 people could get lung cancer	About 7 people could get lung cancer
2	About 32 people could get lung cancer	About 4 people could get lung cancer
1.3	About 20 people could get lung cancer	About 2 people could get lung cancer
0.4	About 3 people could get lung cancer	

Source: (USEPA Report 2012)

1.6. WHAT ARE THE APPLICABLE CONTROLLING MEASURES FOR INDOOR ^{222}RN?

The good news about ^{222}Rn is that it is relatively easy to control. There are a wide variety of remediation techniques available, and the identification of the most appropriate technique for a given building will depend on the ^{222}Rn sources, concentrations, and transportation mechanisms. In-house ^{222}Rn concentrations can be reduced by (i) reducing the ^{222}Rn entry from soil through depressurization of subfloor spaces, (ii) increasing the building ventilation rate with a consequent increase of ^{222}Rn removal, (iii) increasing the resistance of building to ^{222}Rn entry by sealing the floor (or the walls, in case of building materials with high ^{222}Rn exhalation rate), and (iv) not using the ^{222}Rn affected water supply system.

Tables **10** and **11** present a summary of the available ^{222}Rn control techniques for new and existing constructions, respectively. One may choose an optimal performing ^{222}Rn control technique (listed in Table **10**) on the basis of cost efficiency and ^{222}Rn reduction potential to ensure that the new construction occupants are not exposed to ^{222}Rn levels exceeding the regulated action limits. From the early 1990s, attempts were made to analyze the performance or cost effectiveness of various control techniques in mitigating indoor ^{222}Rn concentrations in the USA (Osborne and Harrison 1992; Prill *et al.* 1990), the UK (Denman *et al.* 2000; Coskeran *et al.* 2001), Finland (Korhonen *et al.* 2000; Arvela 2001), Belgium (Paridaens *et al.* 2005), Czech Republic (Jiranek and Neznal 2008), Austria (Maringer *et al.* 2008; Ringer *et al.* 2008), and Japan (Kranrod *et al.* 2009). However, these studies were largely limited by the number of tests performed (due to limited number of selected houses or control techniques adopted) to draw valid statistical conclusions for comparing the performance of available mitigation systems. For additional details on the performance of ^{222}Rn control techniques, refer to the specialized publications on ^{222}Rn prevention and mitigation strategies (USEPA Report 1993b, 2003; WHO Report 2009).

Table 10: ^{222}Rn control techniques for new constructions

Technique	^{222}**Rn Reduction Potential**	**Long-term Performance**	**Monitoring Ease**	**Quiet and Unobtrusive**	**Cost**	
					Installation	**Operating**
Sealing soil contacted surfaces	None to low or moderate	Usually poor to fair	Repeated random testing required	Usually very good	Moderate	Very low
Soil gas barriers	Highly variable	Stable, but often limited ^{222}Rn reduction	Repeated random testing required	Very good	Depends on care and quality	None
Passive ventilation unoccupied lower space	Moderate to good	Very good	Repeated random testing required	Very good	Low	Low

Technique	^{222}Rn Reduction Potential	Long-term Performance	Monitoring Ease	Quiet and Unobtrusive	Cost	
					Installation	Operating
Active ventilation unoccupied lower space	Good	Very good	Repeated random testing required	good	Moderate	Moderate
Passive soil depressurization	Low to moderate	Good if sealing is maintained	Repeated random testing required	Usually very good	Low	Very low
Active soil depressurization	Moderate to greatest	Very good	Pressure and/or radon testing needed	Usually very good	Low	Moderate
Balanced ventilation	Low to moderate	Good if operated and maintained	Repeated random testing required	Usually very good	Low to high	Moderate to high

Source: (WHO Report 2009)

Table 11: ^{222}Rn control techniques for existing constructions

Technique	Typical ^{222}Rn Reduction Potential	Comments
Passive sub-slab suction	30 to 70 percent	Can be more effective in cold climates; not as effective as active sub-slab suction
Drain tile suction, also known as drain tile depressurization (DTD)	50 to 99 percent	Can work with either partial or complete drain tile loops
Block-wall suction, also known as block-wall depressurization (BWD)	50 to 99 percent	Only in homes with hollow block-walls; requires sealing of major openings
Sump-hole suction, also known as sump-pit perimeter depressurization (SUMP)	50 to 99 percent	Works best if air moves easily to sump from under the slab
Submembrane depressurization (SMD) in a crawlspace	50 to 99 percent	Less heat loss than natural ventilation in cold winter climates
Natural ventilation in a crawlspace	0 to 50 percent	Costs variable
Sealing of radon entry routes	See comments	Normally used only with other techniques; proper materials and installation are required
House (basement) pressurization	50 to 99 percent	Works best with tight basement isolated from outdoors and upper floors

Technique	Typical ^{222}Rn Reduction Potential	Comments
Natural ventilation	Variable and temporary	Significant heated or cooled air loss; operating costs depend on the utility rates and the amount of ventilation
Heat recovery ventilation (HRV)	Variable	Best applied in limited-space areas like basements and has only limited use; effectiveness limited by the ^{222}Rn concentration or the amount of ventilation air for dilution by the HRV
Private well water systems: granular activated carbon (GAC)	85 to 95 percent	Less efficient for higher levels than aeration; use for moderate levels, around 50,000 pCi/L or less in water: radioactive ^{222}Rn by-products can build on carbon; may need radiation shield around tank and care in disposal
Private well water systems: aeration	95 to 99 percent	Generally more efficient than GAC; requires annual cleaning to maintain effectiveness and to prevent contamination; requires venting ^{222}Rn to outdoors

Source: (USEPA Report 2010)

CHAPTER 2

Radon Problem in the State of Ohio, USA

Abstract: This chapter provides extensive details on the problem of radon in Ohio, USA by developing and analyzing the associated geospatial maps and statistical metrics. The geospatial and statistical analyses are examined at both the county level and the zip code level to have a thorough understanding of the radon problem that assists in accurately assessing the radon problem in reference to a specific location.

Keywords: Radon, radon in Ohio, Ohio shale, ORIS, GIS, geospatial radon maps, radon statistical metrics, USEPA action limit for radon, WHO action limit for radon, number of radon homes tested statistics, minimum radon concentration statistics, maximum radon concentration statistics, arithmetic mean radon concentration statistics, geometric mean radon concentration statistics, standard deviation in radon measurements statistics, coefficient of variation in radon measurements statistics, median radon concentration statistics, first quartile radon concentration statistics, third quartile radon concentration statistics, radon distributions in Ohio counties and zip codes.

2.1. BACKGROUND ON ^{222}RN PROBLEM IN OHIO

After the Limerick Nuclear Generating Station construction worker home tested for significant indoor ^{222}Rn concentrations in 1984 (Pennsylvania Bureau of Radiation Protection 2012), the Ohio Department of Health (ODH) initiated a ^{222}Rn program in 1988 to examine the extent of ^{222}Rn problem in Ohio homes. As already mentioned in Chapter 1, the main contributor to in-house ^{222}Rn concentration is the geological formation beneath the house. An adequate understanding of the Ohio geological formations can help in understanding the ^{222}Rn problem in Ohio homes. Fig. **1** illustrates the bedrock geology of Ohio. The geological formation in Ohio comprises of Ohio Shale, a geologic rock unit of upper Devonian age and glacial deposits that are fairly widespread as can be seen from Fig. **1**. The Ohio Shale is enriched with ^{238}U concentrations ranging between 10 ppm to 40 ppm (Bates and Strahl 1958; Swanson 1960; Tracy 1983;

Ashok Kumar and Akhil Kadiyala

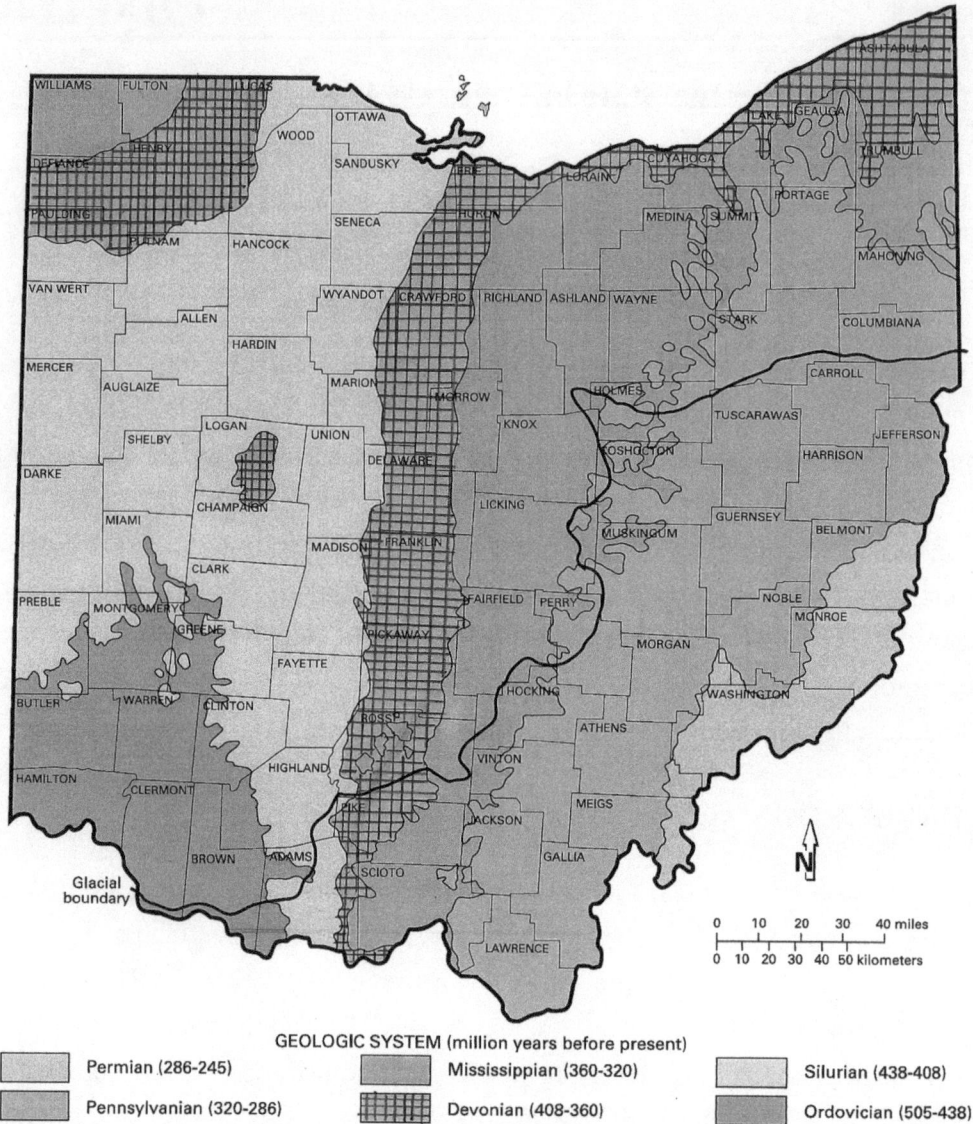

Figure 1: Bedrock geological map of Ohio (Source: Ohio Department of Geological Survey, 2006, Geologic map and cross section of Ohio: Ohio Department of Natural Resources, Division of Geological Survey, page-size map, 1p., 1:2,000,000.). Note: Glacial boundary represents the southern boundary of glacial deposits.

Harrell and Kumar 1988). The Ohio Shale underlies the entire eastern half of the state and also the extreme northwest corner; however, it is only when the Ohio Shale rises to the surface (as represented by Devonian formation in Fig. **1**) that it poses a potential health threat (Harrell and Kumar 1989). It was observed that the Ohio Shale outcrop was mostly covered with up to 60 meters of river, lake, and glacial sediments (Soller 1986); when the sediment overburden is thick, or is of low permeability, the sediment overburden acts as a barrier to the ^{222}Rn migration upwards from Ohio Shale. It was noted that in some parts of Ohio, the sediment overburden itself was enriched in ^{238}U, which acts as a potent ^{222}Rn source; especially true of the glacial outwash, kame and esker gravel deposits (that consist in part granitic rock fragments), and the glacial tills when they contain abundant fragments of Ohio Shale (Harrell and Kumar 1989). The geological setting of Ohio, in reference to the ^{238}U and ^{222}Rn, is similar to that of Sweden with Alum Shale and its glacial deposits (Akerblom 1986).

Initial research on examining the ^{222}Rn concentrations in Ohio homes was restricted to the areas of Ohio Shale outcrop, as the geological unit represented the single largest concentration of ^{238}U. A total of 125 houses within six 100 square kilometer areas in Ashtabula, Erie-Huron, Franklin, Pike, and Logan counties in the state of Ohio were examined for indoor ^{222}Rn concentrations with simultaneous measurements of 14 independent variables that included penetration factor, air exchange factor, temperature difference between indoors and outdoors, age of the house, volume of the basement, condition of the sump pump (if present), condition of the crawl space (if present), frequency of the use of the fire place (if present), type of heating fuel used, type of cooking fuel used, source of indoor water, type of soil surrounding the house, number of people living in the house, and number of smokers in the house. Only the first four factors showed a significant correlation with indoor ^{222}Rn in one or more of the study areas; only the penetration factor and air exchange factor had consistently high correlations with indoor ^{222}Rn, while the depressurization effect was also noted to significantly influence the in-house ^{222}Rn concentrations (Harrell and Kumar 1988). The study was extended further by developing a ^{222}Rn predictive model for each of the six test areas, on the basis of performing multivariate stepwise regression analysis of the indoor ^{222}Rn concentrations as a function of the associated geological and house parameters (Harrell and Kumar 1989). A careful examination of the thickness and lithology of the sediment overburden above the Ohio Shale revealed that (i) when the average thickness of the sediment overburden exceeded 27m (Franklin and Logan counties), much of the ^{222}Rn in houses was noted to be a result of Ohio Shale clasts in the glacial till, and (ii) when the average thickness of the sediment overburden is 6m or less (Ashtabula, Cuyahoga, Erie-Huron, and Pike counties), the indoor ^{222}Rn levels were noted to be directly controlled by the emanations from the underlying Ohio Shale bedrock (Harrell *et al*. 1991).

In context of the findings from initial research studies on ^{222}Rn in Ohio (Harrell and Kumar 1988; Harrell and Kumar 1989; Harrell *et al*. 1991), the ODH in co-ordination with the USEPA aggressively promoted for the ^{222}Rn testing in Ohio homes, and awarded a grant for maintaining the Ohio homes database to the Air pollution Research Group (APRG) of the Civil Engineering Department at The University of Toledo. ^{222}Rn concentrations in Ohio homes were collected from various commercial testing services, health departments, and university researchers. The initial homes database of 50,000 ^{222}Rn concentrations was prepared by Kumar *et al*. (1990) and was organized in the form of Ohio Radon Information System (ORIS) by Heydinger *et al*. (1991). More data were added every year to the Ohio homes ^{222}Rn database that resulted in development and analysis of 80,436 data points in 1988 (Sud 1998), 121,959 data points in 2005 (Kumar and Varadarajan 2005), 133,343 data points in 2009 (Manthena *et al*. 2009), and 145,849 data points in 2011 (Kumar *et al*. 2010). This book provides a detailed explanation of the ^{222}Rn problem in Ohio by computing different statistical metrics and developing Geographical Information Systems (GIS) based maps using 219,114 Ohio homes ^{222}Rn observations.

The statistical metrics computed and presented in this book are (i) the number of ^{222}Rn tests performed (No.), (ii) the minimum ^{222}Rn concentration (Min.), (iii) the maximum ^{222}Rn concentration (Min.), (iv) the arithmetic mean ^{222}Rn concentration (AM), (v) the geometric mean ^{222}Rn concentration (GM), (vi) the standard deviation in ^{222}Rn measurements (SD), (vii) the coefficient of variation (CV), (viii) the median, also known as the second quartile ^{222}Rn concentration (Md), (ix) the first quartile ^{222}Rn concentration (Q1), and (x) the third quartile

^{222}Rn concentration (Q3). Three different versions of GIS based ^{222}Rn maps are presented in this book:

i).   the ^{222}Rn maps categorized on basis of the WHO action level of 2.7 pCi/L and the USEPA action levels of 4 pCi/L (hereon, referred as the WHO-USEPA classification) that produced three classes (< 2.7 pCi/L, 2.7 – 4 pCi/L, > 4 pCi/L); the ^{222}Rn maps generated using these classes were denoted with suffix 'a' in the figure caption,

ii). the ^{222}Rn maps categorized on basis of the USEPA recommended zonal classification (Zone 3: < 2 pCi/L indicating low potential, Zone 2: 2 – 4 pCi/L indicating moderate potential, and Zone 1: > 4 pCi/L indicating highest potential); the ^{222}Rn maps generated using these classes were denoted with suffix 'b' in the figure caption, and

iii). the ^{222}Rn maps detailed classification in intervals of 2 pCi/L (hereon, referred as the 2 pCi/L breakdown classification) (< 2 pCi/L, 2 - 4 pCi/L, 4 - 6 pCi/L, 6 – 8 pCi/L, 8 - 10 pCi/L, > 10 pCi/L); the ^{222}Rn maps generated using these classes were denoted with suffix 'c' in the figure caption.

2.2. DISTRIBUTION OF INDOOR ^{222}RN CONCENTRATIONS IN OHIO – A COUNTY BASED ANALYSIS

Table **1** presents a summary of the computed ^{222}Rn statistics for all the 88 counties in Ohio. There are a total of 29 counties with GM ^{222}Rn concentrations less than 2.7 pCi/L, 30 counties with GM ^{222}Rn concentrations between 2.7 and 4 pCi/L, and 29 counties with ^{222}Rn concentrations greater than 4 pCi/L. These statistics indicate that 67% and 33% of the Ohio counties exhibited the property of GM ^{222}Rn concentrations exceeding the WHO and the USEPA action levels, respectively. Licking county had the highest GM ^{222}Rn concentration (8.13 pCi/L), followed by Pickaway (7.57 pCi/L), Knox (6.95 pCi/L), and Harrison (6.55 pCi/L). The county with the maximum number of homes tested for ^{222}Rn in Ohio is Franklin (31,909 observations). Fig. **2** illustrates the graphical presentation of the number of Ohio homes tested for ^{222}Rn in each county. Figs.

3a, **3b**, and **3c** illustrate the graphical representations of ^{222}Rn GM concentrations in Ohio counties based on the WHO-USEPA, the USEPA, and the 2 pCi/L breakdown classifications, respectively. Table **2** presents a summary of the ^{222}Rn distribution variation in the 88 Ohio counties. Nearly half of the Ohio homes tested for ^{222}Rn had the GM ^{222}Rn concentrations exceeding the USEPA action level.

Table 1: ^{222}Rn statistics for Ohio counties

County Name	No.	Min.	Max.	AM	GM	SD	CV	Md	Q1	Q3
Adams	147	0.10	42.50	4.76	2.14	7.18	1.51	2.20	0.90	5.00
Allen	623	0.10	95.80	6.17	3.57	7.81	1.27	4.20	1.70	7.50
Ashland	683	0.10	260.00	9.15	3.89	20.82	2.28	4.10	1.90	8.10
Ashtabula	856	0.10	210.00	4.01	1.62	10.68	2.66	1.40	0.70	3.20
Athens	976	0.10	82.10	4.83	3.20	5.17	1.07	3.60	1.70	6.50
Auglaize	430	0.10	64.50	7.09	4.01	7.64	1.08	4.50	2.10	9.30
Belmont	840	0.10	288.55	7.12	3.14	17.19	2.42	3.00	1.60	6.00
Brown	118	0.10	20.10	2.92	1.67	3.42	1.17	1.85	0.80	3.70
Butler	7033	0.10	470.00	8.55	3.07	35.10	4.11	3.30	1.50	6.40
Carroll	485	0.10	220.20	12.18	4.44	25.45	2.09	4.00	1.90	10.10
Champaign	1756	0.10	860.00	9.77	4.59	27.92	2.86	5.10	2.20	10.45
Clark	3434	0.10	750.10	8.33	4.42	20.47	2.46	4.80	2.30	9.60
Clermont	3191	0.10	470.80	4.86	2.97	9.91	2.04	3.30	1.60	6.00
Clinton	512	0.10	55.60	5.10	2.94	6.80	1.33	3.20	1.60	6.00
Columbiana	1157	0.10	820.00	10.62	4.08	30.43	2.87	3.80	1.90	8.20
Coshocton	314	0.10	180.00	13.48	5.83	21.60	1.60	5.30	2.30	13.00
Crawford	924	0.10	163.00	6.39	3.20	11.41	1.79	3.80	1.70	7.20
Cuyahoga	14661	0.05	656.20	2.99	1.71	9.33	3.13	1.80	0.90	3.30
Darke	797	0.10	1400.00	10.13	4.44	50.53	4.99	5.30	2.30	10.10
Defiance	252	0.10	21.20	3.47	2.29	3.26	0.94	2.60	1.30	4.60
Delaware	6603	0.10	735.00	7.29	4.32	14.91	2.05	5.10	2.60	8.90
Erie	1755	0.10	279.20	6.99	3.68	14.35	2.05	4.00	1.90	7.80
Fairfield	2164	0.10	340.50	9.21	4.56	16.64	1.81	5.50	2.20	11.00
Fayette	145	0.10	300.00	7.10	3.11	25.14	3.54	3.50	1.70	6.50

County Name	No.	Min.	Max.	AM	GM	SD	CV	Md	Q1	Q3
Franklin	31909	0.05	939.00	9.44	5.56	13.20	1.40	6.20	3.10	11.50
Fulton	315	0.10	21.90	4.06	2.46	4.05	1.00	2.80	1.20	5.40
Gallia	158	0.20	10.60	2.53	1.89	1.97	0.78	1.85	1.10	3.50
Geauga	1670	0.10	733.80	3.25	1.80	19.31	5.94	1.90	1.00	3.20
Greene	8231	0.10	163.00	7.32	4.18	9.44	1.29	4.60	2.20	8.80
Guernsey	280	0.10	115.10	5.62	2.66	11.12	1.98	2.55	1.30	6.40
Hamilton	14584	0.10	316.65	3.45	2.20	4.72	1.37	2.30	1.20	4.20
Hancock	1197	0.10	130.00	5.55	3.14	8.11	1.46	3.40	1.60	6.70
Hardin	184	0.10	57.10	4.52	2.77	5.91	1.31	2.90	1.40	5.70
Harrison	135	0.10	210.00	17.23	6.55	29.92	1.74	4.90	2.30	18.70
Henry	155	0.10	22.90	3.87	2.32	3.72	0.96	2.80	1.50	5.40
Highland	189	0.10	43.00	4.53	2.64	5.37	1.19	3.15	1.40	6.00
Hocking	257	0.10	238.00	10.20	4.27	23.46	2.30	4.50	1.90	10.60
Holmes	150	0.10	140.30	9.64	4.21	18.45	1.91	4.00	1.70	9.50
Huron	887	0.10	159.00	11.39	4.58	22.01	1.93	4.30	2.10	10.00
Jackson	146	0.10	24.30	3.61	2.05	4.12	1.14	2.30	1.00	4.50
Jefferson	465	0.10	927.60	10.85	3.96	46.33	4.27	3.50	1.90	7.50
Knox	4165	0.10	844.20	24.30	6.95	52.55	2.16	6.20	2.30	20.40
Lake	2903	0.10	139.30	3.57	1.93	5.81	1.63	2.00	1.00	3.80
Lawrence	178	0.10	333.40	4.95	1.78	25.16	5.08	1.70	0.90	3.30
Licking	10798	0.10	684.50	18.11	8.13	32.77	1.81	8.30	3.70	19.30
Logan	887	0.10	319.20	10.99	5.80	17.24	1.57	6.40	2.80	12.90
Lorain	4926	0.10	618.20	4.66	2.58	10.37	2.23	3.00	1.40	5.70
Lucas	4960	0.10	260.00	3.74	2.11	6.44	1.72	2.30	1.10	4.50
Madison	899	0.10	82.10	8.05	5.03	9.43	1.17	5.30	2.80	9.70
Mahoning	2367	0.10	220.00	3.17	1.90	8.24	2.60	2.10	1.10	3.60
Marion	1113	0.10	67.10	6.41	3.81	6.99	1.09	4.30	1.80	8.40
Medina	3168	0.10	169.30	4.15	2.72	5.62	1.35	2.90	1.70	4.90
Meigs	77	0.10	6.40	2.03	1.44	1.45	0.72	1.70	0.90	2.90
Mercer	472	0.10	56.10	6.92	4.11	8.07	1.17	4.50	2.10	8.40
Miami	3188	0.10	252.00	8.40	4.95	12.11	1.44	5.20	2.60	10.00
Monroe	88	0.10	39.30	5.36	2.86	7.21	1.34	3.10	1.30	6.10
Montgomery	21202	0.10	440.70	5.65	3.23	10.11	1.79	3.50	1.70	6.70

County Name	No.	Min.	Max.	AM	GM	SD	CV	Md	Q1	Q3
Morgan	56	0.10	33.90	4.96	3.09	6.00	1.21	3.33	1.70	5.10
Morrow	430	0.10	77.20	7.31	3.86	10.51	1.44	4.00	2.00	7.80
Muskingum	996	0.10	131.10	7.25	3.78	12.81	1.77	3.90	2.00	7.10
Noble	38	0.10	10.30	2.63	1.65	2.36	0.90	2.05	1.00	3.40
Ottawa	531	0.10	320.00	5.04	2.62	14.61	2.90	3.10	1.20	5.70
Paulding	72	0.10	13.20	3.52	2.07	3.58	1.02	2.40	1.10	3.50
Perry	263	0.10	274.50	9.34	3.69	27.89	2.99	3.70	2.00	6.90
Pickaway	664	0.10	102.50	12.06	7.57	13.52	1.12	7.60	4.00	14.50
Pike	1008	0.10	74.90	6.57	2.97	8.94	1.36	4.10	1.30	8.20
Portage	1810	0.10	70.10	3.98	2.38	4.88	1.23	2.50	1.25	4.80
Preble	664	0.10	90.30	7.49	4.26	9.01	1.20	4.60	2.20	9.50
Putnam	265	0.10	52.50	6.09	3.86	6.61	1.09	4.10	2.20	6.80
Richland	2790	0.10	318.80	9.07	4.34	19.28	2.13	4.35	2.20	8.40
Ross	1236	0.10	223.40	9.84	4.94	17.34	1.76	5.70	2.50	10.40
Sandusky	600	0.10	116.10	6.57	3.78	9.62	1.46	4.20	2.00	7.50
Scioto	314	0.10	29.50	3.47	1.97	4.29	1.24	2.00	1.10	3.80
Seneca	1823	0.10	187.90	7.77	4.14	11.72	1.51	4.80	2.20	9.20
Shelby	805	0.10	111.10	7.68	4.61	8.71	1.13	5.50	2.50	9.60
Stark	5067	0.10	580.00	7.58	3.92	16.12	2.13	4.20	2.00	8.40
Summit	9702	0.08	908.10	6.26	2.58	34.36	5.49	2.70	1.40	5.00
Trumbull	1348	0.10	71.70	2.85	1.82	3.88	1.36	1.90	1.10	3.50
Tuscarawas	1086	0.10	208.70	9.37	4.60	16.36	1.75	4.80	2.30	10.00
Union	1373	0.10	63.50	5.36	3.30	6.20	1.16	3.70	1.90	6.60
Van Wert	109	0.10	24.50	5.77	3.84	5.07	0.88	4.70	2.20	7.30
Vinton	32	0.10	16.35	2.97	1.65	3.69	1.24	1.75	0.70	3.20
Warren	6328	0.10	219.00	4.89	3.10	5.99	1.22	3.45	1.70	6.10
Washington	748	0.10	91.90	4.74	2.65	7.39	1.56	2.80	1.40	5.00
Wayne	1507	0.10	498.40	11.16	4.22	30.24	2.71	3.90	2.00	8.80
Williams	173	0.10	67.80	5.79	3.01	8.59	1.48	2.90	1.40	6.50
Wood	3883	0.10	145.50	4.19	2.28	5.73	1.37	2.60	1.10	5.30
Wyandot	486	0.10	49.17	5.42	3.21	7.23	1.33	3.35	1.80	5.50
Unknown	748	182.30	10.55	2.70	0.10	2.90	1.07	1.30	6.20	4.90
All of Ohio	219114	0.10	1400.00	7.34	3.49	18.68	2.55	3.7	1.7	7.5

Figure 2: Number of Ohio homes tested for ^{222}Rn in each county.

Figure 3a: GM ^{222}Rn concentrations in Ohio counties based on the WHO-USEPA classification.

Figure 3b: GM ^{222}Rn concentrations in Ohio counties based on the USEPA classification.

Figure 3c: GM ^{222}Rn concentration in Ohio counties based on the 2 pCi/L breakdown classifications.

Table 2: ^{222}Rn concentration distribution (% variation) in Ohio counties

County Name	≤ 2.7 pCi/L	2.7 - 4 pCi/L	4 - 10 pCi/L	10 - 100 pCi/L	≥ 100 pCi/L	Count
Adams	55.78	11.56	17.69	14.97	0.00	147
Allen	35.47	12.68	35.63	16.21	0.00	623
Ashland	34.70	13.91	33.38	16.25	1.76	683
Ashtabula	70.44	8.53	13.08	7.71	0.23	856
Athens	36.27	17.42	36.99	9.32	0.00	976
Auglaize	30.93	12.56	32.79	23.72	0.00	430
Belmont	43.21	16.90	25.83	12.98	1.07	840
Brown	66.10	10.17	20.34	3.39	0.00	118
Butler	41.80	14.83	30.81	11.36	1.19	7033
Carroll	36.49	13.40	24.95	23.30	1.86	485
Champaign	29.67	12.19	31.32	26.31	0.51	1756
Clark	29.53	13.05	33.52	23.79	0.12	3434
Clermont	40.96	17.77	31.06	10.15	0.06	3191
Clinton	43.16	16.02	30.66	10.16	0.00	512
Columbiana	36.47	14.69	26.79	21.43	0.61	1157
Coshocton	28.66	10.51	28.66	31.21	0.96	314
Crawford	35.50	15.15	35.06	13.96	0.32	924
Cuyahoga	67.61	13.36	15.46	3.47	0.10	14661
Darke	29.23	11.79	33.38	25.35	0.25	797
Defiance	50.79	17.46	26.98	4.76	0.00	252
Delaware	25.40	14.01	40.10	20.37	0.12	6603
Erie	35.10	14.76	32.31	17.55	0.28	1755
Fairfield	29.39	10.07	31.33	28.74	0.46	2164
Fayette	39.31	13.79	36.55	9.66	0.69	145
Franklin	20.92	11.29	37.22	30.41	0.18	31909
Fulton	47.94	16.19	27.62	8.25	0.00	315
Gallia	61.39	18.99	18.99	0.63	0.00	158
Geauga	67.13	15.39	14.91	2.40	0.18	1670
Greene	29.94	14.43	34.52	21.05	0.06	8231
Guernsey	51.43	10.00	26.43	11.43	0.71	280
Hamilton	56.03	16.36	22.38	5.22	0.01	14584
Hancock	40.60	15.20	30.74	13.28	0.17	1197

County Name	≤ 2.7 pCi/L	2.7 - 4 pCi/L	4 - 10 pCi/L	10 - 100 pCi/L	≥ 100 pCi/L	Count
Hardin	46.74	16.30	28.26	8.70	0.00	184
Harrison	28.89	13.33	20.00	34.07	3.70	135
Henry	47.10	18.06	27.74	7.10	0.00	155
Highland	46.03	13.23	33.33	7.41	0.00	189
Hocking	36.19	9.34	26.85	26.46	1.17	257
Holmes	38.00	11.33	26.67	22.67	1.33	150
Huron	31.23	14.88	28.75	22.77	2.37	887
Jackson	56.85	15.07	19.18	8.90	0.00	146
Jefferson	37.20	16.99	27.31	16.77	1.72	465
Knox	27.68	9.36	24.95	31.74	6.27	4165
Lake	61.87	14.67	16.64	6.79	0.03	2903
Lawrence	66.85	12.92	15.17	4.49	0.56	178
Licking	17.10	9.78	28.96	41.85	2.32	10798
Logan	23.68	10.94	34.27	30.89	0.23	887
Lorain	45.35	16.22	28.83	9.58	0.02	4926
Lucas	56.53	14.23	22.72	6.45	0.06	4960
Madison	24.25	13.35	37.93	24.47	0.00	899
Mahoning	61.13	18.25	18.29	2.15	0.17	2367
Marion	36.39	11.77	32.52	19.32	0.00	1113
Medina	45.49	21.81	26.55	6.12	0.03	3168
Meigs	66.23	23.38	10.39	0.00	0.00	77
Mercer	30.72	11.65	38.56	19.07	0.00	472
Miami	25.28	13.43	36.17	24.91	0.22	3188
Monroe	47.73	10.23	30.68	11.36	0.00	88
Montgomery	39.35	15.53	31.35	13.65	0.13	21202
Morgan	39.29	23.21	26.79	10.71	0.00	56
Morrow	34.42	14.65	32.79	18.14	0.00	430
Muskingum	34.44	15.76	34.34	15.06	0.40	996
Noble	57.89	23.68	15.79	2.63	0.00	38
Ottawa	44.63	14.69	31.07	9.42	0.19	531
Paulding	55.56	19.44	13.89	11.11	0.00	72
Perry	35.36	20.91	31.18	9.89	2.66	263
Pickaway	13.86	10.54	36.30	39.16	0.15	664

County Name	≤ 2.7 pCi/L	2.7 - 4 pCi/L	4 - 10 pCi/L	10 - 100 pCi/L	≥ 100 pCi/L	Count
Pike	37.80	11.01	32.84	18.35	0.00	1008
Portage	52.21	15.69	24.75	7.35	0.00	1810
Preble	31.17	13.40	32.08	23.34	0.00	664
Putnam	32.45	15.85	35.85	15.85	0.00	265
Richland	31.22	15.30	32.37	20.36	0.75	2790
Ross	26.13	10.28	37.06	25.97	0.57	1236
Sandusky	32.67	13.33	36.67	17.00	0.33	600
Scioto	58.60	18.47	17.20	5.73	0.00	314
Seneca	31.27	12.34	33.41	22.71	0.27	1823
Shelby	25.96	14.29	35.53	24.10	0.12	805
Stark	32.41	15.16	32.76	19.30	0.37	5067
Summit	49.27	17.24	25.22	7.78	0.48	9702
Trumbull	64.91	14.76	18.10	2.23	0.00	1348
Tuscarawas	29.47	14.64	30.85	24.22	0.83	1086
Union	36.49	16.68	34.60	12.24	0.00	1373
Van Wert	34.86	11.01	40.37	13.76	0.00	109
Vinton	65.63	15.63	12.50	6.25	0.00	32
Warren	40.23	15.77	33.75	10.21	0.03	6328
Washington	47.99	17.65	24.06	10.29	0.00	748
Wayne	33.51	17.25	27.41	20.24	1.59	1507
Williams	48.55	15.03	20.23	16.18	0.00	173
Wood	50.66	13.73	26.40	9.17	0.05	3883
Wyandot	40.12	19.34	28.81	11.73	0.00	486
All of Ohio	*38.69*	*14.10*	*29.54*	*17.20*	*0.46*	*218366*

Note: The 748 data points for unknown counties were not considered.

2.3. DISTRIBUTION OF INDOOR ^{222}RN CONCENTRATIONS IN OHIO – A ZIP CODE BASED ANALYSIS

^{222}Rn concentrations were observed in 1,450 Ohio zip codes. Table **3** presents a summary of the computed ^{222}Rn statistics for the 1,450 zip codes in Ohio. There are a total of 641 zip codes with GM ^{222}Rn concentrations less than 2.7 pCi/L, 352 zip codes with GM ^{222}Rn concentrations between 2.7 and 4 pCi/L, 457 zip codes with ^{222}Rn concentrations greater than 4 pCi/L; thereby, indicating that 56% and

32% of the monitored Ohio zip codes had the GM ^{222}Rn concentrations exceeding the WHO and the USEPA action levels, respectively. Of the 457 zip codes that had the GM ^{222}Rn concentrations greater than 4 pCi/L, 96 zip codes had the GM ^{222}Rn concentrations exceeding 8 pCi/L and 13 zip codes had the GM ^{222}Rn concentrations exceeding 20 pCi/L. Of the monitored zip codes with at least five observations, zip code 43916 had the highest GM ^{222}Rn concentration (25.91 pCi/L), followed by the zip codes 43988 (21.87 pCi/L), 43216 (19.5 pCi/L), and 43033 (13.89 pCi/L). Figs. **4a**, **4b**, and **4c** illustrate the graphical representations of ^{222}Rn GM concentrations in Ohio zip codes based on the WHO-USEPA, the USEPA, and the 2 pCi/L breakdown classifications, respectively.

Table 3: ^{222}Rn statistics for Ohio zip codes

Zip Code	No.	Min.	Max.	AM	GM	SD	CV	Md	Q1	Q3
43001	168	0.10	317.70	19.53	6.47	52.61	2.69	5.75	2.80	11.80
43002	11	2.70	40.10	12.21	7.82	12.41	0.64	6.40	3.20	21.20
43003	44	0.10	50.10	6.10	2.83	9.28	0.48	2.25	1.40	7.20
43004	469	0.10	126.50	11.50	6.10	16.93	0.87	6.00	3.20	12.20
43005	6	0.90	14.00	7.73	5.46	5.38	0.28	8.50	2.90	11.60
43006	4	1.00	15.90	10.35	6.83	7.12	0.36	12.25	1.00	15.90
43007	5	1.10	8.30	5.06	4.05	3.16	0.16	3.80	3.80	8.30
43008	34	0.10	14.30	2.76	1.55	3.04	0.16	2.05	1.00	3.90
43009	63	0.60	106.50	14.13	6.68	21.32	1.09	7.50	2.90	12.90
43010	1	3.20	3.20	3.20	3.20	0.00	0.00	3.20	3.20	3.20
43011	154	0.10	61.70	8.61	5.53	9.87	0.51	5.80	3.50	9.20
43013	41	0.40	25.40	7.33	4.73	6.65	0.34	5.60	2.50	10.30
43014	120	0.10	239.30	19.33	6.06	39.09	2.00	5.50	2.20	19.50
43015	1518	0.10	116.50	7.57	4.65	9.40	0.48	5.20	2.70	9.10
43016	1225	0.10	137.20	7.83	4.57	8.55	0.44	5.50	2.30	10.40
43017	2855	0.10	193.70	10.78	5.93	13.25	0.68	6.60	3.20	13.40
43018	5	0.10	19.90	6.82	1.64	8.61	0.44	2.70	0.20	11.20
43019	410	0.10	189.50	9.45	3.38	19.71	1.01	3.00	1.10	9.60
43020	2	3.60	42.50	23.05	12.37	27.51	1.41	23.05	3.60	42.50
43021	591	0.10	31.30	6.47	4.34	5.38	0.28	4.80	2.90	8.60
43022	254	0.10	553.70	29.08	7.58	67.88	3.48	6.35	2.60	23.10
43023	3339	0.10	1400.00	20.24	8.16	44.70	2.29	8.20	3.60	18.60

Zip Code	No.	Min.	Max.	AM	GM	SD	CV	Md	Q1	Q3
43025	283	0.10	684.50	13.08	5.65	50.58	2.59	5.40	3.10	9.60
43026	1847	0.10	151.00	9.33	5.69	9.99	0.51	6.70	3.40	12.30
43027	2	4.90	5.10	5.00	5.00	0.14	0.01	5.00	4.90	5.10
43028	1568	0.10	844.20	39.53	11.00	69.26	3.55	9.35	3.20	43.90
43029	18	0.20	13.30	5.46	3.63	3.47	0.18	5.80	3.90	7.30
43030	9	3.60	40.80	14.72	10.44	12.78	0.65	8.70	4.70	18.20
43031	398	0.10	49.90	6.72	4.21	7.01	0.36	4.60	2.40	8.00
43032	9	0.10	9.00	3.49	1.95	2.64	0.13	3.30	2.70	4.00
43033	11	1.20	145.00	30.15	13.89	40.76	2.09	19.80	3.40	39.00
43035	719	0.10	101.90	6.59	4.16	7.24	0.37	4.90	2.80	8.40
43036	6	1.40	10.20	6.83	5.86	2.98	0.15	7.55	6.10	8.20
43037	4	1.00	1.20	1.10	1.10	0.12	0.01	1.10	1.00	1.20
43040	1063	0.10	63.50	5.28	3.28	6.13	0.31	3.70	2.00	6.50
43042	2	0.20	2.70	1.45	0.73	1.77	0.09	1.45	0.20	2.70
43044	110	0.10	41.50	7.51	4.32	7.59	0.39	4.90	2.00	10.00
43045	87	0.10	26.60	6.05	4.00	5.17	0.26	5.00	2.40	7.40
43046	53	0.90	29.00	5.24	3.60	5.27	0.27	3.60	1.80	6.00
43047	1	15.30	15.30	15.30	15.30	0.00	0.00	15.30	15.30	15.30
43050	1677	0.10	742.70	14.53	5.48	32.72	1.68	5.20	2.10	14.30
43051	1	8.50	8.50	8.50	8.50	0.00	0.00	8.50	8.50	8.50
43052	1	19.20	19.20	19.20	19.20	0.00	0.00	19.20	19.20	19.20
43053	1	7.30	7.30	7.30	7.30	0.00	0.00	7.30	7.30	7.30
43054	1061	0.10	60.10	6.34	4.00	6.70	0.34	4.70	2.40	8.10
43055	4170	0.10	559.00	20.63	10.09	32.49	1.66	11.10	4.70	24.60
43056	1060	0.10	191.00	18.28	8.92	23.76	1.22	9.90	3.90	23.20
43057	2	10.20	10.20	10.20	10.20	0.00	0.00	10.20	10.20	10.20
43058	46	1.50	125.10	17.45	9.00	27.63	1.41	6.50	4.70	12.60
43060	51	0.20	62.70	8.24	4.51	11.03	0.56	5.00	2.20	9.40
43061	133	0.10	29.80	6.10	3.97	5.38	0.28	4.40	2.30	8.10
43062	910	0.10	470.00	9.39	5.00	21.67	1.11	5.20	2.90	9.90
43063	1	8.20	8.20	8.20	8.20	0.00	0.00	8.20	8.20	8.20
43064	329	0.50	70.40	8.28	5.12	10.09	0.52	4.90	2.80	9.70
43065	1938	0.10	735.00	7.62	4.65	18.64	0.95	5.38	2.70	8.90
43066	75	0.10	80.70	6.83	2.45	13.38	0.68	3.30	1.00	5.30

Zip Code	No.	Min.	Max.	AM	GM	SD	CV	Md	Q1	Q3
43067	29	0.30	18.20	3.94	2.46	4.47	0.23	2.60	1.50	5.40
43068	951	0.10	180.50	8.60	5.03	12.12	0.62	5.70	3.00	9.70
43069	1	1.40	1.40	1.40	1.40	0.00	0.00	1.40	1.40	1.40
43070	34	0.10	4.70	1.26	0.83	1.10	0.06	1.05	0.40	1.60
43071	95	0.10	160.80	14.57	6.83	22.37	1.15	5.50	2.90	15.40
43072	217	0.10	46.00	4.33	2.01	6.95	0.36	2.00	1.00	4.40
43073	3	2.50	17.50	7.50	4.78	8.66	0.44	2.50	2.50	17.50
43074	547	0.10	75.70	6.52	3.49	8.28	0.42	4.10	2.10	7.40
43075	3	1.30	15.30	6.93	4.37	7.39	0.38	4.20	1.30	15.30
43076	168	0.20	274.50	13.41	4.40	36.30	1.86	4.15	2.50	7.60
43077	4	0.10	5.40	2.23	1.12	2.25	0.12	1.70	0.10	1.70
43078	719	0.10	265.30	9.30	5.34	14.18	0.73	5.70	2.90	11.10
43080	204	0.10	146.30	14.43	6.40	24.59	1.26	6.50	2.70	15.70
43081	2239	0.10	114.80	8.13	4.92	8.99	0.46	6.00	2.90	10.10
43082	1068	0.10	58.80	7.37	4.58	6.90	0.35	5.70	2.80	9.50
43084	12	0.10	18.60	5.54	2.53	5.78	0.30	4.20	1.10	6.40
43085	2323	0.10	369.90	12.36	7.48	15.83	0.81	8.60	4.20	15.90
43086	9	1.00	5.70	2.82	2.39	1.61	0.08	2.40	1.40	3.60
43087	2	0.40	6.90	3.65	1.66	4.60	0.24	3.65	0.40	6.90
43088	2	5.90	11.30	8.60	8.17	3.82	0.20	8.60	5.90	11.30
43089	1	2.90	2.90	2.90	2.90	0.00	0.00	2.90	2.90	2.90
43093	5	0.90	1.20	1.08	1.07	0.13	0.01	1.10	1.00	1.20
43101	1	2.10	2.10	2.10	2.10	0.00	0.00	2.10	2.10	2.10
43102	18	0.20	33.40	8.13	3.79	9.88	0.51	4.80	1.20	13.80
43103	159	0.10	102.50	15.65	8.55	18.32	0.94	9.90	3.80	21.00
43105	261	0.10	44.50	4.63	1.80	7.08	0.36	1.80	0.70	5.40
43106	7	0.10	36.70	11.60	4.23	13.47	0.69	5.00	1.80	21.60
43107	32	0.80	260.00	16.24	5.71	46.20	2.37	4.80	2.60	9.80
43109	4	3.40	20.50	9.33	7.48	7.61	0.39	6.70	3.40	6.80
43110	438	0.10	112.00	10.56	6.45	11.49	0.59	7.40	3.80	13.20
43111	1	21.20	21.20	21.20	21.20	0.00	0.00	21.20	21.20	21.20
43112	130	0.10	238.50	12.25	6.33	22.96	1.18	6.85	3.40	12.80
43113	342	0.20	86.60	11.96	7.89	12.53	0.64	7.80	4.50	14.20
43114	2	15.30	15.30	15.30	15.30	0.00	0.00	15.30	15.30	15.30

Zip Code	No.	Min.	Max.	AM	GM	SD	CV	Md	Q1	Q3
43115	28	0.10	16.10	4.16	2.58	3.78	0.19	3.25	1.70	4.90
43116	10	3.30	12.50	7.39	6.75	3.16	0.16	7.80	4.00	8.50
43117	5	2.50	8.60	5.58	5.15	2.31	0.12	5.20	4.70	6.90
43119	393	0.10	63.90	8.33	5.44	7.39	0.38	6.10	3.40	11.20
43123	920	0.10	91.10	8.02	5.57	7.61	0.39	6.25	3.50	10.30
43124	1	9.00	9.00	9.00	9.00	0.00	0.00	9.00	9.00	9.00
43125	253	0.10	94.80	13.06	8.20	13.75	0.70	10.00	4.50	16.20
43126	6	3.10	18.50	10.08	8.36	6.17	0.32	9.45	5.20	14.80
43127	2	1.60	15.20	8.40	4.93	9.62	0.49	8.40	1.60	15.20
43128	17	0.50	300.00	21.38	3.50	71.87	3.68	3.10	1.90	7.00
43129	2	1.00	10.40	5.70	3.22	6.65	0.34	5.70	1.00	10.40
43130	700	0.10	340.50	10.97	5.97	21.42	1.10	7.00	3.20	12.00
43132	1	7.20	7.20	7.20	7.20	0.00	0.00	7.20	7.20	7.20
43135	26	0.80	93.20	11.76	6.68	18.43	0.94	5.25	3.40	10.20
43136	9	3.80	23.80	14.68	13.15	6.00	0.31	15.70	12.00	18.10
43137	21	0.90	100.50	24.05	13.41	27.44	1.41	12.90	5.70	27.80
43138	169	0.10	99.50	6.96	3.63	9.68	0.50	4.20	1.50	10.10
43140	397	0.10	82.10	7.87	4.88	9.39	0.48	5.20	2.50	9.90
43141	4	2.90	8.80	6.20	5.57	3.06	0.16	6.55	2.90	8.80
43142	1	9.60	9.60	9.60	9.60	0.00	0.00	9.60	9.60	9.60
43143	58	0.10	28.50	5.21	3.37	4.88	0.25	4.20	1.60	7.00
43144	1	23.30	23.30	23.30	23.30	0.00	0.00	23.30	23.30	23.30
43145	4	2.90	41.10	16.88	11.32	16.68	0.85	11.75	2.90	12.00
43146	143	0.50	43.10	9.05	6.17	8.39	0.43	5.70	3.40	12.30
43147	1316	0.10	110.10	9.34	5.41	9.88	0.51	6.70	3.10	12.10
43148	24	0.10	45.70	8.63	5.28	9.74	0.50	5.65	3.10	7.70
43149	49	0.30	238.00	21.55	6.55	47.35	2.42	5.90	2.50	17.90
43150	6	0.30	5.90	2.45	1.49	2.39	0.12	1.45	0.60	5.00
43151	4	1.00	9.30	6.30	4.75	3.72	0.19	7.45	1.00	8.40
43152	10	0.10	10.20	3.48	2.01	3.02	0.15	3.40	0.90	4.30
43153	8	0.90	10.90	5.10	4.10	3.25	0.17	4.80	2.40	5.00
43154	20	0.30	19.10	7.31	4.86	5.79	0.30	6.15	2.10	12.50
43155	21	0.80	40.20	9.25	4.28	12.24	0.63	3.20	1.60	7.90
43158	1	7.50	7.50	7.50	7.50	0.00	0.00	7.50	7.50	7.50

Zip Code	No.	Min.	Max.	AM	GM	SD	CV	Md	Q1	Q3
43160	115	0.20	38.30	4.71	2.89	5.09	0.26	3.50	1.60	6.20
43162	142	0.50	54.40	8.54	6.20	7.90	0.40	6.10	3.80	10.70
43164	9	2.30	45.80	9.80	5.82	13.87	0.71	3.70	3.40	10.30
43170	2	0.30	0.40	0.35	0.35	0.07	0.00	0.35	0.30	0.40
43173	1	3.80	3.80	3.80	3.80	0.00	0.00	3.80	3.80	3.80
43188	2	1.50	6.20	3.85	3.05	3.32	0.17	3.85	1.50	6.20
43201	357	0.10	73.60	8.37	4.67	9.78	0.50	5.80	2.20	10.10
43202	445	0.10	152.90	7.21	4.66	9.91	0.51	4.90	2.70	8.20
43203	48	0.10	47.40	5.91	1.45	9.84	0.50	2.40	0.10	5.50
43204	395	0.10	114.60	6.91	4.10	9.37	0.48	4.60	2.30	8.40
43205	59	0.40	50.40	9.50	5.41	11.53	0.59	4.70	2.20	11.50
43206	359	0.20	35.30	6.43	4.62	5.81	0.30	4.90	2.70	7.80
43207	255	0.20	125.30	11.24	6.62	15.06	0.77	6.60	3.80	11.80
43208	3	9.50	39.50	20.90	17.26	16.24	0.83	13.70	9.50	39.50
43209	1302	0.10	65.10	7.07	4.66	6.92	0.35	5.20	2.60	9.30
43210	59	0.10	30.70	5.73	2.51	6.52	0.33	2.70	1.00	9.70
43211	90	0.10	25.90	5.17	3.39	4.54	0.23	4.00	2.30	6.60
43212	589	0.10	178.20	6.17	4.09	10.38	0.53	4.40	2.40	7.20
43213	591	0.10	79.60	9.77	5.20	11.63	0.60	6.20	2.70	12.50
43214	1307	0.10	939.00	10.85	6.36	28.21	1.44	7.20	3.70	12.40
43215	318	0.10	174.80	8.95	4.33	16.92	0.87	4.70	2.30	9.20
43216	25	1.10	232.00	42.25	19.50	57.33	2.94	23.10	7.70	40.30
43217	16	0.90	16.00	4.66	3.46	3.93	0.20	2.90	1.80	5.90
43218	4	3.80	10.10	6.93	6.50	2.70	0.14	6.90	3.80	7.90
43219	129	0.10	112.20	10.80	5.67	17.05	0.87	5.30	3.20	10.70
43220	1513	0.10	68.10	9.40	6.53	8.07	0.41	7.10	3.90	12.20
43221	1877	0.10	165.90	9.36	6.30	10.31	0.53	6.60	3.70	11.70
43222	17	0.40	16.40	6.49	4.54	4.97	0.25	5.80	2.30	9.50
43223	157	0.10	25.40	5.34	3.58	4.61	0.24	4.50	2.00	6.70
43224	356	0.10	333.90	8.16	5.34	18.27	0.94	5.90	3.40	9.20
43225	3	5.60	16.00	12.30	11.11	5.81	0.30	15.30	5.60	16.00
43226	6	2.10	14.70	8.22	6.40	5.40	0.28	8.00	2.50	14.00
43227	309	0.10	76.70	9.32	6.14	8.73	0.45	7.10	3.50	11.70
43228	716	0.10	73.50	8.17	4.48	9.03	0.46	5.50	2.50	10.50

Zip Code	No.	Min.	Max.	AM	GM	SD	CV	Md	Q1	Q3
43229	854	0.10	102.80	7.75	4.95	7.69	0.39	5.80	2.80	10.40
43230	1508	0.10	417.70	11.09	5.90	17.28	0.88	6.30	3.20	13.00
43231	278	0.10	52.00	7.92	4.74	8.78	0.45	5.70	2.30	9.00
43232	579	0.10	142.80	16.55	8.08	17.83	0.91	10.10	4.10	23.50
43234	4	2.40	18.10	9.28	7.40	6.51	0.33	8.30	2.40	8.40
43235	1657	0.10	720.00	10.69	5.97	21.06	1.08	7.20	3.30	13.00
43236	3	3.90	18.80	9.60	7.65	8.04	0.41	6.10	3.90	18.80
43239	2	5.20	48.60	26.90	15.90	30.69	1.57	26.90	5.20	48.60
43240	41	0.10	16.10	5.13	3.50	3.96	0.20	4.60	2.20	6.60
43266	99	0.10	35.00	3.79	1.23	8.56	0.44	1.10	0.60	1.80
43270	1	3.70	3.70	3.70	3.70	0.00	0.00	3.70	3.70	3.70
43271	1	5.20	5.20	5.20	5.20	0.00	0.00	5.20	5.20	5.20
43279	1	10.40	10.40	10.40	10.40	0.00	0.00	10.40	10.40	10.40
43290	1	8.40	8.40	8.40	8.40	0.00	0.00	8.40	8.40	8.40
43301	2	5.30	5.30	5.30	5.30	0.00	0.00	5.30	5.30	5.30
43302	911	0.10	67.10	6.25	3.47	7.20	0.37	4.00	1.60	8.30
43303	1	1.50	1.50	1.50	1.50	0.00	0.00	1.50	1.50	1.50
43304	2	7.00	7.10	7.05	7.05	0.07	0.00	7.05	7.00	7.10
43305	2	10.20	60.80	35.50	24.90	35.78	1.83	35.50	10.20	60.80
43306	3	0.50	6.80	2.77	1.50	3.50	0.18	1.00	0.50	6.80
43310	32	0.50	29.80	7.51	5.19	6.84	0.35	4.50	2.50	9.50
43311	448	0.10	319.20	10.90	6.21	18.72	0.96	6.85	3.10	12.90
43313	3	2.90	3.70	3.23	3.22	0.42	0.02	3.10	2.90	3.70
43314	83	0.10	26.30	4.65	3.38	4.12	0.21	3.00	2.20	6.70
43315	116	0.10	77.20	7.48	3.85	12.18	0.62	4.25	1.70	7.40
43316	63	0.60	42.10	4.70	3.02	6.45	0.33	3.30	1.50	5.70
43317	4	1.40	14.80	9.68	7.13	5.87	0.30	11.25	1.40	12.60
43318	59	0.80	150.30	9.64	5.33	20.02	1.03	6.20	3.30	8.30
43319	20	0.60	20.40	3.97	2.81	4.37	0.22	2.90	1.60	4.10
43320	24	0.60	32.40	6.84	3.68	8.43	0.43	3.25	1.40	7.80
43321	8	0.20	14.60	4.96	2.25	5.10	0.26	3.30	0.20	6.20
43322	2	2.50	7.80	5.15	4.42	3.75	0.19	5.15	2.50	7.80
43323	15	1.40	31.90	5.78	3.50	8.09	0.41	2.70	1.90	4.60
43324	32	0.20	34.10	8.36	4.52	7.91	0.41	5.55	2.00	12.10

Zip Code	No.	Min.	Max.	AM	GM	SD	CV	Md	Q1	Q3
43325	2	4.90	6.40	5.65	5.60	1.06	0.05	5.65	4.90	6.40
43326	101	0.10	16.10	3.71	2.40	3.51	0.18	2.10	1.20	5.20
43327	1	2.60	2.60	2.60	2.60	0.00	0.00	2.60	2.60	2.60
43328	4	2.20	13.40	6.00	4.77	5.03	0.26	4.20	2.20	4.50
43330	2	0.90	3.10	2.00	1.67	1.56	0.08	2.00	0.90	3.10
43331	19	0.50	8.70	2.06	1.47	2.08	0.11	1.20	0.80	2.50
43332	18	2.00	30.30	12.28	8.78	8.91	0.46	11.65	4.90	19.00
43333	13	0.30	14.60	7.36	5.48	4.05	0.21	7.50	5.90	10.40
43334	94	0.10	62.90	9.01	4.68	12.88	0.66	4.30	2.30	9.40
43335	1	6.10	6.10	6.10	6.10	0.00	0.00	6.10	6.10	6.10
43336	3	1.40	3.80	2.77	2.55	1.23	0.06	3.10	1.40	3.80
43337	7	0.80	8.60	4.06	3.05	2.96	0.15	3.20	1.40	7.00
43338	168	0.10	53.10	6.52	3.51	8.51	0.44	4.05	2.00	6.70
43340	9	1.10	13.80	3.52	2.41	4.02	0.21	2.50	1.10	3.60
43341	12	1.40	19.30	7.21	5.50	5.96	0.31	4.75	4.20	5.80
43342	58	0.60	21.30	9.15	6.79	5.92	0.30	7.55	3.90	14.00
43343	18	0.10	38.80	14.47	6.01	14.88	0.76	6.50	2.40	27.00
43344	153	0.10	61.80	5.54	3.14	7.70	0.39	2.80	1.60	6.50
43345	6	1.20	6.40	3.08	2.37	2.39	0.12	2.00	1.20	5.70
43347	7	0.70	4.90	2.60	2.07	1.71	0.09	2.50	0.80	4.80
43348	9	0.70	11.00	4.33	3.01	3.61	0.18	2.80	2.70	5.80
43349	6	0.70	12.90	4.93	3.11	4.93	0.25	2.50	1.80	9.20
43350	2	5.10	10.70	7.90	7.39	3.96	0.20	7.90	5.10	10.70
43351	294	0.10	49.20	6.02	3.47	8.10	0.41	3.60	2.00	5.70
43356	33	0.70	28.80	6.16	3.94	6.15	0.32	3.90	1.80	8.00
43357	207	0.10	75.00	16.00	8.17	17.07	0.87	10.10	3.90	21.60
43358	24	0.50	12.60	4.75	3.47	3.49	0.18	4.00	1.90	7.70
43359	9	0.50	10.80	3.00	2.04	3.16	0.16	1.90	1.10	3.60
43360	66	0.40	90.20	11.66	6.23	15.94	0.82	6.20	3.10	12.70
43376	1	3.20	3.20	3.20	3.20	0.00	0.00	3.20	3.20	3.20
43397	1	4.70	4.70	4.70	4.70	0.00	0.00	4.70	4.70	4.70
43401	2	2.30	2.30	2.30	2.30	0.00	0.00	2.30	2.30	2.30
43402	1188	0.10	100.00	3.72	1.96	5.07	0.26	2.20	1.00	4.70
43403	13	0.10	4.40	1.17	0.81	1.11	0.06	0.80	0.70	1.00

Zip Code	No.	Min.	Max.	AM	GM	SD	CV	Md	Q1	Q3
43406	11	0.10	5.90	3.10	2.12	2.11	0.11	2.10	1.80	5.30
43407	6	0.10	5.10	2.55	1.57	1.87	0.10	2.10	1.50	4.40
43410	126	0.10	116.10	8.53	4.02	16.11	0.82	3.95	1.70	9.20
43412	30	0.10	260.00	13.30	3.40	46.76	2.39	4.00	1.70	7.00
43413	31	0.10	35.00	4.97	2.40	6.87	0.35	2.50	1.50	6.40
43414	8	0.50	3.00	2.15	1.76	1.05	0.05	2.55	0.50	2.80
43416	23	1.10	17.20	5.96	4.75	4.07	0.21	4.90	3.20	8.00
43417	1	5.00	5.00	5.00	5.00	0.00	0.00	5.00	5.00	5.00
43420	354	0.10	47.70	6.29	4.00	7.06	0.36	4.20	2.30	7.50
43425	1	1.80	1.80	1.80	1.80	0.00	0.00	1.80	1.80	1.80
43430	54	0.40	27.80	4.84	3.22	4.99	0.26	3.25	1.50	6.70
43431	28	0.40	17.00	5.13	3.61	4.14	0.21	4.35	1.90	6.30
43432	13	0.80	6.50	2.08	1.65	1.67	0.09	1.20	1.00	2.60
43434	1	2.80	2.80	2.80	2.80	0.00	0.00	2.80	2.80	2.80
43435	18	0.10	11.00	3.71	2.74	2.47	0.13	3.95	1.60	4.60
43437	6	0.10	31.20	6.32	1.07	12.32	0.63	0.75	0.20	4.90
43438	7	1.10	3.90	2.16	1.92	1.16	0.06	1.80	1.10	3.70
43439	1	4.50	4.50	4.50	4.50	0.00	0.00	4.50	4.50	4.50
43440	84	0.10	30.20	5.13	2.74	5.52	0.28	3.00	1.10	8.10
43441	3	0.70	1.60	1.30	1.21	0.52	0.03	1.60	0.70	1.60
43442	9	0.60	18.20	6.79	4.14	5.98	0.31	5.70	2.10	11.20
43443	40	0.20	15.30	3.85	2.58	3.26	0.17	3.20	1.30	5.10
43445	10	0.10	7.80	3.06	1.93	2.31	0.12	3.05	0.80	4.30
43446	2	12.40	12.80	12.60	12.60	0.28	0.01	12.60	12.40	12.80
43447	42	0.20	17.20	4.98	2.96	4.80	0.25	4.30	1.30	6.80
43449	59	0.10	14.10	3.08	2.00	3.00	0.15	2.20	1.00	4.20
43450	146	0.10	33.00	3.77	2.21	4.52	0.23	2.55	1.20	4.90
43451	14	0.10	7.80	2.10	1.20	2.04	0.10	1.60	0.60	3.10
43452	265	0.10	320.00	5.44	2.43	20.18	1.03	3.00	1.30	5.20
43457	10	0.10	9.13	3.91	2.26	3.52	0.18	2.45	1.10	8.70
43459	1	2.40	2.40	2.40	2.40	0.00	0.00	2.40	2.40	2.40
43460	118	0.10	25.00	4.24	2.51	4.49	0.23	2.50	1.40	5.00
43462	27	0.10	11.80	1.86	0.86	2.73	0.14	0.90	0.50	1.40
43464	7	0.20	9.00	2.14	0.98	3.17	0.16	0.70	0.30	3.00

Zip Code	No.	Min.	Max.	AM	GM	SD	CV	Md	Q1	Q3
43465	66	0.10	18.80	4.54	2.44	4.70	0.24	2.60	1.10	7.30
43466	23	0.10	9.00	1.69	1.12	1.93	0.10	1.10	0.60	1.70
43468	5	2.20	14.80	6.14	4.95	4.99	0.26	5.10	3.50	5.10
43469	28	0.10	46.10	6.34	3.65	8.48	0.43	5.05	2.00	6.90
43500	3	2.50	7.10	4.03	3.54	2.66	0.14	2.50	2.50	7.10
43501	1	1.20	1.20	1.20	1.20	0.00	0.00	1.20	1.20	1.20
43502	91	0.20	19.40	5.06	3.40	4.41	0.23	3.90	1.90	7.10
43504	14	0.50	16.60	5.57	3.72	5.06	0.26	4.20	2.40	5.40
43505	1	1.40	1.40	1.40	1.40	0.00	0.00	1.40	1.40	1.40
43506	94	0.10	53.40	6.67	3.41	8.72	0.45	3.20	1.70	7.90
43511	15	0.10	24.10	5.45	2.12	7.84	0.40	2.70	0.70	4.90
43512	211	0.10	21.20	3.26	2.12	3.24	0.17	2.40	1.20	4.30
43515	54	0.10	20.20	4.36	2.75	4.25	0.22	2.95	1.40	6.10
43516	13	0.70	6.10	2.35	2.04	1.39	0.07	2.10	1.40	2.80
43517	11	0.40	67.80	8.99	2.51	19.89	1.02	2.10	0.90	4.80
43518	9	1.00	5.50	2.70	2.29	1.52	0.08	3.00	1.10	3.30
43521	13	0.60	12.90	4.51	3.40	3.32	0.17	4.20	2.20	5.50
43522	60	0.10	27.50	4.06	2.18	5.33	0.27	2.50	1.30	4.40
43523	1	2.00	2.00	2.00	2.00	0.00	0.00	2.00	2.00	2.00
43524	2	0.10	2.10	1.10	0.46	1.41	0.07	1.10	0.10	2.10
43525	35	0.10	14.10	3.52	1.84	3.94	0.20	2.40	0.90	4.20
43526	12	0.10	5.60	3.32	2.52	1.62	0.08	3.15	2.20	4.60
43527	12	0.60	29.40	6.14	3.11	8.26	0.42	2.70	1.20	7.80
43528	306	0.10	17.50	2.63	1.85	2.57	0.13	1.95	1.10	3.00
43531	1	1.60	1.60	1.60	1.60	0.00	0.00	1.60	1.60	1.60
43532	28	0.10	10.80	2.24	0.91	2.79	0.14	1.45	0.10	2.50
43533	9	1.20	10.00	4.74	3.58	3.43	0.18	2.70	2.00	7.40
43534	8	1.90	13.50	5.75	4.78	3.87	0.20	4.65	2.50	6.10
43535	3	3.60	5.80	5.00	4.89	1.22	0.06	5.60	3.60	5.80
43537	846	0.10	125.50	5.19	2.93	6.98	0.36	3.70	1.50	6.60
43540	9	0.10	5.20	2.00	1.23	1.64	0.08	1.50	0.90	2.80
43542	138	0.10	19.60	4.08	2.40	3.53	0.18	3.35	1.30	6.10
43543	23	0.70	9.40	3.72	2.94	2.76	0.14	2.60	2.30	3.80
43545	79	0.10	19.50	3.81	2.58	3.24	0.17	3.10	1.70	5.50

Zip Code	No.	Min.	Max.	AM	GM	SD	CV	Md	Q1	Q3
43547	2	0.60	0.80	0.70	0.69	0.14	0.01	0.70	0.60	0.80
43548	10	1.50	22.90	8.33	6.26	6.81	0.35	5.45	4.20	11.60
43549	2	4.40	4.60	4.50	4.50	0.14	0.01	4.50	4.40	4.60
43550	3	5.60	12.90	10.47	9.77	4.21	0.22	12.90	5.60	12.90
43551	1875	0.10	145.50	4.68	2.62	6.87	0.35	3.00	1.30	5.80
43552	3	1.00	7.80	3.27	1.98	3.93	0.20	1.00	1.00	7.80
43553	5	1.50	21.90	8.00	4.77	8.67	0.44	3.80	1.80	11.00
43554	10	0.50	2.50	1.26	1.12	0.60	0.03	1.30	0.70	1.50
43556	9	0.40	5.30	2.50	1.77	1.89	0.10	2.20	0.90	4.10
43557	14	0.40	18.70	5.40	2.92	5.64	0.29	4.15	1.00	7.10
43558	85	0.10	20.40	2.75	1.61	3.26	0.17	1.50	0.90	3.30
43560	816	0.10	70.30	3.30	1.99	5.13	0.26	2.10	1.10	3.80
43565	6	0.50	15.90	7.62	4.52	6.76	0.35	5.50	2.40	15.90
43566	268	0.10	71.20	5.36	2.94	6.55	0.34	3.75	1.40	6.80
43567	59	0.10	16.20	3.79	2.20	3.56	0.18	2.70	1.20	5.20
43569	44	0.10	18.10	2.57	1.31	3.77	0.19	1.90	0.60	2.70
43570	16	0.70	22.90	6.53	3.81	6.70	0.34	4.50	1.10	7.50
43571	107	0.10	18.80	3.28	1.97	3.45	0.18	1.90	1.10	4.60
43585	3	6.90	9.60	8.70	8.60	1.56	0.08	9.60	6.90	9.60
43590	1	1.10	1.10	1.10	1.10	0.00	0.00	1.10	1.10	1.10
43595	1	4.70	4.70	4.70	4.70	0.00	0.00	4.70	4.70	4.70
43601	2	0.60	11.20	5.90	2.59	7.50	0.38	5.90	0.60	11.20
43602	4	0.60	5.70	2.33	1.59	2.34	0.12	1.50	0.60	2.10
43604	42	0.10	13.80	2.76	1.29	3.36	0.17	1.15	0.50	3.50
43605	58	0.10	12.10	3.29	2.09	2.80	0.14	2.55	1.00	5.00
43606	310	0.10	25.50	2.38	1.58	2.61	0.13	1.75	0.90	2.90
43607	46	0.10	17.70	3.39	1.96	4.09	0.21	2.35	1.00	3.60
43608	93	0.10	18.60	4.48	3.28	3.33	0.17	4.00	2.30	5.80
43609	49	0.30	26.20	2.79	1.77	4.13	0.21	1.80	0.90	2.80
43610	11	0.20	6.00	2.29	1.57	1.84	0.09	1.80	1.10	3.10
43611	113	0.10	24.40	4.67	2.63	4.70	0.24	3.20	1.30	6.90
43612	154	0.10	15.50	2.54	1.75	2.37	0.12	1.80	1.00	3.20
43613	209	0.10	19.20	1.86	1.23	2.20	0.11	1.30	0.70	2.20
43614	461	0.10	103.90	5.04	2.65	8.20	0.42	3.00	1.50	5.80

Zip Code	No.	Min.	Max.	AM	GM	SD	CV	Md	Q1	Q3
43615	362	0.10	16.40	2.61	1.64	2.60	0.13	1.60	0.90	3.40
43616	139	0.10	46.30	4.81	2.87	5.52	0.28	3.30	1.50	6.80
43617	156	0.10	26.10	2.10	1.64	2.28	0.12	1.70	1.20	2.50
43618	13	0.90	10.80	6.25	4.86	3.55	0.18	7.20	4.00	8.80
43619	88	0.10	18.70	4.96	2.81	4.54	0.23	3.65	1.30	6.80
43620	10	0.40	9.40	3.38	2.32	2.99	0.15	2.25	1.20	6.20
43621	1	3.50	3.50	3.50	3.50	0.00	0.00	3.50	3.50	3.50
43623	193	0.10	50.80	2.08	1.37	3.80	0.19	1.40	1.00	2.50
43624	11	1.40	4.00	2.45	2.35	0.72	0.04	2.40	1.80	2.80
43628	1	0.50	0.50	0.50	0.50	0.00	0.00	0.50	0.50	0.50
43635	1	6.60	6.60	6.60	6.60	0.00	0.00	6.60	6.60	6.60
43637	1	0.70	0.70	0.70	0.70	0.00	0.00	0.70	0.70	0.70
43641	1	0.10	0.10	0.10	0.10	0.00	0.00	0.10	0.10	0.10
43650	2	1.30	2.90	2.10	1.94	1.13	0.06	2.10	1.30	2.90
43652	1	3.10	3.10	3.10	3.10	0.00	0.00	3.10	3.10	3.10
43655	1	27.90	27.90	27.90	27.90	0.00	0.00	27.90	27.90	27.90
43662	1	3.80	3.80	3.80	3.80	0.00	0.00	3.80	3.80	3.80
43690	1	2.80	2.80	2.80	2.80	0.00	0.00	2.80	2.80	2.80
43694	1	1.30	1.30	1.30	1.30	0.00	0.00	1.30	1.30	1.30
43697	1	2.50	2.50	2.50	2.50	0.00	0.00	2.50	2.50	2.50
43699	7	0.10	1.00	0.29	0.20	0.32	0.02	0.20	0.10	0.30
43701	584	0.10	118.30	6.54	3.73	10.79	0.55	3.80	2.00	6.90
43702	13	0.20	19.10	4.56	2.96	4.78	0.24	3.70	2.30	4.50
43713	48	0.50	17.90	5.44	3.48	4.91	0.25	3.95	1.40	7.40
43716	39	0.60	51.20	8.27	4.27	10.97	0.56	4.70	1.50	9.40
43718	62	0.10	43.20	5.62	2.82	8.11	0.42	2.90	1.10	5.40
43719	9	1.30	14.30	5.62	4.28	4.70	0.24	4.00	2.60	5.00
43720	4	1.10	5.30	3.70	3.15	1.95	0.10	4.20	1.10	5.10
43721	7	2.00	5.90	3.60	3.32	1.57	0.08	3.00	2.30	5.00
43723	24	0.10	28.10	5.68	2.73	7.63	0.39	2.15	1.30	6.10
43724	32	0.10	10.30	2.79	1.68	2.51	0.13	2.20	1.00	3.80
43725	187	0.10	108.10	6.04	3.16	10.31	0.53	2.90	1.40	7.30
43727	10	0.50	7.10	3.68	2.67	2.60	0.13	3.65	1.10	6.70
43728	3	1.70	5.00	2.90	2.57	1.82	0.09	2.00	1.70	5.00

Zip Code	No.	Min.	Max.	AM	GM	SD	CV	Md	Q1	Q3
43730	8	0.60	15.30	4.09	2.30	4.88	0.25	3.00	0.70	4.60
43731	15	0.60	6.50	3.37	2.82	1.86	0.10	2.40	2.10	5.00
43732	13	0.10	8.60	2.93	1.66	3.04	0.16	1.70	0.90	3.90
43734	8	0.80	13.90	4.96	3.38	4.74	0.24	3.45	1.70	3.70
43735	5	2.60	14.20	8.82	7.47	5.00	0.26	7.20	6.40	13.70
43738	2	1.50	11.70	6.60	4.19	7.21	0.37	6.60	1.50	11.70
43739	19	2.00	19.40	7.67	5.78	5.56	0.28	5.50	2.50	12.30
43740	4	12.20	12.30	12.25	12.25	0.06	0.00	12.25	12.20	12.30
43745	1	17.00	17.00	17.00	17.00	0.00	0.00	17.00	17.00	17.00
43746	7	1.30	27.00	6.43	3.58	9.23	0.47	2.60	1.70	5.70
43747	5	0.40	6.10	3.74	2.11	3.01	0.15	5.60	0.50	6.10
43748	20	0.10	11.10	2.66	1.68	2.51	0.13	1.90	1.00	3.70
43749	12	0.10	115.10	12.85	2.11	32.85	1.68	1.40	1.00	2.60
43750	2	3.80	9.10	6.45	5.88	3.75	0.19	6.45	3.80	9.10
43752	1	1.10	1.10	1.10	1.10	0.00	0.00	1.10	1.10	1.10
43754	4	0.10	3.90	1.73	0.90	1.67	0.09	1.45	0.10	2.10
43755	6	0.10	5.70	2.32	1.17	2.19	0.11	2.15	0.40	3.40
43756	24	0.10	12.40	3.33	2.44	2.57	0.13	3.00	1.30	3.80
43758	12	0.20	11.60	4.67	3.07	3.74	0.19	3.23	1.50	6.20
43759	5	2.00	2.60	2.40	2.39	0.24	0.01	2.40	2.40	2.60
43760	10	0.50	3.10	1.89	1.59	0.97	0.05	2.10	0.70	2.40
43761	2	1.30	3.80	2.55	2.22	1.77	0.09	2.55	1.30	3.80
43762	112	0.10	129.00	6.70	3.21	13.62	0.70	3.45	1.60	5.70
43764	43	0.80	15.80	4.95	3.73	3.80	0.19	3.50	1.90	7.50
43766	4	1.30	2.40	1.80	1.75	0.47	0.02	1.75	1.30	1.90
43767	12	1.20	58.50	12.53	5.21	19.61	1.00	4.05	1.40	8.70
43768	4	1.90	2.90	2.20	2.17	0.48	0.02	2.00	1.90	2.10
43771	8	0.20	15.40	6.56	3.35	6.33	0.32	3.30	1.10	13.40
43772	13	0.10	8.40	2.91	1.37	3.02	0.15	1.30	0.90	4.90
43773	5	0.60	9.20	3.28	2.21	3.42	0.18	2.10	1.50	3.00
43777	5	0.60	7.50	4.80	3.56	2.94	0.15	5.10	3.30	7.50
43778	3	1.70	5.90	3.70	3.27	2.11	0.11	3.50	1.70	5.90
43779	2	1.50	2.70	2.10	2.01	0.85	0.04	2.10	1.50	2.70
43780	9	0.50	4.60	1.58	1.04	1.73	0.09	0.70	0.60	1.30

Zip Code	No.	Min.	Max.	AM	GM	SD	CV	Md	Q1	Q3
43782	3	0.60	9.10	6.27	3.68	4.91	0.25	9.10	0.60	9.10
43783	25	1.00	42.00	6.28	4.12	8.35	0.43	3.70	2.70	6.20
43784	1	0.20	0.20	0.20	0.20	0.00	0.00	0.20	0.20	0.20
43787	11	0.70	33.90	7.81	4.02	10.14	0.52	3.40	1.70	11.60
43788	2	1.00	1.30	1.15	1.14	0.21	0.01	1.15	1.00	1.30
43789	1	3.80	3.80	3.80	3.80	0.00	0.00	3.80	3.80	3.80
43793	24	0.10	8.60	3.33	2.38	2.36	0.12	2.80	1.30	4.50
43802	4	0.80	6.90	3.40	2.62	2.57	0.13	2.95	0.80	3.40
43804	2	0.80	2.50	1.65	1.41	1.20	0.06	1.65	0.80	2.50
43805	1	2.30	2.30	2.30	2.30	0.00	0.00	2.30	2.30	2.30
43811	12	0.40	51.70	15.86	7.25	18.56	0.95	6.60	2.50	23.60
43812	180	0.10	95.60	11.78	5.17	18.64	0.95	4.90	2.20	12.10
43820	2	4.00	6.50	5.25	5.10	1.77	0.09	5.25	4.00	6.50
43821	30	0.20	14.60	4.16	2.90	3.15	0.16	3.65	1.70	6.40
43822	84	0.20	131.10	11.75	5.70	18.43	0.94	6.55	3.10	12.90
43824	27	0.70	62.50	25.70	11.75	24.91	1.28	10.30	4.00	55.30
43827	2	3.60	3.60	3.60	3.60	0.00	0.00	3.60	3.60	3.60
43828	1	8.80	8.80	8.80	8.80	0.00	0.00	8.80	8.80	8.80
43830	129	0.20	129.70	9.20	4.00	17.16	0.88	4.00	1.60	9.30
43832	26	0.20	20.40	6.77	3.22	7.41	0.38	2.90	2.40	11.50
43835	1	0.70	0.70	0.70	0.70	0.00	0.00	0.70	0.70	0.70
43837	16	0.80	40.30	9.74	5.88	10.43	0.53	6.35	2.60	12.20
43840	9	1.10	10.50	5.31	3.97	3.71	0.19	3.80	2.40	7.30
43842	1	1.90	1.90	1.90	1.90	0.00	0.00	1.90	1.90	1.90
43843	11	1.60	60.50	10.39	4.74	17.77	0.91	3.10	2.00	8.60
43844	17	0.60	117.70	25.59	9.82	35.85	1.84	14.50	3.60	22.20
43845	29	0.20	40.80	10.46	6.68	8.42	0.43	10.10	4.40	13.20
43885	2	12.50	12.50	12.50	12.50	0.00	0.00	12.50	12.50	12.50
43901	5	1.10	8.90	4.74	3.46	3.60	0.18	4.20	1.50	8.00
43902	2	0.80	3.10	1.95	1.57	1.63	0.08	1.95	0.80	3.10
43903	19	0.10	23.00	3.58	1.85	5.22	0.27	1.60	1.00	4.10
43905	5	0.90	13.30	4.80	3.28	4.90	0.25	3.80	2.20	3.80
43906	57	0.80	91.30	8.52	5.06	13.51	0.69	5.00	2.70	9.10
43907	43	1.20	42.60	9.52	5.37	11.78	0.60	4.60	2.70	10.10

Zip Code	No.	Min.	Max.	AM	GM	SD	CV	Md	Q1	Q3
43908	3	1.90	3.50	2.47	2.37	0.90	0.05	2.00	1.90	3.50
43909	2	2.30	2.30	2.30	2.30	0.00	0.00	2.30	2.30	2.30
43910	23	0.60	44.70	9.69	3.73	14.14	0.72	2.10	1.50	19.00
43912	41	0.10	23.40	3.97	2.53	4.47	0.23	2.50	1.50	4.50
43913	5	1.50	5.80	3.92	3.31	2.19	0.11	4.90	1.60	5.80
43915	9	0.10	39.30	8.21	3.28	12.19	0.62	3.70	1.90	8.20
43916	8	2.70	108.50	60.89	25.91	49.20	2.52	79.60	3.00	105.20
43917	26	2.30	65.30	11.88	7.91	13.93	0.71	7.90	3.60	12.70
43920	286	0.10	83.40	12.14	5.57	15.93	0.82	5.85	2.60	13.70
43921	1	3.00	3.00	3.00	3.00	0.00	0.00	3.00	3.00	3.00
43925	1	10.50	10.50	10.50	10.50	0.00	0.00	10.50	10.50	10.50
43926	2	2.00	6.10	4.05	3.49	2.90	0.15	4.05	2.00	6.10
43927	1	2.40	2.40	2.40	2.40	0.00	0.00	2.40	2.40	2.40
43930	4	3.50	39.00	15.05	9.08	16.74	0.86	8.85	3.50	14.20
43931	1	0.80	0.80	0.80	0.80	0.00	0.00	0.80	0.80	0.80
43932	4	1.90	36.70	14.93	9.39	15.08	0.77	10.55	1.90	10.60
43933	13	0.50	54.00	17.93	6.04	21.79	1.12	7.40	1.90	43.10
43934	1	1.50	1.50	1.50	1.50	0.00	0.00	1.50	1.50	1.50
43935	64	0.10	13.40	3.28	2.39	2.57	0.13	2.30	1.40	4.50
43937	4	3.10	24.30	9.95	7.31	9.69	0.50	6.20	3.10	6.90
43938	17	0.50	21.40	5.19	3.18	5.58	0.29	3.00	1.70	7.30
43940	4	1.00	20.50	11.43	6.92	9.34	0.48	12.10	1.00	18.00
43941	1	5.00	5.00	5.00	5.00	0.00	0.00	5.00	5.00	5.00
43942	18	0.50	15.30	5.57	3.06	5.32	0.27	2.60	1.00	9.10
43943	36	0.20	180.00	29.74	9.69	40.64	2.08	11.75	2.70	45.70
43944	11	0.50	9.30	4.05	2.53	3.35	0.17	3.80	0.70	7.50
43945	29	0.60	177.80	20.52	5.57	41.09	2.10	3.80	1.60	14.50
43946	2	0.90	1.70	1.30	1.24	0.57	0.03	1.30	0.90	1.70
43947	56	0.10	288.60	10.40	3.40	38.47	1.97	2.70	2.00	6.00
43948	1	0.30	0.30	0.30	0.30	0.00	0.00	0.30	0.30	0.30
43950	338	0.10	119.40	7.15	3.26	16.31	0.84	3.10	1.80	5.60
43952	186	0.10	927.60	10.17	3.50	67.92	3.48	3.45	1.90	6.40
43953	61	0.20	61.20	8.35	4.16	11.45	0.59	3.90	1.80	8.60
43961	1	1.90	1.90	1.90	1.90	0.00	0.00	1.90	1.90	1.90

Zip Code	No.	Min.	Max.	AM	GM	SD	CV	Md	Q1	Q3
43962	2	3.70	3.70	3.70	3.70	0.00	0.00	3.70	3.70	3.70
43963	7	0.90	11.40	3.87	2.84	3.63	0.19	2.10	2.00	5.60
43964	51	0.50	132.20	13.79	4.64	31.68	1.62	3.40	2.50	7.50
43967	2	0.90	0.90	0.90	0.90	0.00	0.00	0.90	0.90	0.90
43968	43	0.10	14.60	3.66	2.39	3.41	0.17	2.60	1.30	5.20
43971	3	0.80	6.70	3.37	2.41	3.02	0.15	2.60	0.80	6.70
43972	1	2.20	2.20	2.20	2.20	0.00	0.00	2.20	2.20	2.20
43973	2	0.60	1.40	1.00	0.92	0.57	0.03	1.00	0.60	1.40
43974	3	0.80	2.10	1.47	1.36	0.65	0.03	1.50	0.80	2.10
43976	10	1.10	20.70	4.67	3.02	5.94	0.30	2.00	1.90	4.10
43977	15	0.90	26.90	5.99	3.35	7.28	0.37	3.70	1.20	11.10
43983	6	0.60	10.40	4.22	2.94	3.66	0.19	2.60	2.30	6.80
43985	2	0.50	5.90	3.20	1.72	3.82	0.20	3.20	0.50	5.90
43986	20	0.70	23.90	7.56	3.75	8.25	0.42	2.90	1.00	12.70
43988	28	0.80	124.50	43.55	21.87	40.86	2.09	29.20	8.90	60.70
44001	594	0.10	42.80	2.96	1.20	4.87	0.25	1.30	0.50	3.20
44003	32	0.20	4.70	1.27	1.00	0.94	0.05	1.05	0.60	1.70
44004	209	0.10	51.70	3.73	1.51	7.12	0.36	1.20	0.70	2.70
44005	1	1.50	1.50	1.50	1.50	0.00	0.00	1.50	1.50	1.50
44007	3	1.10	1.60	1.43	1.41	0.29	0.01	1.60	1.10	1.60
44010	25	0.40	5.30	2.05	1.54	1.49	0.08	1.60	0.70	3.00
44011	738	0.10	52.70	5.02	3.04	6.22	0.32	3.50	1.90	5.70
44012	1910	0.10	618.20	5.33	3.12	15.05	0.77	3.40	1.70	6.50
44017	297	0.10	30.30	3.79	2.43	3.66	0.19	2.60	1.20	5.40
44021	82	0.10	13.30	2.71	1.94	2.30	0.12	2.05	1.20	3.70
44022	698	0.10	733.80	3.66	1.95	27.79	1.42	2.00	1.20	3.20
44023	441	0.10	46.60	3.21	2.32	3.33	0.17	2.30	1.40	3.90
44024	319	0.10	19.60	2.83	2.07	2.51	0.13	2.30	1.25	3.60
44026	194	0.10	22.80	2.23	1.61	2.19	0.11	1.80	1.00	2.70
44027	3	0.90	3.10	1.63	1.36	1.27	0.07	0.90	0.90	3.10
44028	75	0.20	16.60	4.94	3.68	3.79	0.19	3.40	2.60	7.80
44029	2	4.10	4.20	4.15	4.15	0.07	0.00	4.15	4.10	4.20
44030	142	0.10	122.20	7.70	2.91	16.31	0.83	2.95	1.30	7.50
44032	11	0.30	3.90	1.30	0.95	1.14	0.06	0.90	0.40	1.40

Zip Code	No.	Min.	Max.	AM	GM	SD	CV	Md	Q1	Q3
44033	2	0.80	2.70	1.75	1.47	1.34	0.07	1.75	0.80	2.70
44035	384	0.10	70.10	3.30	1.93	4.84	0.25	2.10	1.00	4.20
44036	8	0.20	4.70	2.49	1.79	1.66	0.09	2.15	1.10	3.90
44037	2	1.80	2.60	2.20	2.16	0.57	0.03	2.20	1.80	2.60
44038	2	0.60	0.70	0.65	0.65	0.07	0.00	0.65	0.60	0.70
44039	465	0.10	40.20	4.12	2.47	4.72	0.24	2.80	1.40	5.20
44040	133	0.30	22.40	2.95	2.09	3.33	0.17	2.00	1.20	3.10
44041	87	0.10	14.00	2.18	1.16	2.66	0.14	1.20	0.50	2.50
44044	127	0.20	28.70	4.02	2.69	3.82	0.20	3.10	1.60	5.30
44045	6	0.10	5.10	1.65	0.56	2.14	0.11	0.55	0.10	3.50
44046	18	0.80	260.00	18.49	4.27	60.33	3.09	4.15	1.50	7.10
44047	92	0.10	210.00	3.67	1.11	21.78	1.12	1.10	0.70	1.70
44048	26	0.40	37.40	7.47	3.76	9.89	0.51	3.25	1.60	11.40
44049	1	1.30	1.30	1.30	1.30	0.00	0.00	1.30	1.30	1.30
44050	33	0.10	51.20	7.02	4.03	9.23	0.47	4.90	2.20	9.20
44051	1	1.30	1.30	1.30	1.30	0.00	0.00	1.30	1.30	1.30
44052	130	0.10	20.20	3.01	1.47	3.55	0.18	1.80	0.90	3.80
44053	156	0.10	32.80	5.16	3.09	5.22	0.27	4.00	1.70	6.60
44054	117	0.10	27.20	5.19	3.13	4.99	0.26	3.70	1.60	7.00
44055	42	0.10	23.40	5.05	2.88	5.02	0.26	3.65	1.40	7.40
44056	274	0.10	32.30	3.49	2.27	3.91	0.20	2.60	1.20	4.20
44057	152	0.10	37.80	3.97	1.86	6.77	0.35	1.70	0.90	3.20
44058	3	0.60	2.20	1.23	1.06	0.85	0.04	0.90	0.60	2.20
44060	899	0.10	49.90	3.78	2.06	5.86	0.30	2.20	1.10	3.80
44061	3	0.50	20.00	7.10	2.00	11.17	0.57	0.80	0.50	20.00
44062	38	0.10	5.80	1.83	1.10	1.72	0.09	1.10	0.50	2.80
44064	12	0.40	12.50	3.98	2.69	3.77	0.19	2.85	1.30	3.60
44065	41	0.10	16.50	3.73	2.12	4.18	0.21	2.00	1.10	4.00
44067	471	0.10	47.50	4.02	2.30	5.85	0.30	2.40	1.20	4.60
44068	24	0.10	51.00	12.02	4.17	14.82	0.76	4.10	2.10	18.70
44069	2	0.80	7.70	4.25	2.48	4.88	0.25	4.25	0.80	7.70
44070	465	0.10	95.80	4.85	2.86	7.29	0.37	3.10	1.60	5.40
44072	123	0.10	20.20	2.39	1.54	2.70	0.14	1.80	1.10	2.40
44073	1	1.20	1.20	1.20	1.20	0.00	0.00	1.20	1.20	1.20

Zip Code	No.	Min.	Max.	AM	GM	SD	CV	Md	Q1	Q3
44074	120	0.10	54.10	3.87	2.47	5.59	0.29	2.50	1.60	4.20
44076	39	0.10	9.30	1.56	1.08	1.62	0.08	1.10	0.60	1.90
44077	870	0.10	139.30	4.38	2.38	7.51	0.38	2.50	1.30	4.60
44080	4	1.60	12.50	5.85	4.01	5.25	0.27	4.65	1.60	7.60
44081	124	0.10	22.60	4.43	2.72	4.52	0.23	2.90	1.40	5.90
44082	19	0.30	8.40	2.12	1.42	2.07	0.11	1.60	0.70	2.60
44084	32	0.20	5.60	2.08	1.56	1.57	0.08	1.40	1.00	2.50
44085	52	0.20	14.10	2.63	1.66	2.88	0.15	1.50	1.00	3.00
44086	25	0.10	27.50	3.72	2.07	5.59	0.29	2.20	1.00	3.40
44087	610	0.10	908.10	35.11	3.48	129.15	6.61	2.50	1.30	5.10
44088	3	12.70	28.30	22.10	20.87	8.28	0.42	25.30	12.70	28.30
44089	270	0.10	89.60	7.21	4.03	10.70	0.55	4.50	1.90	8.20
44090	75	0.10	28.30	5.56	3.41	5.72	0.29	3.80	1.80	6.70
44092	137	0.10	24.70	2.32	1.42	2.94	0.15	1.40	0.80	3.00
44093	13	0.30	8.50	2.23	1.61	2.10	0.11	1.70	1.00	2.50
44094	528	0.10	33.30	2.76	1.55	3.61	0.18	1.70	0.80	3.10
44095	203	0.10	9.30	1.56	1.07	1.52	0.08	1.10	0.60	2.00
44096	1	2.50	2.50	2.50	2.50	0.00	0.00	2.50	2.50	2.50
44097	2	1.00	3.40	2.20	1.84	1.70	0.09	2.20	1.00	3.40
44099	16	0.30	5.30	1.31	0.89	1.35	0.07	0.95	0.40	1.30
44100	1	5.40	5.40	5.40	5.40	0.00	0.00	5.40	5.40	5.40
44101	15	0.20	4.10	1.43	0.95	1.22	0.06	1.20	0.50	2.30
44102	107	0.10	13.30	2.14	1.26	2.34	0.12	1.50	0.60	2.60
44103	36	0.10	10.30	1.66	0.87	2.41	0.12	0.65	0.40	2.10
44104	23	0.20	10.10	2.08	1.08	2.75	0.14	0.70	0.40	2.90
44105	85	0.10	13.70	2.31	1.34	2.33	0.12	1.60	0.80	3.20
44106	223	0.10	24.90	1.96	1.30	2.33	0.12	1.40	0.70	2.60
44107	397	0.10	17.50	1.63	1.00	2.04	0.10	1.00	0.50	1.90
44108	50	0.40	8.50	2.64	1.83	2.34	0.12	1.80	1.00	4.20
44109	134	0.10	14.40	2.29	1.32	2.56	0.13	1.30	0.70	3.20
44110	30	0.10	7.90	1.32	0.88	1.49	0.08	0.90	0.50	1.40
44111	217	0.10	30.00	2.32	1.36	3.09	0.16	1.40	0.80	2.60
44112	76	0.10	10.00	1.26	0.71	1.49	0.08	0.90	0.20	1.80
44113	115	0.10	51.80	3.30	1.66	5.50	0.28	2.10	0.70	3.90

Zip Code	No.	Min.	Max.	AM	GM	SD	CV	Md	Q1	Q3
44114	33	0.10	20.00	2.03	0.79	3.73	0.19	0.80	0.30	2.50
44115	10	0.50	14.30	2.82	1.61	4.18	0.21	1.50	0.80	1.80
44116	444	0.10	112.40	2.68	1.71	6.10	0.31	1.80	1.00	2.90
44117	58	0.10	6.30	1.26	0.87	1.23	0.06	0.90	0.50	1.60
44118	844	0.10	161.80	1.78	0.94	6.00	0.31	1.00	0.50	1.80
44119	82	0.10	17.10	1.53	0.98	2.11	0.11	1.00	0.60	1.60
44120	369	0.10	10.50	1.64	1.12	1.55	0.08	1.20	0.70	2.10
44121	372	0.10	114.60	1.87	1.04	6.13	0.31	1.10	0.60	1.90
44122	802	0.10	114.60	1.81	1.24	4.28	0.22	1.20	0.80	2.00
44123	83	0.10	6.30	1.31	0.79	1.42	0.07	0.90	0.40	1.70
44124	678	0.10	19.00	2.13	1.54	2.18	0.11	1.50	1.00	2.50
44125	138	0.10	16.00	2.20	1.35	2.95	0.15	1.40	0.70	2.30
44126	303	0.10	89.90	3.35	1.94	7.25	0.37	1.90	1.10	3.30
44127	16	0.60	7.40	2.41	1.93	1.81	0.09	2.10	1.20	2.60
44128	83	0.10	14.50	1.17	0.76	1.77	0.09	0.70	0.50	1.30
44129	224	0.10	29.30	2.07	1.24	3.16	0.16	1.30	0.70	2.20
44130	457	0.10	105.00	2.61	1.49	6.01	0.31	1.55	0.90	2.70
44131	288	0.10	79.20	3.28	2.17	5.85	0.30	2.20	1.30	3.40
44132	63	0.10	14.80	1.30	0.79	2.13	0.11	0.80	0.50	1.10
44133	345	0.10	15.20	2.87	2.09	2.34	0.12	2.30	1.30	3.60
44134	263	0.10	23.10	1.85	1.27	2.05	0.11	1.30	0.80	2.30
44135	120	0.10	14.90	2.33	1.46	2.66	0.14	1.50	0.80	2.70
44136	560	0.10	656.20	5.12	2.16	38.94	1.99	2.20	1.40	3.30
44137	92	0.10	6.90	1.32	0.96	1.19	0.06	1.00	0.60	1.70
44138	328	0.10	68.40	4.01	2.56	4.85	0.25	3.00	1.50	5.10
44139	718	0.10	156.40	4.12	1.87	14.70	0.75	1.80	1.10	3.00
44140	859	0.10	37.00	5.06	3.34	4.73	0.24	3.60	1.90	6.80
44141	524	0.10	72.90	3.53	2.23	4.67	0.24	2.20	1.30	4.10
44142	108	0.10	6.50	1.52	1.01	1.43	0.07	1.00	0.60	1.80
44143	391	0.10	23.30	3.04	2.22	2.61	0.13	2.40	1.40	3.90
44144	117	0.10	8.70	1.70	1.09	1.79	0.09	1.00	0.70	1.80
44145	1063	0.10	68.70	4.69	2.99	5.36	0.27	3.10	1.70	5.80
44146	169	0.10	23.70	2.07	1.31	2.60	0.13	1.30	0.80	2.50
44147	331	0.10	19.60	2.72	2.04	2.24	0.11	2.10	1.30	3.40

Zip Code	No.	Min.	Max.	AM	GM	SD	CV	Md	Q1	Q3
44148	2	2.30	2.90	2.60	2.58	0.42	0.02	2.60	2.30	2.90
44149	194	0.10	12.70	3.09	2.42	2.32	0.12	2.40	1.70	3.80
44190	1	1.10	1.10	1.10	1.10	0.00	0.00	1.10	1.10	1.10
44194	1	6.10	6.10	6.10	6.10	0.00	0.00	6.10	6.10	6.10
44195	1	5.00	5.00	5.00	5.00	0.00	0.00	5.00	5.00	5.00
44199	7	1.80	4.60	2.93	2.71	1.24	0.06	2.10	1.80	4.10
44201	38	0.20	8.90	3.11	2.25	2.41	0.12	2.50	1.30	3.60
44202	465	0.10	22.40	2.43	1.65	2.35	0.12	1.80	0.90	3.30
44203	482	0.10	30.00	4.21	2.74	4.16	0.21	3.00	1.50	5.20
44205	2	1.40	1.50	1.45	1.45	0.07	0.00	1.45	1.40	1.50
44208	3	2.30	3.50	2.80	2.76	0.62	0.03	2.60	2.30	3.50
44209	1	1.90	1.90	1.90	1.90	0.00	0.00	1.90	1.90	1.90
44210	44	0.10	12.10	3.50	2.72	2.25	0.12	3.35	2.00	4.40
44211	1	2.10	2.10	2.10	2.10	0.00	0.00	2.10	2.10	2.10
44212	545	0.10	169.30	3.38	2.22	7.54	0.39	2.50	1.40	4.00
44214	24	0.30	5.60	2.90	2.29	1.67	0.09	2.60	1.70	4.40
44215	21	0.90	21.20	4.00	2.69	4.55	0.23	2.20	1.40	4.70
44216	175	0.30	68.10	6.85	4.27	8.30	0.42	4.60	2.40	7.60
44217	28	0.10	90.40	9.45	2.88	18.33	0.94	2.70	1.10	5.50
44221	346	0.10	22.50	3.01	1.93	3.09	0.16	2.10	1.10	3.70
44222	3	0.70	2.70	1.97	1.68	1.10	0.06	2.50	0.70	2.70
44223	418	0.10	20.90	3.45	2.23	3.01	0.15	2.80	1.40	4.70
44224	1231	0.10	58.50	4.05	2.47	4.97	0.25	2.70	1.40	4.90
44230	99	0.30	77.50	7.22	4.34	9.86	0.50	4.10	2.40	7.20
44231	63	0.40	14.10	3.46	2.64	2.65	0.14	3.00	1.60	4.40
44232	14	0.60	18.70	4.07	2.81	4.52	0.23	3.05	1.40	4.90
44233	172	0.10	22.10	3.11	2.26	2.67	0.14	2.45	1.40	4.00
44234	36	0.50	29.80	3.48	2.20	5.37	0.27	2.10	1.40	2.90
44235	21	0.80	44.20	6.23	3.89	9.32	0.48	3.30	2.60	4.40
44236	1270	0.10	319.00	3.76	2.21	10.53	0.54	2.40	1.20	4.10
44237	3	0.60	5.90	2.50	1.52	2.95	0.15	1.00	0.60	5.90
44238	14	0.40	9.10	4.79	3.89	2.37	0.12	5.10	3.60	6.20
44240	503	0.10	70.10	4.33	2.56	5.53	0.28	2.70	1.30	5.40
44241	162	0.10	31.20	4.75	2.93	5.13	0.26	3.05	1.50	6.30

Zip Code	No.	Min.	Max.	AM	GM	SD	CV	Md	Q1	Q3
44242	8	0.10	4.05	1.96	1.12	1.58	0.08	1.55	0.20	3.20
44243	4	0.10	5.80	3.50	1.71	2.65	0.14	4.05	0.10	5.40
44244	2	2.00	2.90	2.45	2.41	0.64	0.03	2.45	2.00	2.90
44247	3	1.50	4.30	2.60	2.35	1.49	0.08	2.00	1.50	4.30
44249	2	2.20	20.60	11.40	6.73	13.01	0.67	11.40	2.20	20.60
44250	8	0.30	5.30	2.61	2.07	1.48	0.08	2.75	1.40	3.20
44251	38	0.60	11.60	4.22	3.59	2.50	0.13	3.85	2.50	5.00
44252	3	2.80	11.60	6.67	5.67	4.50	0.23	5.60	2.80	11.60
44253	33	1.00	23.90	4.69	3.33	4.80	0.25	3.30	1.80	5.60
44254	50	0.50	31.50	5.18	3.31	5.90	0.30	3.35	1.90	5.50
44255	65	0.20	41.20	4.69	2.72	6.07	0.31	3.00	1.50	5.70
44256	1446	0.10	60.30	3.93	2.73	4.35	0.22	2.90	1.75	4.70
44258	9	1.60	13.60	4.33	3.23	4.16	0.21	2.70	1.90	3.00
44260	197	0.10	44.90	5.71	3.38	6.09	0.31	3.70	1.80	7.90
44262	124	0.10	27.00	4.17	2.64	4.37	0.22	2.80	1.50	4.80
44264	49	0.10	13.20	2.21	1.43	2.32	0.12	1.50	0.90	2.40
44265	7	0.10	21.80	11.01	4.87	8.98	0.46	14.80	2.20	19.30
44266	386	0.10	25.60	2.62	1.48	3.34	0.17	1.30	0.60	2.90
44270	43	0.10	14.70	3.90	2.60	3.37	0.17	3.20	1.80	4.70
44272	52	0.10	37.40	5.83	3.38	7.98	0.41	3.70	1.70	5.40
44273	93	0.40	48.40	6.49	4.08	8.96	0.46	3.50	2.70	6.10
44274	8	0.30	6.40	2.65	1.78	2.19	0.11	2.40	0.70	2.80
44275	11	0.60	14.10	5.28	3.31	4.93	0.25	3.20	1.55	10.20
44276	17	1.50	55.00	7.93	4.69	12.58	0.64	2.90	2.60	9.80
44278	293	0.10	55.30	3.44	2.30	4.08	0.21	2.50	1.50	4.40
44280	52	0.50	36.10	7.66	5.01	8.40	0.43	4.60	2.40	8.20
44281	688	0.10	52.50	4.71	2.86	5.69	0.29	3.10	1.80	5.70
44282	3	0.80	9.10	3.97	2.44	4.49	0.23	2.00	0.80	9.10
44285	2	4.70	5.50	5.10	5.08	0.57	0.03	5.10	4.70	5.50
44286	195	0.10	27.60	3.39	2.50	3.12	0.16	2.60	1.70	3.80
44287	39	0.10	17.00	4.78	2.85	4.49	0.23	3.40	1.90	7.50
44288	9	0.20	5.30	2.21	1.49	1.81	0.09	2.10	0.80	2.70
44290	1	3.30	3.30	3.30	3.30	0.00	0.00	3.30	3.30	3.30
44296	1	2.80	2.80	2.80	2.80	0.00	0.00	2.80	2.80	2.80

Zip Code	No.	Min.	Max.	AM	GM	SD	CV	Md	Q1	Q3
44301	122	0.10	14.20	2.37	1.62	2.09	0.11	1.55	1.00	3.30
44302	34	0.10	11.50	2.23	1.67	2.20	0.11	1.75	1.10	2.30
44303	133	0.10	7.40	2.67	2.09	1.74	0.09	2.10	1.30	3.70
44304	11	0.40	6.80	2.62	1.56	2.53	0.13	1.40	0.50	5.40
44305	122	0.10	24.20	3.82	2.52	3.61	0.18	2.80	1.60	4.60
44306	66	0.10	44.50	3.44	1.66	6.06	0.31	1.65	0.70	3.70
44307	18	0.20	5.40	2.48	1.98	1.52	0.08	2.10	1.60	3.20
44308	4	1.20	7.50	4.93	3.92	3.12	0.16	5.50	1.20	7.50
44309	1	4.80	4.80	4.80	4.80	0.00	0.00	4.80	4.80	4.80
44310	127	0.10	20.80	3.52	2.33	3.18	0.16	2.80	1.40	4.80
44311	19	0.10	10.60	2.12	1.04	2.70	0.14	1.00	0.50	2.70
44312	374	0.10	48.20	5.09	3.10	5.91	0.30	3.55	1.70	6.40
44313	708	0.10	46.70	3.87	2.54	4.02	0.21	2.50	1.50	4.90
44314	118	0.10	14.70	2.68	1.75	2.63	0.13	2.00	1.10	3.30
44315	1	3.50	3.50	3.50	3.50	0.00	0.00	3.50	3.50	3.50
44316	2	1.50	3.20	2.35	2.19	1.20	0.06	2.35	1.50	3.20
44318	8	0.70	4.20	1.61	1.38	1.12	0.06	1.20	0.90	1.60
44319	402	0.10	77.30	4.59	2.97	5.62	0.29	3.00	1.70	5.70
44320	163	0.10	131.90	4.57	1.51	14.93	0.76	1.10	0.50	3.70
44321	554	0.10	591.60	7.19	3.26	26.23	1.34	3.50	1.60	6.60
44322	2	1.00	1.20	1.10	1.10	0.14	0.01	1.10	1.00	1.20
44323	1	12.60	12.60	12.60	12.60	0.00	0.00	12.60	12.60	12.60
44324	1	1.50	1.50	1.50	1.50	0.00	0.00	1.50	1.50	1.50
44326	7	0.70	7.50	3.10	2.13	2.76	0.14	2.40	0.70	6.40
44333	621	0.10	50.80	6.67	3.75	8.28	0.42	3.50	1.90	7.70
44398	1	210.00	210.00	210.00	210.00	0.00	0.00	210.00	210.00	210.00
44400	1	2.70	2.70	2.70	2.70	0.00	0.00	2.70	2.70	2.70
44401	33	0.10	7.80	2.19	1.40	1.72	0.09	1.70	1.00	3.30
44402	24	0.10	11.60	3.26	2.22	3.17	0.16	2.05	1.50	3.30
44403	35	0.50	11.00	2.87	2.22	2.27	0.12	2.00	1.40	3.70
44404	12	0.20	8.50	2.20	1.28	2.29	0.12	1.70	0.30	3.10
44405	43	0.10	16.90	2.93	1.68	3.12	0.16	2.20	0.70	3.60
44406	415	0.10	16.30	2.99	2.13	2.51	0.13	2.20	1.30	3.90
44408	157	0.10	234.20	6.31	3.07	19.41	0.99	3.00	1.90	5.50

Zip Code	No.	Min.	Max.	AM	GM	SD	CV	Md	Q1	Q3
44410	253	0.10	53.00	3.69	2.41	4.83	0.25	2.40	1.40	4.50
44411	8	0.60	9.70	3.51	2.27	3.40	0.17	1.65	1.00	5.20
44412	27	0.70	11.30	3.50	2.40	3.10	0.16	2.20	1.00	5.10
44413	60	0.10	21.30	3.15	2.26	3.22	0.16	2.25	1.50	3.80
44415	3	0.90	9.20	3.93	2.41	4.58	0.23	1.70	0.90	9.20
44416	2	1.60	1.90	1.75	1.74	0.21	0.01	1.75	1.60	1.90
44417	17	0.30	3.60	2.06	1.63	1.22	0.06	1.80	0.90	3.10
44418	10	0.40	3.30	1.81	1.48	1.03	0.05	1.80	1.10	2.40
44420	100	0.10	17.10	2.47	1.54	2.44	0.12	1.70	1.00	3.40
44422	2	0.40	1.70	1.05	0.82	0.92	0.05	1.05	0.40	1.70
44423	26	0.10	41.00	12.72	6.13	14.05	0.72	5.75	2.90	23.30
44424	2	0.50	1.20	0.85	0.77	0.49	0.03	0.85	0.50	1.20
44425	152	0.10	40.10	3.11	2.05	3.96	0.20	1.95	1.20	3.60
44426	1	9.60	9.60	9.60	9.60	0.00	0.00	9.60	9.60	9.60
44427	19	0.50	219.10	31.13	8.10	57.50	2.94	10.40	2.20	25.60
44428	21	0.30	10.70	2.73	1.74	2.79	0.14	1.20	1.10	3.40
44429	14	0.30	4.90	2.39	1.93	1.41	0.07	1.85	1.40	3.80
44430	15	0.10	94.90	13.53	1.59	32.17	1.65	0.80	0.50	2.50
44431	56	0.20	15.30	4.00	3.13	2.94	0.15	3.25	1.80	5.30
44432	123	0.10	136.50	17.72	6.29	27.84	1.43	5.10	2.40	18.00
44435	1	1.20	1.20	1.20	1.20	0.00	0.00	1.20	1.20	1.20
44436	21	0.40	15.80	3.06	2.20	3.16	0.16	2.70	1.90	3.10
44437	14	0.10	6.20	2.21	1.72	1.41	0.07	1.80	1.60	3.10
44438	31	0.10	5.50	2.13	1.63	1.32	0.07	1.70	1.10	3.10
44439	2	0.90	1.30	1.10	1.08	0.28	0.01	1.10	0.90	1.30
44440	32	0.10	57.00	6.63	2.44	12.56	0.64	2.45	1.30	3.80
44441	46	0.20	41.60	14.62	8.44	11.77	0.60	11.85	5.70	20.80
44442	35	1.00	12.20	3.40	2.65	2.75	0.14	2.10	1.70	4.10
44443	31	0.10	207.00	27.88	6.65	50.99	2.61	3.80	2.20	21.40
44444	64	0.10	13.90	2.47	1.74	2.22	0.11	1.85	0.90	3.50
44445	27	0.20	39.00	5.23	2.84	8.09	0.41	2.90	1.50	5.30
44446	72	0.10	12.80	2.12	1.35	2.06	0.11	1.55	0.80	2.80
44447	1	1.10	1.10	1.10	1.10	0.00	0.00	1.10	1.10	1.10
44449	10	0.30	4.80	2.07	1.52	1.40	0.07	2.00	1.00	2.90

Zip Code	No.	Min.	Max.	AM	GM	SD	CV	Md	Q1	Q3
44450	14	0.40	5.90	2.71	1.89	1.93	0.10	2.60	0.60	3.80
44451	46	0.20	12.90	2.07	1.47	2.02	0.10	1.85	0.80	2.60
44452	46	0.30	29.10	4.43	2.96	5.27	0.27	3.00	2.00	4.30
44453	1	2.30	2.30	2.30	2.30	0.00	0.00	2.30	2.30	2.30
44454	10	1.10	26.80	6.03	3.82	7.60	0.39	3.80	1.80	6.90
44455	11	0.60	41.10	13.93	5.79	15.42	0.79	4.80	1.40	32.50
44460	226	0.10	76.70	6.56	2.95	12.28	0.63	2.90	1.50	4.90
44470	14	0.10	3.20	1.39	1.11	0.74	0.04	1.50	0.90	1.70
44471	46	0.30	12.90	2.17	1.64	2.09	0.11	1.70	1.10	2.30
44472	1	4.50	4.50	4.50	4.50	0.00	0.00	4.50	4.50	4.50
44473	28	0.30	6.40	2.20	1.84	1.38	0.07	1.80	1.30	2.50
44481	87	0.10	11.10	2.63	1.53	2.48	0.13	1.70	0.70	3.60
44482	2	1.50	1.60	1.55	1.55	0.07	0.00	1.55	1.50	1.60
44483	110	0.10	13.60	2.10	1.50	2.01	0.10	1.50	0.80	2.80
44484	192	0.10	71.70	3.17	2.06	5.48	0.28	2.00	1.40	3.80
44485	35	0.10	8.80	1.41	0.81	1.58	0.08	1.20	0.20	1.80
44486	1	2.50	2.50	2.50	2.50	0.00	0.00	2.50	2.50	2.50
44487	2	5.50	8.00	6.75	6.63	1.77	0.09	6.75	5.50	8.00
44490	2	3.50	4.00	3.75	3.74	0.35	0.02	3.75	3.50	4.00
44491	14	0.10	5.10	1.59	1.16	1.27	0.07	1.15	0.80	2.50
44493	3	2.30	6.20	4.00	3.68	2.00	0.10	3.50	2.30	6.20
44500	1	2.10	2.10	2.10	2.10	0.00	0.00	2.10	2.10	2.10
44501	11	0.60	4.90	2.17	1.81	1.43	0.07	2.00	1.30	2.30
44502	30	0.10	9.90	1.81	1.19	1.82	0.09	1.40	0.90	2.20
44503	17	0.40	50.00	8.95	4.75	11.92	0.61	3.70	3.30	12.20
44504	16	0.50	6.80	1.56	1.26	1.46	0.07	1.30	0.90	1.60
44505	113	0.10	9.60	2.62	2.02	1.75	0.09	2.30	1.50	3.40
44506	6	0.70	8.40	4.23	2.98	2.98	0.15	4.75	0.90	5.90
44507	16	0.30	3.60	1.03	0.77	0.91	0.05	0.80	0.40	1.20
44508	2	1.80	1.80	1.80	1.80	0.00	0.00	1.80	1.80	1.80
44509	49	0.10	5.70	1.88	1.27	1.48	0.08	1.50	0.80	2.75
44510	8	0.50	3.80	2.26	1.83	1.27	0.07	2.30	0.60	3.10
44511	166	0.10	32.60	2.81	1.74	3.32	0.17	2.00	0.90	3.30
44512	406	0.10	20.00	2.65	2.00	1.99	0.10	2.20	1.30	3.50

Zip Code	No.	Min.	Max.	AM	GM	SD	CV	Md	Q1	Q3
44513	4	1.40	3.90	2.83	2.65	1.04	0.05	3.00	1.40	3.00
44514	369	0.10	12.10	2.84	2.03	2.20	0.11	2.20	1.30	3.90
44515	251	0.10	11.70	2.52	1.85	1.98	0.10	1.90	1.10	3.30
44555	113	0.10	16.60	2.58	1.00	3.89	0.20	0.90	0.30	2.90
44572	2	0.50	8.30	4.40	2.04	5.52	0.28	4.40	0.50	8.30
44575	2	3.40	4.10	3.75	3.73	0.49	0.03	3.75	3.40	4.10
44587	2	8.00	8.00	8.00	8.00	0.00	0.00	8.00	8.00	8.00
44601	214	0.10	16.30	2.62	1.89	2.21	0.11	2.00	1.20	3.50
44602	3	1.10	7.30	3.17	2.07	3.58	0.18	1.10	1.10	7.30
44606	33	0.40	83.00	8.14	4.53	13.99	0.72	5.80	2.60	9.50
44607	5	0.30	18.20	5.42	2.46	7.33	0.38	2.30	1.50	4.80
44608	10	0.10	19.00	7.39	3.32	7.26	0.37	6.20	0.90	13.00
44609	17	0.30	5.30	1.43	1.06	1.31	0.07	1.10	0.60	1.60
44610	5	1.40	7.80	4.44	3.67	2.80	0.14	3.70	2.40	6.90
44611	8	1.80	9.20	4.33	3.81	2.42	0.12	3.80	2.40	4.60
44612	84	0.50	165.40	12.81	5.44	26.02	1.33	4.70	2.30	10.70
44613	4	2.10	4.30	3.45	3.31	1.06	0.05	3.70	2.10	4.30
44614	249	0.10	49.30	5.59	2.99	6.21	0.32	3.80	1.70	7.00
44615	258	0.10	220.20	13.50	3.95	30.90	1.58	3.40	1.60	8.20
44617	1	0.60	0.60	0.60	0.60	0.00	0.00	0.60	0.60	0.60
44618	47	0.30	17.90	5.59	3.43	5.31	0.27	3.30	1.90	8.40
44619	5	0.10	10.10	3.04	0.76	4.32	0.22	0.60	0.10	4.30
44620	29	0.90	15.20	4.84	3.65	3.82	0.20	3.50	2.60	6.00
44621	25	0.10	6.70	2.72	2.10	1.57	0.08	2.40	1.80	3.40
44622	341	0.10	56.00	6.70	4.18	7.08	0.36	4.60	2.30	8.00
44623	1	9.80	9.80	9.80	9.80	0.00	0.00	9.80	9.80	9.80
44624	16	1.10	58.00	8.66	4.16	15.11	0.77	3.65	1.80	5.70
44625	21	0.60	48.30	7.29	3.70	11.59	0.59	3.30	1.90	6.60
44626	26	0.10	245.10	33.31	7.82	60.41	3.09	6.65	1.90	23.40
44627	3	2.20	17.50	7.30	4.39	8.83	0.45	2.20	2.20	17.50
44628	16	0.80	23.60	6.16	3.09	7.64	0.39	1.90	1.20	8.90
44629	22	0.10	68.10	9.64	4.95	13.84	0.71	5.60	2.80	11.80
44630	2	0.60	6.60	3.60	1.99	4.24	0.22	3.60	0.60	6.60
44631	3	0.80	9.50	6.07	3.92	4.63	0.24	7.90	0.80	9.50

Zip Code	No.	Min.	Max.	AM	GM	SD	CV	Md	Q1	Q3
44632	128	0.10	23.60	4.67	2.92	4.71	0.24	3.05	1.50	6.50
44633	7	0.10	35.20	8.43	2.06	12.55	0.64	3.60	0.10	12.30
44634	17	0.20	10.20	2.59	1.78	2.56	0.13	1.90	1.10	3.30
44636	2	1.00	1.00	1.00	1.00	0.00	0.00	1.00	1.00	1.00
44637	19	0.60	140.30	27.02	9.77	43.48	2.23	9.50	2.90	17.00
44638	7	0.90	11.80	4.79	3.75	3.56	0.18	3.50	3.10	6.90
44639	3	7.90	9.50	8.47	8.44	0.90	0.05	8.00	7.90	9.50
44640	2	1.20	2.10	1.65	1.59	0.64	0.03	1.65	1.20	2.10
44641	167	0.10	65.20	5.56	3.53	6.69	0.34	3.80	2.00	6.90
44643	64	0.20	95.70	14.50	6.82	21.45	1.10	5.80	3.10	12.60
44644	103	0.10	69.50	10.72	4.73	14.08	0.72	4.40	1.50	12.60
44645	34	0.10	24.20	4.13	2.33	4.92	0.25	2.40	1.50	4.00
44646	603	0.10	125.20	7.17	3.69	12.40	0.63	3.70	1.90	7.70
44647	73	0.30	33.50	4.43	2.93	5.10	0.26	2.70	1.80	5.20
44648	2	0.10	3.30	1.70	0.57	2.26	0.12	1.70	0.10	3.30
44651	9	0.40	9.50	3.98	2.59	3.25	0.17	4.10	1.30	6.90
44653	1	9.50	9.50	9.50	9.50	0.00	0.00	9.50	9.50	9.50
44654	77	0.10	43.70	7.24	3.74	9.54	0.49	3.40	1.70	8.40
44655	2	0.10	0.10	0.10	0.10	0.00	0.00	0.10	0.10	0.10
44656	36	0.20	47.50	11.71	5.44	12.81	0.66	6.10	1.60	17.00
44657	111	0.10	580.00	17.61	6.09	56.16	2.88	7.10	2.40	16.70
44659	1	1.50	1.50	1.50	1.50	0.00	0.00	1.50	1.50	1.50
44660	2	1.10	1.70	1.40	1.37	0.42	0.02	1.40	1.10	1.70
44661	1	11.10	11.10	11.10	11.10	0.00	0.00	11.10	11.10	11.10
44662	93	0.10	110.00	12.29	5.59	19.36	0.99	6.00	2.10	14.70
44663	357	0.10	208.70	11.14	4.98	21.51	1.10	5.40	2.60	11.10
44665	2	2.00	2.60	2.30	2.28	0.42	0.02	2.30	2.00	2.60
44666	32	0.40	16.20	5.98	4.66	3.76	0.19	5.30	2.90	8.00
44667	224	0.10	368.70	17.30	4.51	49.80	2.55	4.00	1.70	9.70
44669	13	0.60	17.10	7.80	5.07	6.05	0.31	9.50	2.50	11.30
44671	4	1.70	43.90	24.10	12.26	22.96	1.18	25.40	1.70	43.90
44672	10	0.10	2.00	0.95	0.72	0.62	0.03	0.85	0.50	1.40
44675	12	0.50	55.60	10.60	4.68	16.79	0.86	4.05	2.00	6.40
44676	18	1.10	164.20	20.16	5.05	45.49	2.33	3.55	1.60	8.00

Zip Code	No.	Min.	Max.	AM	GM	SD	CV	Md	Q1	Q3
44677	43	0.80	259.00	14.89	5.11	40.29	2.06	4.10	2.80	6.80
44679	1	0.90	0.90	0.90	0.90	0.00	0.00	0.90	0.90	0.90
44680	53	0.20	84.10	15.00	7.59	15.39	0.79	13.00	2.20	22.30
44681	29	0.10	88.20	8.70	3.76	16.84	0.86	3.80	2.70	7.10
44682	8	0.10	18.80	8.26	4.08	6.78	0.35	7.20	1.10	12.60
44683	38	0.20	46.60	8.02	3.45	11.44	0.59	3.10	1.40	7.50
44685	724	0.10	120.00	8.46	4.27	12.74	0.65	4.40	2.20	9.50
44686	1	0.60	0.60	0.60	0.60	0.00	0.00	0.60	0.60	0.60
44687	1	3.80	3.80	3.80	3.80	0.00	0.00	3.80	3.80	3.80
44688	34	0.60	55.70	6.47	3.95	9.53	0.49	3.45	2.20	9.40
44691	867	0.10	498.40	11.58	4.47	29.06	1.49	4.30	2.10	9.60
44695	18	0.10	33.00	9.53	5.77	8.54	0.44	5.55	3.60	15.50
44697	5	0.90	2.30	1.44	1.37	0.52	0.03	1.30	1.30	1.40
44699	7	0.70	28.70	10.80	5.10	12.18	0.62	4.90	1.20	27.40
44701	2	3.10	9.40	6.25	5.40	4.45	0.23	6.25	3.10	9.40
44702	5	1.70	5.60	4.06	3.78	1.46	0.07	4.40	3.90	4.70
44703	35	0.30	10.70	4.48	3.66	2.67	0.14	3.80	2.50	7.30
44704	4	2.00	7.50	5.03	4.41	2.66	0.14	5.30	2.00	7.00
44705	99	0.10	31.60	5.83	3.32	5.79	0.30	4.00	1.60	8.40
44706	118	0.10	163.90	10.11	4.84	21.08	1.08	4.85	2.30	8.60
44707	52	0.40	55.30	11.14	5.91	12.75	0.65	5.50	3.30	13.80
44708	277	0.10	94.70	7.09	4.40	8.48	0.43	5.10	2.50	9.00
44709	242	0.20	49.90	6.07	3.99	6.29	0.32	4.45	2.30	7.60
44710	38	0.10	31.00	8.54	5.10	7.44	0.38	6.80	2.50	12.30
44711	4	1.70	10.90	5.60	4.48	4.03	0.21	4.90	1.70	6.40
44714	120	0.10	34.90	5.91	3.60	6.35	0.33	3.80	2.10	6.30
44718	408	0.10	550.60	8.98	4.04	28.74	1.47	4.90	2.20	9.60
44720	996	0.10	144.30	6.92	4.00	9.81	0.50	4.40	2.20	8.30
44721	266	0.20	220.00	9.91	5.19	17.81	0.91	5.20	2.70	11.20
44726	2	2.80	9.00	5.90	5.02	4.38	0.22	5.90	2.80	9.00
44728	3	1.00	5.50	3.43	2.75	2.27	0.12	3.80	1.00	5.50
44730	47	0.10	63.60	6.88	3.51	11.13	0.57	3.10	2.00	7.50
44735	2	8.30	13.60	10.95	10.62	3.75	0.19	10.95	8.30	13.60
44799	3	3.10	8.00	6.07	5.60	2.61	0.13	7.10	3.10	8.00

Zip Code	No.	Min.	Max.	AM	GM	SD	CV	Md	Q1	Q3
44802	19	0.10	120.00	11.21	4.05	26.65	1.36	5.60	1.90	8.10
44803	1	3.70	3.70	3.70	3.70	0.00	0.00	3.70	3.70	3.70
44804	15	0.80	19.90	8.59	6.18	6.45	0.33	6.20	4.70	17.40
44805	451	0.10	155.80	8.17	3.62	17.70	0.91	3.90	1.70	7.30
44806	1	2.40	2.40	2.40	2.40	0.00	0.00	2.40	2.40	2.40
44807	25	0.20	44.70	7.13	3.10	10.74	0.55	3.00	1.30	5.90
44809	2	7.00	14.90	10.95	10.21	5.59	0.29	10.95	7.00	14.90
44810	1	1.00	1.00	1.00	1.00	0.00	0.00	1.00	1.00	1.00
44811	262	0.10	159.00	15.21	6.34	24.33	1.25	5.75	2.80	17.90
44813	209	0.20	318.80	20.72	8.99	38.36	1.96	8.30	3.90	24.00
44814	77	0.10	247.50	13.11	5.86	28.79	1.47	6.20	2.40	16.70
44815	8	2.20	15.80	7.83	6.35	5.16	0.26	6.20	3.30	10.00
44816	6	0.50	11.00	3.87	2.71	3.64	0.19	2.95	2.50	3.30
44817	30	0.10	18.40	4.31	2.59	3.91	0.20	3.45	1.80	5.10
44818	21	0.10	16.40	4.30	2.72	3.99	0.20	2.80	2.00	5.80
44820	223	0.10	163.00	6.61	4.11	11.93	0.61	4.20	2.50	7.80
44822	56	1.20	264.00	36.45	9.52	63.76	3.26	6.20	2.70	20.00
44823	1	0.60	0.60	0.60	0.60	0.00	0.00	0.60	0.60	0.60
44824	129	0.10	279.20	13.43	5.23	36.71	1.88	5.80	2.50	11.20
44826	15	0.60	15.80	5.37	3.42	5.12	0.26	3.20	2.20	6.10
44827	181	0.10	56.10	3.18	1.25	5.67	0.29	1.20	0.50	3.70
44828	2	2.50	4.50	3.50	3.35	1.41	0.07	3.50	2.50	4.50
44829	1	6.20	6.20	6.20	6.20	0.00	0.00	6.20	6.20	6.20
44830	272	0.10	86.20	6.32	3.21	8.51	0.44	3.60	1.60	8.40
44833	493	0.10	140.20	7.75	4.41	12.72	0.65	4.70	2.60	8.40
44834	1	4.00	4.00	4.00	4.00	0.00	0.00	4.00	4.00	4.00
44835	1	13.30	13.30	13.30	13.30	0.00	0.00	13.30	13.30	13.30
44836	26	0.10	20.50	5.16	2.04	6.04	0.31	2.20	0.80	9.10
44837	17	0.70	50.80	8.52	4.20	13.38	0.69	3.50	2.10	7.30
44838	6	1.00	23.70	6.15	3.23	8.74	0.45	2.90	1.10	5.30
44839	464	0.10	70.10	6.40	3.72	7.78	0.40	4.30	2.00	7.90
44840	26	0.30	185.30	17.34	5.93	37.57	1.92	4.20	2.40	16.30
44841	10	0.10	5.20	1.69	1.01	1.59	0.08	1.25	0.70	2.50
44842	52	0.10	29.90	6.49	3.53	6.88	0.35	3.75	1.70	9.60

Zip Code	No.	Min.	Max.	AM	GM	SD	CV	Md	Q1	Q3
44843	35	0.10	179.00	24.30	8.47	42.04	2.15	9.00	3.50	26.20
44844	13	0.20	26.10	7.32	3.32	8.30	0.42	4.30	1.80	7.10
44845	1	10.20	10.20	10.20	10.20	0.00	0.00	10.20	10.20	10.20
44846	101	0.10	81.60	9.22	4.50	13.68	0.70	4.70	2.30	8.70
44847	132	0.10	149.70	23.18	9.05	35.72	1.83	8.20	3.80	20.80
44848	2	6.80	10.90	8.85	8.61	2.90	0.15	8.85	6.80	10.90
44849	52	0.20	20.90	4.15	2.88	4.36	0.22	3.15	1.60	4.40
44850	1	1.00	1.00	1.00	1.00	0.00	0.00	1.00	1.00	1.00
44851	32	0.90	13.70	5.01	4.14	3.28	0.17	4.20	2.50	6.40
44853	53	0.20	23.00	6.86	4.26	6.43	0.33	3.90	1.90	10.90
44854	15	0.20	7.10	2.96	2.21	1.78	0.09	2.50	2.00	4.40
44855	6	3.50	22.40	8.22	6.47	7.21	0.37	5.75	3.70	8.20
44856	3	1.70	5.60	3.00	2.53	2.25	0.12	1.70	1.70	5.60
44857	335	0.10	104.10	6.41	3.23	10.19	0.52	3.40	1.80	7.20
44859	27	0.10	17.40	5.75	4.58	3.40	0.17	5.10	4.00	6.50
44860	3	3.40	7.30	4.70	4.39	2.25	0.12	3.40	3.40	7.30
44861	2	2.20	3.70	2.95	2.85	1.06	0.05	2.95	2.20	3.70
44862	2	1.90	4.40	3.15	2.89	1.77	0.09	3.15	1.90	4.40
44864	42	0.90	115.00	15.46	7.68	23.56	1.21	7.60	3.10	18.50
44865	15	2.30	38.10	10.89	7.47	11.62	0.60	6.60	5.00	10.15
44866	7	1.10	11.90	3.98	2.85	3.91	0.20	2.30	1.50	6.45
44867	34	0.40	10.50	3.70	2.96	2.39	0.12	3.20	1.90	5.40
44870	674	0.10	134.90	4.89	2.89	7.31	0.37	3.20	1.60	5.60
44875	169	0.10	32.60	5.68	4.02	5.21	0.27	4.30	2.50	6.90
44878	14	0.40	16.60	4.61	3.35	4.13	0.21	2.90	2.80	5.30
44879	1	1.90	1.90	1.90	1.90	0.00	0.00	1.90	1.90	1.90
44880	13	0.60	17.70	5.42	3.18	5.48	0.28	3.90	0.90	5.10
44881	1	3.90	3.90	3.90	3.90	0.00	0.00	3.90	3.90	3.90
44882	32	0.50	9.10	4.19	2.91	3.00	0.15	3.60	1.00	7.30
44883	1352	0.10	187.90	8.30	4.50	12.43	0.64	5.30	2.30	9.80
44887	7	1.80	7.80	4.16	3.57	2.57	0.13	3.40	2.00	7.80
44888	4	1.20	7.60	4.55	3.76	2.64	0.14	4.70	1.20	5.10
44889	44	0.10	24.80	4.74	2.88	4.77	0.24	3.25	1.30	5.90
44890	80	0.30	139.30	8.69	3.92	21.85	1.12	3.60	2.10	6.20

Zip Code	No.	Min.	Max.	AM	GM	SD	CV	Md	Q1	Q3
44891	1	1.40	1.40	1.40	1.40	0.00	0.00	1.40	1.40	1.40
44892	1	1.70	1.70	1.70	1.70	0.00	0.00	1.70	1.70	1.70
44901	57	0.10	43.40	4.60	2.39	7.81	0.40	2.20	1.30	4.20
44902	37	0.50	10.20	2.58	1.81	2.52	0.13	1.50	1.00	3.20
44903	494	0.10	150.40	8.02	4.43	12.43	0.64	4.70	2.30	8.60
44904	798	0.10	292.30	9.75	4.76	17.45	0.89	4.90	2.50	9.80
44905	123	0.10	24.60	5.65	3.70	4.97	0.25	4.00	2.10	8.00
44906	447	0.10	44.00	5.51	3.40	6.03	0.31	3.60	1.90	7.10
44907	341	0.10	64.90	5.55	3.44	7.54	0.39	3.50	2.10	6.00
45001	5	1.60	3.60	2.74	2.63	0.84	0.04	2.70	2.30	3.50
45002	125	0.10	28.00	3.78	2.66	3.56	0.18	2.70	1.20	5.10
45003	10	0.90	14.00	6.01	4.40	4.43	0.23	4.00	3.10	10.20
45005	580	0.10	70.70	6.82	3.56	8.76	0.45	3.90	1.80	8.30
45011	1111	0.10	85.90	4.90	3.18	5.37	0.27	3.60	1.70	6.30
45012	1	0.90	0.90	0.90	0.90	0.00	0.00	0.90	0.90	0.90
45013	562	0.10	470.00	49.54	5.07	114.91	5.88	3.10	1.50	8.20
45014	776	0.10	61.10	6.45	3.44	7.75	0.40	4.05	1.60	8.10
45015	72	0.10	49.70	5.60	3.05	7.34	0.38	3.10	1.60	6.60
45017	1	5.40	5.40	5.40	5.40	0.00	0.00	5.40	5.40	5.40
45018	5	1.30	19.70	6.66	3.93	7.72	0.40	3.20	1.50	7.60
45020	1	1.30	1.30	1.30	1.30	0.00	0.00	1.30	1.30	1.30
45022	1	1.80	1.80	1.80	1.80	0.00	0.00	1.80	1.80	1.80
45023	2	3.70	3.80	3.75	3.75	0.07	0.00	3.75	3.70	3.80
45026	1	5.40	5.40	5.40	5.40	0.00	0.00	5.40	5.40	5.40
45028	1	2.60	2.60	2.60	2.60	0.00	0.00	2.60	2.60	2.60
45029	3	1.80	4.20	2.60	2.39	1.39	0.07	1.80	1.80	4.20
45030	261	0.10	59.00	7.21	4.35	7.40	0.38	4.90	2.50	9.40
45031	1	1.70	1.70	1.70	1.70	0.00	0.00	1.70	1.70	1.70
45033	4	1.00	6.00	3.38	2.78	2.14	0.11	3.25	1.00	4.00
45034	44	0.10	27.40	3.52	1.95	4.65	0.24	2.05	1.00	4.20
45036	870	0.10	34.80	4.49	2.97	4.38	0.22	3.20	1.70	5.60
45037	1	3.40	3.40	3.40	3.40	0.00	0.00	3.40	3.40	3.40
45039	875	0.10	219.00	4.87	3.08	9.13	0.47	3.50	1.70	5.90
45040	2066	0.10	44.10	4.43	2.89	4.44	0.23	3.25	1.60	5.70

Zip Code	No.	Min.	Max.	AM	GM	SD	CV	Md	Q1	Q3
45041	3	1.20	3.20	2.10	1.94	1.01	0.05	1.90	1.20	3.20
45042	640	0.10	143.80	4.51	1.96	7.76	0.40	2.10	0.50	6.20
45043	3	3.20	11.40	6.30	5.39	4.45	0.23	4.30	3.20	11.40
45044	963	0.10	50.20	4.83	3.14	4.94	0.25	3.40	1.70	6.20
45045	2	3.50	7.50	5.50	5.12	2.83	0.14	5.50	3.50	7.50
45046	1	3.80	3.80	3.80	3.80	0.00	0.00	3.80	3.80	3.80
45049	2	3.40	4.50	3.95	3.91	0.78	0.04	3.95	3.40	4.50
45050	153	0.10	30.60	4.75	2.78	5.09	0.26	2.80	1.20	6.10
45051	1	7.40	7.40	7.40	7.40	0.00	0.00	7.40	7.40	7.40
45052	36	0.40	21.80	3.87	2.74	3.84	0.20	2.85	1.20	5.10
45053	34	0.10	11.00	3.93	2.77	2.69	0.14	3.95	1.60	5.70
45054	45	0.10	30.20	5.23	2.87	5.55	0.28	3.60	1.90	6.50
45056	366	0.10	61.10	4.93	2.95	5.99	0.31	3.40	1.80	6.30
45057	1	3.80	3.80	3.80	3.80	0.00	0.00	3.80	3.80	3.80
45059	1	0.50	0.50	0.50	0.50	0.00	0.00	0.50	0.50	0.50
45060	5	1.00	6.80	3.66	2.65	2.88	0.15	2.90	1.00	6.60
45061	5	2.70	19.40	7.10	5.41	6.92	0.35	4.60	4.00	4.80
45062	11	2.60	15.20	8.10	6.99	4.36	0.22	7.80	4.50	13.30
45063	3	2.90	5.70	3.97	3.79	1.51	0.08	3.30	2.90	5.70
45064	18	0.50	30.90	6.14	2.77	9.39	0.48	2.40	1.00	5.40
45065	118	0.20	34.00	5.40	3.79	4.82	0.25	4.55	2.50	6.30
45066	1044	0.10	58.10	4.50	2.89	4.82	0.25	3.15	1.50	5.80
45067	126	0.10	48.90	9.89	5.71	10.16	0.52	7.70	2.70	14.20
45068	335	0.10	44.90	5.63	3.58	6.00	0.31	3.90	1.90	7.10
45069	1942	0.10	69.80	4.52	2.92	4.71	0.24	3.30	1.60	5.60
45070	2	1.70	2.20	1.95	1.93	0.35	0.02	1.95	1.70	2.20
45071	4	1.00	1.80	1.38	1.34	0.35	0.02	1.35	1.00	1.50
45072	1	1.90	1.90	1.90	1.90	0.00	0.00	1.90	1.90	1.90
45076	1	12.10	12.10	12.10	12.10	0.00	0.00	12.10	12.10	12.10
45081	3	0.40	3.70	2.60	1.76	1.91	0.10	3.70	0.40	3.70
45085	1	7.20	7.20	7.20	7.20	0.00	0.00	7.20	7.20	7.20
45101	16	0.90	9.80	3.88	2.92	3.01	0.15	2.65	1.40	5.20
45102	190	0.10	26.30	4.42	2.70	4.16	0.21	3.00	1.40	6.50
45103	325	0.10	37.20	4.72	3.06	4.61	0.24	3.20	1.70	5.90

Zip Code	No.	Min.	Max.	AM	GM	SD	CV	Md	Q1	Q3
45104	3	6.60	12.60	10.60	10.16	3.46	0.18	12.60	6.60	12.60
45105	8	1.30	7.80	3.35	2.75	2.33	0.12	2.20	1.40	5.00
45106	59	0.10	13.60	3.39	2.08	2.99	0.15	2.70	0.90	5.00
45107	66	0.10	9.80	3.17	2.23	2.39	0.12	2.55	1.30	5.20
45109	1	4.30	4.30	4.30	4.30	0.00	0.00	4.30	4.30	4.30
45110	2	0.70	7.60	4.15	2.31	4.88	0.25	4.15	0.70	7.60
45111	7	2.60	15.60	9.06	7.79	4.55	0.23	10.50	3.90	11.80
45113	44	0.60	54.20	5.94	3.36	8.98	0.46	3.25	1.50	5.70
45118	15	0.10	8.00	2.93	1.86	2.39	0.12	2.40	1.20	4.10
45120	20	0.50	18.20	3.22	1.96	4.18	0.21	1.70	1.10	3.20
45121	42	0.20	20.10	3.13	1.50	4.16	0.21	1.90	0.50	3.30
45122	59	0.30	15.20	3.77	2.83	3.03	0.16	2.70	1.90	4.80
45123	20	0.20	43.00	8.27	4.41	10.50	0.54	4.60	2.70	8.50
45125	1	7.20	7.20	7.20	7.20	0.00	0.00	7.20	7.20	7.20
45129	2	0.90	3.20	2.05	1.70	1.63	0.08	2.05	0.90	3.20
45130	3	0.30	1.60	0.77	0.58	0.72	0.04	0.40	0.30	1.60
45131	2	1.10	1.10	1.10	1.10	0.00	0.00	1.10	1.10	1.10
45132	5	0.90	9.20	3.48	2.38	3.40	0.17	3.20	0.90	3.20
45133	126	0.10	31.10	4.14	2.49	4.51	0.23	2.60	1.50	5.10
45135	10	0.50	15.40	3.54	1.94	4.58	0.23	1.90	0.80	3.50
45138	9	1.90	42.10	11.28	6.18	14.07	0.72	3.80	2.80	13.70
45139	2	0.10	6.70	3.40	0.82	4.67	0.24	3.40	0.10	6.70
45140	1852	0.10	470.80	4.95	2.99	12.09	0.62	3.40	1.60	6.20
45141	3	0.30	7.20	2.73	1.15	3.87	0.20	0.70	0.30	7.20
45142	25	0.50	9.40	4.15	3.05	2.78	0.14	4.20	1.40	5.90
45144	8	0.60	16.80	5.41	3.23	5.81	0.30	3.30	0.90	3.70
45146	10	0.20	40.30	7.98	3.49	11.83	0.61	3.90	2.20	8.60
45147	4	2.30	4.90	3.28	3.11	1.24	0.06	2.95	2.30	3.60
45148	7	0.80	7.30	4.50	3.59	2.64	0.13	4.90	2.40	7.10
45150	644	0.10	83.50	5.14	3.30	6.53	0.33	3.40	1.90	6.30
45152	236	0.10	30.10	5.59	3.68	4.90	0.25	4.20	2.10	7.30
45153	9	0.60	2.90	1.09	0.91	0.81	0.04	0.70	0.60	0.90
45154	19	0.50	18.30	3.42	1.88	4.35	0.22	1.90	0.70	5.90
45155	2	5.50	6.50	6.00	5.98	0.71	0.04	6.00	5.50	6.50

Zip Code	No.	Min.	Max.	AM	GM	SD	CV	Md	Q1	Q3
45157	80	0.10	25.20	3.85	2.54	4.23	0.22	2.75	1.40	4.50
45159	37	0.10	30.90	6.73	2.94	7.69	0.39	3.10	1.60	9.90
45160	3	1.00	1.40	1.20	1.19	0.20	0.01	1.20	1.00	1.40
45162	35	0.20	6.40	2.05	1.51	1.61	0.08	1.20	0.90	3.40
45165	1	14.30	14.30	14.30	14.30	0.00	0.00	14.30	14.30	14.30
45166	1	2.00	2.00	2.00	2.00	0.00	0.00	2.00	2.00	2.00
45167	10	0.60	8.20	3.09	2.10	2.80	0.14	1.75	1.00	4.60
45168	2	1.80	2.50	2.15	2.12	0.49	0.03	2.15	1.80	2.50
45169	23	0.40	12.00	4.24	3.11	3.46	0.18	3.50	1.90	5.40
45170	1	4.10	4.10	4.10	4.10	0.00	0.00	4.10	4.10	4.10
45171	27	0.40	9.80	3.01	2.05	2.64	0.14	1.70	0.90	4.50
45172	1	0.10	0.10	0.10	0.10	0.00	0.00	0.10	0.10	0.10
45174	212	0.10	74.60	6.27	3.98	7.14	0.37	4.40	2.60	7.70
45175	1	2.40	2.40	2.40	2.40	0.00	0.00	2.40	2.40	2.40
45176	19	0.50	11.10	4.02	2.86	3.17	0.16	3.30	1.50	5.90
45177	317	0.10	55.60	4.80	2.85	6.47	0.33	3.20	1.50	5.90
45200	20	0.40	3.80	1.43	1.14	0.98	0.05	1.15	0.70	2.00
45201	10	0.10	6.30	2.25	1.49	1.79	0.09	2.20	0.70	2.70
45202	105	0.10	27.00	2.29	1.23	3.79	0.19	1.10	0.60	2.50
45203	24	0.10	5.30	1.60	1.22	1.30	0.07	1.10	0.80	1.70
45204	15	0.10	8.60	2.27	1.41	2.37	0.12	1.10	0.80	2.30
45205	76	0.10	14.10	2.24	1.45	2.25	0.12	1.40	0.90	3.10
45206	107	0.10	15.20	2.57	1.69	2.58	0.13	1.80	1.00	3.20
45207	152	0.10	4.90	0.99	0.71	0.97	0.05	0.70	0.40	1.20
45208	868	0.10	35.40	2.79	2.06	2.64	0.14	2.10	1.20	3.50
45209	264	0.10	19.50	2.52	1.75	2.53	0.13	1.70	1.00	3.20
45210	9	0.10	4.70	2.23	1.61	1.34	0.07	2.20	1.20	3.10
45211	315	0.10	25.20	2.69	1.69	3.17	0.16	1.80	0.90	3.60
45212	196	0.10	26.30	2.22	1.46	2.63	0.13	1.40	0.80	2.90
45213	203	0.10	49.50	3.35	2.29	4.40	0.23	2.20	1.40	4.10
45214	27	0.10	6.50	2.70	1.88	1.95	0.10	2.10	0.90	4.80
45215	634	0.10	20.60	3.00	1.96	2.87	0.15	2.05	1.10	4.10
45216	47	0.10	16.80	3.89	2.46	3.79	0.19	3.00	1.00	5.00
45217	55	0.50	30.50	2.99	2.02	4.25	0.22	1.90	1.10	3.00

Zip Code	No.	Min.	Max.	AM	GM	SD	CV	Md	Q1	Q3
45218	56	0.10	17.50	2.92	1.74	3.46	0.18	1.95	1.00	3.20
45219	71	0.10	18.70	2.23	1.15	2.90	0.15	1.30	0.60	2.80
45220	167	0.10	58.50	3.24	1.85	6.57	0.34	1.90	1.00	3.40
45221	7	0.20	13.60	2.90	1.09	4.80	0.25	1.50	0.30	2.20
45223	99	0.10	17.50	3.19	2.11	2.93	0.15	2.30	1.30	4.30
45224	233	0.10	32.40	3.42	2.04	4.06	0.21	2.20	1.00	4.10
45225	12	0.30	5.90	1.83	1.13	1.96	0.10	0.75	0.50	2.50
45226	185	0.20	17.80	2.03	1.46	2.15	0.11	1.40	0.90	2.60
45227	348	0.10	64.80	4.29	2.69	5.67	0.29	2.80	1.50	4.90
45228	3	0.80	8.40	3.77	2.42	4.06	0.21	2.10	0.80	8.40
45229	64	0.30	19.20	2.43	1.71	2.70	0.14	1.70	1.00	2.90
45230	686	0.10	27.00	3.11	2.25	2.78	0.14	2.40	1.40	4.10
45231	607	0.10	40.10	2.99	1.94	3.30	0.17	2.00	1.00	3.60
45232	35	0.30	23.40	3.86	2.23	5.25	0.27	1.80	1.10	3.90
45233	202	0.20	26.70	3.66	2.63	3.34	0.17	2.70	1.60	4.70
45234	6	0.50	26.40	6.18	2.83	9.95	0.51	2.65	1.80	3.10
45235	3	0.10	5.70	2.97	1.21	2.80	0.14	3.10	0.10	5.70
45236	653	0.10	87.00	2.76	1.77	4.32	0.22	1.80	1.00	3.30
45237	151	0.10	19.40	3.00	2.09	2.66	0.14	2.40	1.00	4.00
45238	374	0.10	42.40	2.68	1.60	3.99	0.20	1.80	0.90	3.00
45239	333	0.10	23.50	2.67	1.60	3.16	0.16	1.60	0.90	3.40
45240	285	0.10	18.60	2.57	1.68	2.69	0.14	1.60	0.90	3.10
45241	1001	0.10	42.30	4.07	2.48	4.41	0.23	2.70	1.30	5.20
45242	1039	0.10	33.50	3.91	2.59	4.10	0.21	2.70	1.50	4.80
45243	820	0.10	26.50	2.99	2.21	2.67	0.14	2.30	1.40	3.70
45244	860	0.10	40.50	4.16	2.84	4.19	0.21	3.10	1.60	5.00
45245	338	0.10	23.70	5.03	3.30	4.96	0.25	3.60	1.90	6.30
45246	233	0.10	23.50	3.89	2.50	3.85	0.20	2.50	1.40	4.90
45247	371	0.10	15.90	3.08	2.12	2.67	0.14	2.20	1.30	3.80
45248	263	0.10	316.65	4.28	2.00	19.69	1.01	1.90	1.10	3.60
45249	621	0.10	34.30	4.22	2.75	4.22	0.22	3.00	1.60	5.40
45250	1	3.30	3.30	3.30	3.30	0.00	0.00	3.30	3.30	3.30
45251	188	0.10	45.50	3.21	2.12	4.35	0.22	2.05	1.10	3.80
45252	73	0.10	22.50	3.76	2.71	3.38	0.17	2.70	1.60	5.10

Zip Code	No.	Min.	Max.	AM	GM	SD	CV	Md	Q1	Q3
45253	1	4.00	4.00	4.00	4.00	0.00	0.00	4.00	4.00	4.00
45254	2	2.30	4.30	3.30	3.14	1.41	0.07	3.30	2.30	4.30
45255	557	0.10	42.40	4.19	2.72	4.05	0.21	3.10	1.60	5.30
45258	1	1.30	1.30	1.30	1.30	0.00	0.00	1.30	1.30	1.30
45262	1	3.00	3.00	3.00	3.00	0.00	0.00	3.00	3.00	3.00
45263	1	20.50	20.50	20.50	20.50	0.00	0.00	20.50	20.50	20.50
45264	1	2.80	2.80	2.80	2.80	0.00	0.00	2.80	2.80	2.80
45265	1	0.40	0.40	0.40	0.40	0.00	0.00	0.40	0.40	0.40
45266	3	3.20	5.00	3.80	3.71	1.04	0.05	3.20	3.20	5.00
45271	2	0.50	1.70	1.10	0.92	0.85	0.04	1.10	0.50	1.70
45274	2	1.50	3.10	2.30	2.16	1.13	0.06	2.30	1.50	3.10
45275	3	0.10	4.50	1.83	0.74	2.34	0.12	0.90	0.10	4.50
45280	2	4.30	5.80	5.05	4.99	1.06	0.05	5.05	4.30	5.80
45291	1	4.20	4.20	4.20	4.20	0.00	0.00	4.20	4.20	4.20
45300	1	1.60	1.60	1.60	1.60	0.00	0.00	1.60	1.60	1.60
45301	6	1.80	17.90	13.45	10.69	6.71	0.34	17.15	8.80	17.90
45302	65	0.20	61.30	10.37	5.11	12.75	0.65	5.60	2.40	11.10
45303	15	2.00	125.00	15.37	5.80	32.18	1.65	5.30	2.30	8.20
45304	136	0.10	49.00	7.09	3.95	7.55	0.39	4.60	2.00	9.60
45305	841	0.10	79.50	6.79	4.44	6.93	0.35	4.80	2.50	8.80
45306	32	0.80	18.30	6.79	5.05	5.07	0.26	6.30	2.70	8.30
45307	8	0.80	8.80	3.26	2.65	2.42	0.12	2.85	1.50	3.40
45308	66	0.10	71.00	8.43	4.69	12.25	0.63	4.70	2.70	8.30
45309	428	0.10	267.80	7.08	3.65	14.87	0.76	4.20	1.90	7.80
45310	8	0.70	13.00	5.96	4.16	4.33	0.22	6.20	1.10	6.30
45311	55	0.10	41.50	5.66	3.14	7.16	0.37	3.60	1.90	6.80
45312	43	0.10	59.20	9.64	6.01	10.29	0.53	6.40	3.50	13.80
45313	1	1.60	1.60	1.60	1.60	0.00	0.00	1.60	1.60	1.60
45314	176	0.10	97.70	5.40	2.71	10.73	0.55	3.40	1.50	6.00
45315	183	0.10	67.30	5.36	3.38	6.50	0.33	3.50	2.20	6.90
45316	7	0.20	9.30	3.06	1.87	3.05	0.16	1.80	1.00	4.10
45317	16	0.20	19.60	8.84	6.20	5.70	0.29	9.20	3.30	11.80
45318	88	0.10	252.00	11.42	5.54	27.38	1.40	5.50	2.50	11.20
45319	19	0.20	26.60	5.60	3.48	6.03	0.31	3.50	1.90	7.60

Zip Code	No.	Min.	Max.	AM	GM	SD	CV	Md	Q1	Q3
45320	282	0.10	54.50	6.57	3.97	7.08	0.36	4.10	2.00	8.40
45321	12	1.20	38.10	9.10	5.23	11.11	0.57	3.75	2.80	10.20
45322	655	0.10	69.60	6.06	3.42	7.32	0.37	3.70	1.80	7.80
45323	287	0.10	67.10	9.81	5.70	10.81	0.55	6.40	3.00	12.40
45324	1357	0.10	163.00	7.91	4.13	10.98	0.56	4.60	2.00	9.20
45325	62	0.50	27.70	6.78	4.54	6.61	0.34	4.75	2.20	8.90
45326	27	0.10	20.10	6.49	3.70	5.56	0.28	4.80	2.40	10.00
45327	233	0.10	86.10	9.23	5.43	11.52	0.59	5.30	2.80	11.70
45328	3	0.80	2.70	1.50	1.29	1.04	0.05	1.00	0.80	2.70
45329	3	1.50	4.50	2.70	2.42	1.59	0.08	2.10	1.50	4.50
45330	4	1.00	5.10	2.60	2.18	1.79	0.09	2.15	1.00	2.60
45331	334	0.10	84.50	8.88	4.28	11.47	0.59	5.65	2.50	10.70
45332	5	1.10	9.00	5.68	4.29	3.70	0.19	7.40	2.30	8.60
45333	19	1.30	36.00	8.16	5.12	8.80	0.45	4.80	2.60	11.70
45334	19	0.70	17.00	3.70	2.54	4.05	0.21	2.80	1.30	4.10
45335	195	0.10	39.80	4.88	2.68	6.69	0.34	3.00	1.60	5.40
45336	4	0.60	5.80	2.83	1.95	2.43	0.12	2.45	0.60	3.80
45337	49	0.20	18.10	6.04	4.54	4.44	0.23	4.20	3.10	7.30
45338	117	0.10	57.00	9.57	5.48	10.35	0.53	6.50	3.00	11.50
45339	55	0.50	40.40	11.10	6.91	10.84	0.55	5.80	3.80	15.30
45340	14	0.10	13.10	4.27	1.85	4.68	0.24	2.10	0.60	8.20
45341	151	0.10	750.10	27.41	11.31	85.32	4.37	13.40	5.00	26.40
45342	926	0.10	52.70	4.21	2.52	4.91	0.25	2.60	1.40	5.50
45344	440	0.10	150.10	11.30	5.75	16.58	0.85	5.90	2.80	12.80
45345	145	0.10	40.10	5.92	3.86	5.99	0.31	4.50	1.90	7.00
45346	41	0.40	62.30	10.07	5.54	12.21	0.63	6.90	1.80	12.00
45347	41	0.60	44.40	8.94	5.67	8.97	0.46	6.80	2.50	10.10
45348	11	1.20	33.90	8.80	3.87	12.70	0.65	2.30	1.30	10.30
45349	12	0.80	6.60	3.33	2.75	1.88	0.10	3.20	1.60	4.60
45350	3	0.80	2.20	1.53	1.41	0.70	0.04	1.60	0.80	2.20
45351	10	0.80	38.20	10.30	5.50	11.32	0.58	7.00	2.00	14.30
45352	1	10.80	10.80	10.80	10.80	0.00	0.00	10.80	10.80	10.80
45354	22	0.10	18.30	3.96	1.13	5.58	0.29	2.25	0.10	4.30
45356	454	0.10	160.00	8.59	5.09	12.29	0.63	5.18	3.00	9.90

Zip Code	No.	Min.	Max.	AM	GM	SD	CV	Md	Q1	Q3
45358	3	0.90	2.40	1.70	1.57	0.75	0.04	1.80	0.90	2.40
45359	62	0.50	105.00	12.71	5.96	21.84	1.12	5.35	2.70	11.90
45361	2	6.30	6.30	6.30	6.30	0.00	0.00	6.30	6.30	6.30
45362	9	0.50	47.70	13.72	5.64	19.07	0.98	6.10	2.30	7.60
45363	26	1.10	17.70	5.58	3.96	4.77	0.24	4.20	2.00	7.90
45365	567	0.10	111.10	7.67	4.63	8.64	0.44	5.50	2.60	9.70
45367	1	12.40	12.40	12.40	12.40	0.00	0.00	12.40	12.40	12.40
45368	85	0.20	35.80	5.33	3.18	5.63	0.29	4.70	2.10	7.10
45369	88	0.10	21.30	5.09	3.26	4.74	0.24	3.90	1.50	6.40
45370	346	0.10	129.00	8.17	4.28	13.24	0.68	4.40	2.60	8.80
45371	787	0.10	204.00	7.69	4.54	11.83	0.61	4.70	2.50	8.80
45372	21	0.20	37.50	8.49	5.52	8.28	0.42	6.90	2.90	7.90
45373	1402	0.10	223.00	8.44	5.09	10.72	0.55	5.50	2.50	10.50
45374	1	9.20	9.20	9.20	9.20	0.00	0.00	9.20	9.20	9.20
45375	2	9.40	9.40	9.40	9.40	0.00	0.00	9.40	9.40	9.40
45376	2	14.00	14.00	14.00	14.00	0.00	0.00	14.00	14.00	14.00
45377	538	0.10	40.80	4.58	2.90	4.38	0.22	3.30	1.50	6.20
45378	7	1.00	13.80	6.41	4.38	4.92	0.25	7.00	1.40	9.60
45380	149	0.10	41.60	7.38	4.65	7.48	0.38	4.90	2.50	9.80
45381	123	0.10	90.30	8.12	4.28	11.83	0.61	4.50	2.30	9.00
45382	18	0.10	45.10	10.11	4.19	11.91	0.61	5.20	1.60	15.00
45383	167	0.10	62.90	8.16	4.75	10.13	0.52	5.10	2.70	8.20
45384	35	0.10	26.90	4.70	2.42	6.88	0.35	2.50	0.90	4.00
45385	2910	0.10	142.00	8.21	4.78	9.90	0.51	5.30	2.50	10.10
45387	653	0.10	111.90	6.83	4.02	9.20	0.47	4.30	2.30	8.50
45388	13	1.70	16.20	5.85	4.27	5.00	0.26	2.80	2.10	7.70
45389	12	1.40	30.80	11.13	7.62	8.57	0.44	11.25	2.30	13.40
45390	21	1.30	16.80	6.75	5.38	4.59	0.24	5.50	3.30	8.80
45401	97	0.10	36.50	11.63	6.07	9.59	0.49	9.20	2.50	22.00
45402	127	0.10	84.80	6.57	2.87	10.23	0.52	3.10	1.60	7.20
45403	248	0.10	65.00	5.05	3.31	6.06	0.31	3.55	1.90	6.40
45404	134	0.20	48.20	8.74	6.05	7.92	0.41	6.35	3.80	10.10
45405	462	0.10	49.60	5.78	3.56	5.78	0.30	4.05	1.90	7.90
45406	441	0.10	32.90	3.69	2.44	3.70	0.19	2.60	1.40	4.90

Zip Code	No.	Min.	Max.	AM	GM	SD	CV	Md	Q1	Q3
45407	79	0.10	71.00	10.94	2.85	16.89	0.86	3.00	0.60	16.30
45408	69	0.20	191.00	10.24	3.98	23.73	1.21	4.00	1.70	11.00
45409	462	0.10	39.90	5.97	3.57	5.73	0.29	4.55	2.00	7.80
45410	374	0.10	35.00	3.79	2.55	3.79	0.19	2.55	1.40	5.00
45413	4	2.70	11.40	6.38	5.62	3.64	0.19	5.70	2.70	5.70
45414	720	0.10	410.00	7.41	4.06	16.99	0.87	4.60	2.00	8.30
45415	727	0.10	416.30	7.65	4.35	17.30	0.89	4.60	2.50	8.20
45416	152	0.10	61.20	5.50	2.69	8.82	0.45	2.50	1.20	6.40
45417	95	0.10	60.10	5.46	2.28	8.82	0.45	3.50	1.10	5.50
45418	120	0.10	21.50	3.38	2.09	3.93	0.20	2.20	1.10	3.60
45419	1394	0.10	220.10	5.49	3.28	10.18	0.52	3.40	1.80	6.50
45420	1100	0.10	76.50	4.87	3.06	5.91	0.30	3.30	1.80	5.90
45421	7	1.20	7.20	2.47	2.00	2.14	0.11	1.50	1.20	2.50
45422	8	0.10	10.90	4.05	1.82	3.55	0.18	3.30	0.10	5.40
45423	4	0.30	16.90	5.58	1.92	7.79	0.40	2.55	0.30	4.50
45424	1187	0.10	83.20	5.16	2.89	6.83	0.35	2.90	1.40	6.00
45426	427	0.10	161.00	5.72	2.97	12.79	0.65	3.00	1.50	6.30
45427	96	0.10	18.60	4.44	2.95	3.58	0.18	3.35	1.90	5.90
45428	19	0.10	4.20	1.77	1.16	1.30	0.07	1.60	0.60	3.00
45429	2059	0.10	440.70	5.53	3.30	11.77	0.60	3.60	1.80	6.60
45430	652	0.10	75.90	7.29	4.28	8.39	0.43	4.55	2.20	9.20
45431	975	0.10	117.00	5.65	3.50	6.94	0.36	3.80	1.80	6.70
45432	864	0.10	117.00	5.85	3.70	6.68	0.34	4.10	2.00	7.60
45433	31	0.10	268.80	44.31	3.62	83.01	4.25	2.60	0.50	49.00
45434	1100	0.10	820.00	7.47	4.23	25.62	1.31	4.50	2.40	8.40
45435	23	0.10	12.90	1.77	0.44	3.36	0.17	0.20	0.10	2.20
45437	6	0.90	11.80	3.77	2.52	4.07	0.21	2.95	0.90	3.10
45439	183	0.10	56.10	8.34	5.07	8.23	0.42	5.50	3.00	11.40
45440	1520	0.10	359.00	5.35	3.20	12.18	0.62	3.40	1.80	6.10
45441	5	0.30	13.50	5.36	1.65	6.77	0.35	0.50	0.50	12.00
45448	14	0.50	6.10	1.80	1.26	1.90	0.10	1.10	0.70	1.60
45449	412	0.10	51.60	5.12	2.76	6.49	0.33	2.90	1.40	6.20
45450	6	3.90	7.90	5.63	5.52	1.28	0.07	5.50	5.40	5.60
45451	3	0.80	11.90	5.17	2.99	5.92	0.30	2.80	0.80	11.90

Zip Code	No.	Min.	Max.	AM	GM	SD	CV	Md	Q1	Q3
45453	3	3.10	6.20	5.17	4.92	1.79	0.09	6.20	3.10	6.20
45454	5	1.50	36.90	9.76	4.54	15.23	0.78	3.40	2.10	4.90
45456	2	5.70	15.00	10.35	9.25	6.58	0.34	10.35	5.70	15.00
45457	1	2.10	2.10	2.10	2.10	0.00	0.00	2.10	2.10	2.10
45458	1266	0.10	35.50	3.82	2.27	3.94	0.20	2.60	1.20	4.90
45459	2159	0.10	153.00	5.56	3.22	7.93	0.41	3.40	1.60	6.80
45463	1	5.10	5.10	5.10	5.10	0.00	0.00	5.10	5.10	5.10
45469	6	0.90	7.30	3.10	2.58	2.18	0.11	2.55	2.30	3.00
45475	2	4.30	15.50	9.90	8.16	7.92	0.41	9.90	4.30	15.50
45482	1	20.40	20.40	20.40	20.40	0.00	0.00	20.40	20.40	20.40
45490	3	1.00	2.10	1.37	1.28	0.64	0.03	1.00	1.00	2.10
45495	1	1.40	1.40	1.40	1.40	0.00	0.00	1.40	1.40	1.40
45500	2	1.60	6.00	3.80	3.10	3.11	0.16	3.80	1.60	6.00
45501	37	0.10	88.10	21.56	7.82	29.14	1.49	7.80	3.30	24.20
45502	652	0.10	860.00	10.34	4.52	41.04	2.10	5.10	2.30	9.80
45503	857	0.10	106.00	7.33	4.49	8.07	0.41	5.10	2.50	9.60
45504	567	0.10	100.00	6.05	3.71	7.88	0.40	3.90	2.10	7.40
45505	379	0.10	87.70	6.89	4.25	8.63	0.44	4.80	2.30	8.90
45506	246	0.20	47.10	6.71	4.01	7.37	0.38	4.40	2.10	8.30
45512	1	0.80	0.80	0.80	0.80	0.00	0.00	0.80	0.80	0.80
45601	1058	0.10	223.40	9.61	5.10	15.04	0.77	5.80	2.50	10.70
45612	48	0.20	28.40	5.95	2.28	7.52	0.39	1.80	0.90	8.20
45613	19	0.10	37.90	4.12	1.18	8.54	0.44	1.20	0.20	5.50
45614	21	0.30	10.60	2.17	1.52	2.37	0.12	1.50	1.00	2.20
45616	2	1.60	4.70	3.15	2.74	2.19	0.11	3.15	1.60	4.70
45617	1	2.50	2.50	2.50	2.50	0.00	0.00	2.50	2.50	2.50
45619	19	0.10	7.60	2.16	1.39	1.98	0.10	1.50	0.90	2.80
45620	5	1.60	4.10	2.78	2.65	0.92	0.05	2.70	2.40	3.10
45621	3	0.10	10.50	3.67	0.75	5.92	0.30	0.40	0.10	10.50
45622	2	0.90	2.30	1.60	1.44	0.99	0.05	1.60	0.90	2.30
45623	5	0.60	4.90	1.80	1.32	1.78	0.09	1.00	0.80	1.70
45624	4	0.20	0.20	0.20	0.20	0.00	0.00	0.20	0.20	0.20
45628	61	0.10	188.70	20.13	5.91	44.25	2.27	5.90	4.40	8.30
45629	7	0.40	4.80	2.07	1.57	1.52	0.08	2.10	0.80	2.70

Zip Code	No.	Min.	Max.	AM	GM	SD	CV	Md	Q1	Q3
45631	103	0.20	9.20	2.78	2.11	1.99	0.10	2.30	1.20	3.80
45633	1	15.90	15.90	15.90	15.90	0.00	0.00	15.90	15.90	15.90
45634	8	0.70	16.40	4.98	2.77	5.45	0.28	2.85	0.90	7.50
45636	1	0.60	0.60	0.60	0.60	0.00	0.00	0.60	0.60	0.60
45638	52	0.10	33.00	3.64	2.00	5.33	0.27	1.90	1.20	3.20
45640	99	0.10	20.30	3.43	2.21	3.20	0.16	2.60	1.20	4.50
45644	25	0.60	27.70	7.00	4.47	7.06	0.36	3.60	2.30	9.30
45645	6	0.70	1.90	0.95	0.88	0.47	0.02	0.75	0.70	0.90
45646	20	0.20	73.20	17.08	4.44	22.77	1.17	4.60	0.90	29.40
45647	25	0.70	21.40	6.92	6.09	3.59	0.18	6.40	5.60	7.90
45648	50	0.10	29.50	5.73	2.62	6.58	0.34	3.35	1.10	7.30
45650	2	35.40	35.40	35.40	35.40	0.00	0.00	35.40	35.40	35.40
45651	12	0.10	3.90	1.66	1.09	1.36	0.07	1.05	0.60	3.00
45652	12	0.80	6.30	3.09	2.59	1.80	0.09	2.70	1.30	4.00
45653	103	0.10	7.90	1.54	1.02	1.39	0.07	1.30	0.60	1.80
45654	4	1.10	5.80	2.63	2.13	2.14	0.11	1.80	1.10	1.80
45655	1	0.60	0.60	0.60	0.60	0.00	0.00	0.60	0.60	0.60
45656	14	0.60	12.40	3.62	2.30	3.89	0.20	1.75	1.20	6.40
45658	6	0.30	7.00	2.52	1.56	2.50	0.13	1.75	0.60	3.70
45660	65	0.10	42.50	5.09	1.71	8.41	0.43	1.90	0.50	4.50
45661	140	0.10	43.30	5.36	1.82	8.14	0.42	1.80	0.60	6.70
45662	98	0.10	23.40	3.84	2.65	3.93	0.20	2.60	1.60	4.60
45663	3	3.30	17.15	7.92	5.72	8.00	0.41	3.30	3.30	17.15
45664	1	182.30	182.30	182.30	182.30	0.00	0.00	182.30	182.30	182.30
45665	2	0.10	0.20	0.15	0.14	0.07	0.00	0.15	0.10	0.20
45669	60	0.10	333.40	9.39	2.16	42.88	2.20	1.95	0.90	5.30
45670	3	0.10	2.10	1.23	0.68	1.03	0.05	1.50	0.10	2.10
45672	4	1.90	13.50	5.00	3.39	5.68	0.29	2.30	1.90	2.70
45673	2	1.40	1.90	1.65	1.63	0.35	0.02	1.65	1.40	1.90
45674	6	0.70	5.20	2.17	1.70	1.68	0.09	1.75	0.90	2.70
45678	8	0.10	1.00	0.75	0.59	0.37	0.02	0.90	0.20	1.00
45679	16	0.70	6.90	3.61	3.14	1.81	0.09	3.10	2.20	5.00
45680	26	0.60	8.10	2.28	1.70	2.01	0.10	1.35	0.90	2.70
45681	3	4.50	52.90	24.57	15.71	25.24	1.29	16.30	4.50	52.90

Zip Code	No.	Min.	Max.	AM	GM	SD	CV	Md	Q1	Q3
45682	5	1.80	2.80	2.42	2.39	0.41	0.02	2.40	2.30	2.80
45684	11	0.80	21.60	9.22	6.54	7.21	0.37	7.40	3.30	15.20
45685	2	1.70	1.70	1.70	1.70	0.00	0.00	1.70	1.70	1.70
45686	9	0.50	3.70	1.59	1.41	0.88	0.05	1.40	1.30	1.50
45690	825	0.10	74.90	6.61	3.31	8.36	0.43	4.50	1.80	8.40
45692	30	0.20	24.30	3.93	1.79	6.01	0.31	1.40	0.70	4.00
45693	10	0.10	4.80	1.62	0.96	1.47	0.08	1.35	0.50	2.50
45694	29	0.80	9.30	3.18	2.53	2.35	0.12	2.80	1.60	3.40
45695	1	1.70	1.70	1.70	1.70	0.00	0.00	1.70	1.70	1.70
45697	22	0.10	15.60	4.63	2.41	4.79	0.25	2.40	1.20	10.50
45698	1	0.10	0.10	0.10	0.10	0.00	0.00	0.10	0.10	0.10
45700	1	0.10	0.10	0.10	0.10	0.00	0.00	0.10	0.10	0.10
45701	719	0.10	82.10	5.24	3.68	5.32	0.27	4.10	2.40	7.00
45706	2	1.60	2.40	2.00	1.96	0.57	0.03	2.00	1.60	2.40
45708	1	0.20	0.20	0.20	0.20	0.00	0.00	0.20	0.20	0.20
45710	53	0.30	12.40	3.18	2.06	3.11	0.16	1.90	1.10	4.70
45711	12	0.40	4.50	1.96	1.36	1.67	0.09	1.20	0.50	3.50
45712	2	1.00	3.70	2.35	1.92	1.91	0.10	2.35	1.00	3.70
45714	81	0.10	23.10	2.98	1.93	3.27	0.17	2.20	1.20	3.60
45715	12	2.20	27.00	6.94	5.45	6.62	0.34	5.30	3.00	7.30
45716	2	0.10	1.10	0.60	0.33	0.71	0.04	0.60	0.10	1.10
45717	1	1.60	1.60	1.60	1.60	0.00	0.00	1.60	1.60	1.60
45719	6	1.00	8.40	3.82	2.63	3.38	0.17	2.40	1.00	7.70
45720	1	3.80	3.80	3.80	3.80	0.00	0.00	3.80	3.80	3.80
45723	11	0.30	4.20	2.24	1.66	1.40	0.07	2.70	0.70	3.20
45724	11	0.20	5.80	2.91	2.19	1.83	0.09	2.70	1.70	4.50
45727	3	0.70	3.40	1.70	1.34	1.48	0.08	1.00	0.70	3.40
45729	19	0.80	21.00	3.76	2.51	4.61	0.24	2.40	1.60	4.30
45732	17	0.10	7.90	3.21	2.00	2.69	0.14	2.10	1.20	6.40
45734	2	1.50	1.70	1.60	1.60	0.14	0.01	1.60	1.50	1.70
45735	8	0.70	3.80	2.10	1.76	1.24	0.06	2.00	0.90	2.60
45737	8	3.10	11.20	8.69	8.10	2.85	0.15	9.60	5.60	9.80
45740	4	1.00	9.50	4.85	3.70	3.50	0.18	4.45	1.00	4.50
45741	2	3.20	4.10	3.65	3.62	0.64	0.03	3.65	3.20	4.10

Zip Code	No.	Min.	Max.	AM	GM	SD	CV	Md	Q1	Q3
45742	20	0.30	18.00	2.90	1.65	3.97	0.20	1.45	1.00	2.70
45743	2	1.00	2.80	1.90	1.67	1.27	0.07	1.90	1.00	2.80
45744	25	1.00	10.70	3.89	3.00	3.03	0.16	3.10	1.80	4.30
45745	5	1.10	3.90	2.00	1.74	1.23	0.06	1.30	1.10	2.60
45746	3	0.50	3.40	1.70	1.27	1.51	0.08	1.20	0.50	3.40
45750	496	0.10	91.90	5.45	2.91	8.67	0.44	3.00	1.50	5.50
45760	12	0.50	4.00	1.82	1.48	1.19	0.06	1.50	0.70	2.00
45761	12	0.10	15.50	2.92	1.12	4.33	0.22	1.10	0.30	2.90
45764	22	0.60	9.20	4.13	3.15	2.82	0.14	3.20	2.30	6.30
45766	7	0.80	3.60	2.10	1.75	1.29	0.07	1.40	1.00	3.40
45767	10	0.30	9.35	2.19	1.38	2.67	0.14	1.25	0.80	2.90
45768	6	0.10	5.30	2.32	1.20	2.23	0.11	1.60	0.50	4.80
45769	29	0.10	4.20	1.92	1.32	1.33	0.07	1.70	0.70	2.80
45770	2	0.90	1.00	0.95	0.95	0.07	0.00	0.95	0.90	1.00
45771	13	0.70	6.30	2.60	2.19	1.58	0.08	2.20	1.80	3.20
45772	3	0.10	2.90	1.93	0.93	1.59	0.08	2.80	0.10	2.90
45773	9	0.70	5.40	3.36	2.76	1.80	0.09	4.10	1.90	4.70
45775	3	1.10	1.70	1.43	1.41	0.31	0.02	1.50	1.10	1.70
45776	6	0.20	6.40	2.02	0.96	2.48	0.13	0.85	0.30	3.50
45778	7	0.70	9.40	3.50	2.52	3.08	0.16	2.00	1.50	5.40
45779	4	0.30	3.90	1.45	0.90	1.67	0.09	0.80	0.30	1.10
45780	43	0.60	33.00	7.48	4.47	8.10	0.41	4.20	2.10	12.70
45784	28	0.10	14.20	4.86	3.23	3.63	0.19	3.85	1.50	6.10
45785	1	2.20	2.20	2.20	2.20	0.00	0.00	2.20	2.20	2.20
45786	17	0.40	19.60	3.84	2.48	4.71	0.24	2.40	1.40	3.30
45788	9	0.80	5.50	3.41	2.80	1.93	0.10	4.40	1.60	5.30
45789	1	2.90	2.90	2.90	2.90	0.00	0.00	2.90	2.90	2.90
45800	1	17.50	17.50	17.50	17.50	0.00	0.00	17.50	17.50	17.50
45801	51	0.10	30.80	7.27	3.74	8.27	0.42	4.30	1.70	9.50
45802	4	1.10	4.50	2.73	2.28	1.73	0.09	2.65	1.10	3.90
45804	56	0.30	24.70	4.86	3.12	4.68	0.24	3.50	1.50	6.40
45805	238	0.10	95.80	5.27	3.05	8.30	0.42	3.40	1.50	6.20
45806	75	0.10	38.10	7.64	4.23	7.38	0.38	5.50	2.90	10.10
45807	96	0.10	26.70	5.79	3.77	4.94	0.25	5.00	1.70	7.80

Zip Code	No.	Min.	Max.	AM	GM	SD	CV	Md	Q1	Q3
45809	3	2.00	10.20	4.77	3.50	4.71	0.24	2.10	2.00	10.20
45810	32	0.10	57.10	6.92	3.79	10.47	0.54	4.60	1.90	7.10
45812	10	1.50	8.30	3.50	3.14	1.92	0.10	3.25	2.20	3.90
45813	15	1.00	7.60	3.17	2.64	1.99	0.10	2.50	1.40	4.70
45814	18	0.50	10.50	3.24	2.07	3.18	0.16	1.60	1.00	5.80
45817	38	0.50	62.60	7.51	4.02	12.79	0.65	3.85	2.60	5.50
45819	1	21.50	21.50	21.50	21.50	0.00	0.00	21.50	21.50	21.50
45820	1	2.00	2.00	2.00	2.00	0.00	0.00	2.00	2.00	2.00
45821	1	3.00	3.00	3.00	3.00	0.00	0.00	3.00	3.00	3.00
45822	203	0.10	48.60	7.15	4.25	7.92	0.41	4.70	2.10	9.10
45826	1	5.80	5.80	5.80	5.80	0.00	0.00	5.80	5.80	5.80
45827	10	0.40	11.70	4.77	2.71	4.25	0.22	4.40	0.80	7.10
45828	71	0.70	17.60	5.24	3.98	3.95	0.20	4.20	2.80	6.70
45830	25	0.10	35.90	9.16	5.43	9.14	0.47	6.40	3.20	12.10
45831	12	0.60	12.10	3.78	2.70	3.21	0.16	3.25	1.00	4.70
45832	10	0.50	11.30	4.84	2.89	4.27	0.22	3.80	1.00	10.20
45833	81	0.20	47.50	7.93	4.78	7.51	0.38	6.50	3.10	12.10
45836	4	0.50	6.40	2.73	1.54	2.83	0.14	2.00	0.50	3.50
45838	1	1.50	1.50	1.50	1.50	0.00	0.00	1.50	1.50	1.50
45839	11	0.10	10.60	2.35	1.28	2.96	0.15	1.60	0.60	2.90
45840	1041	0.10	73.90	5.12	3.06	5.93	0.30	3.30	1.60	6.30
45841	8	1.10	14.40	5.98	4.56	4.20	0.21	6.40	1.60	7.20
45843	23	0.10	30.20	6.92	4.35	7.51	0.38	3.90	2.90	7.60
45844	36	1.00	28.70	7.38	4.99	6.97	0.36	4.80	3.10	9.00
45845	60	0.50	48.60	8.24	5.98	8.11	0.42	6.00	4.10	9.70
45846	42	0.10	38.50	5.88	3.37	7.20	0.37	3.45	1.60	7.50
45848	7	0.60	6.30	3.67	2.61	2.39	0.12	4.80	0.60	5.90
45849	4	0.30	0.80	0.50	0.47	0.22	0.01	0.45	0.30	0.50
45850	10	1.20	24.30	10.73	6.74	8.14	0.42	12.20	1.60	15.00
45851	3	0.10	3.00	1.97	0.94	1.62	0.08	2.80	0.10	3.00
45853	9	1.00	9.60	5.28	4.48	2.58	0.13	6.20	3.70	6.50
45854	1	2.20	2.20	2.20	2.20	0.00	0.00	2.20	2.20	2.20
45855	1	4.00	4.00	4.00	4.00	0.00	0.00	4.00	4.00	4.00
45856	50	0.50	52.50	6.57	4.03	8.93	0.46	4.10	2.40	5.70

Zip Code	No.	Min.	Max.	AM	GM	SD	CV	Md	Q1	Q3
45858	40	0.10	62.60	6.09	2.97	10.32	0.53	3.05	1.10	5.30
45860	65	0.50	56.10	9.43	5.43	12.03	0.62	4.90	2.50	11.20
45861	1	0.50	0.50	0.50	0.50	0.00	0.00	0.50	0.50	0.50
45862	7	0.40	16.10	5.23	2.33	6.48	0.33	2.50	0.70	12.90
45863	2	2.60	3.60	3.10	3.06	0.71	0.04	3.10	2.60	3.60
45865	85	0.10	64.50	8.56	4.35	10.44	0.53	5.40	2.00	12.20
45867	7	0.20	80.20	14.73	3.12	29.17	1.49	3.90	0.70	12.50
45868	11	0.80	21.00	6.83	3.97	7.37	0.38	4.80	1.20	7.30
45869	71	0.40	24.90	6.51	4.18	6.03	0.31	5.00	2.00	9.20
45871	33	0.10	38.90	9.63	5.23	9.82	0.50	6.10	2.60	13.00
45872	28	0.10	14.80	3.42	2.11	3.65	0.19	2.40	1.70	3.30
45873	5	0.60	3.10	1.74	1.38	1.25	0.06	1.10	0.80	3.10
45874	10	0.10	7.30	4.73	3.01	2.80	0.14	5.60	1.80	7.30
45875	94	0.10	17.50	4.30	2.92	3.81	0.19	2.85	1.90	5.30
45876	3	2.10	15.00	6.57	4.34	7.31	0.37	2.60	2.10	15.00
45877	20	0.20	28.10	9.70	6.49	7.83	0.40	7.75	3.00	12.00
45879	31	0.10	13.20	5.03	2.91	4.53	0.23	2.70	1.80	11.00
45880	11	0.30	9.50	2.34	1.52	2.63	0.13	1.70	0.60	2.20
45881	7	1.80	130.00	24.51	9.12	46.63	2.39	9.20	3.30	10.20
45882	17	0.10	13.30	3.48	1.83	4.14	0.21	1.40	1.20	4.10
45883	59	0.10	53.00	7.34	4.27	8.33	0.43	4.60	2.60	9.50
45884	1	6.00	6.00	6.00	6.00	0.00	0.00	6.00	6.00	6.00
45885	75	0.10	31.10	6.12	3.35	6.65	0.34	4.00	1.80	6.60
45886	1	0.90	0.90	0.90	0.90	0.00	0.00	0.90	0.90	0.90
45887	14	0.50	14.30	4.41	2.69	4.43	0.23	2.70	1.10	6.20
45889	25	0.10	33.80	7.34	3.72	7.46	0.38	6.80	1.20	9.50
45890	7	2.30	34.70	11.30	8.09	10.96	0.56	9.40	4.40	12.90
45891	70	0.50	24.50	6.17	4.07	5.65	0.29	4.20	2.20	9.20
45892	1	1.80	1.80	1.80	1.80	0.00	0.00	1.80	1.80	1.80
45893	1	7.80	7.80	7.80	7.80	0.00	0.00	7.80	7.80	7.80
45894	2	4.90	7.10	6.00	5.90	1.56	0.08	6.00	4.90	7.10
45895	122	0.10	34.20	6.16	3.81	5.97	0.31	4.45	2.00	8.50
45896	3	0.10	9.10	4.03	1.38	4.61	0.24	2.90	0.10	9.10
45898	8	1.80	7.20	4.61	4.08	2.26	0.12	4.50	2.20	6.90

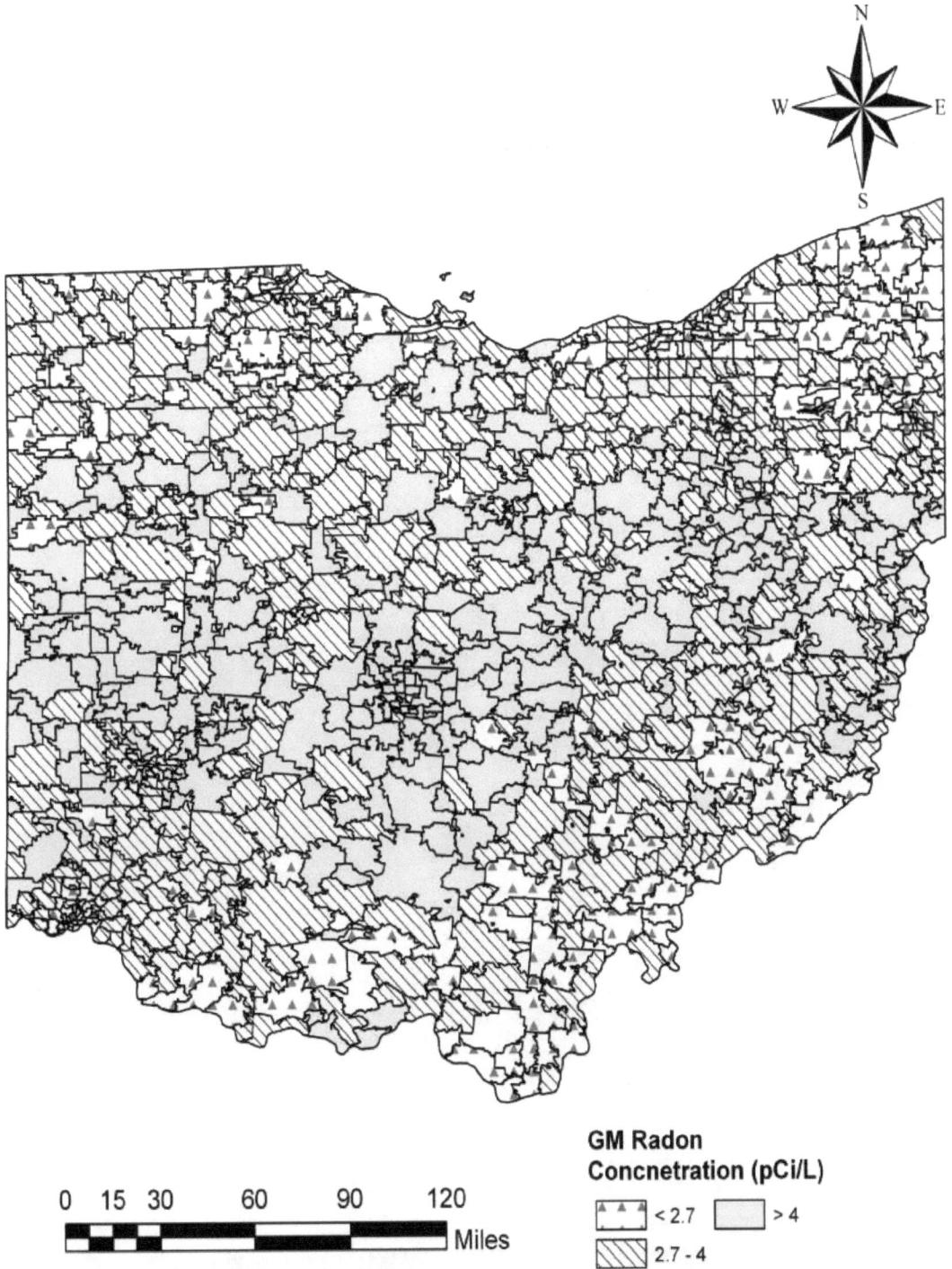

Figure 4a: GM ^{222}Rn concentration in Ohio zip codes based on the WHO-USEPA classification.

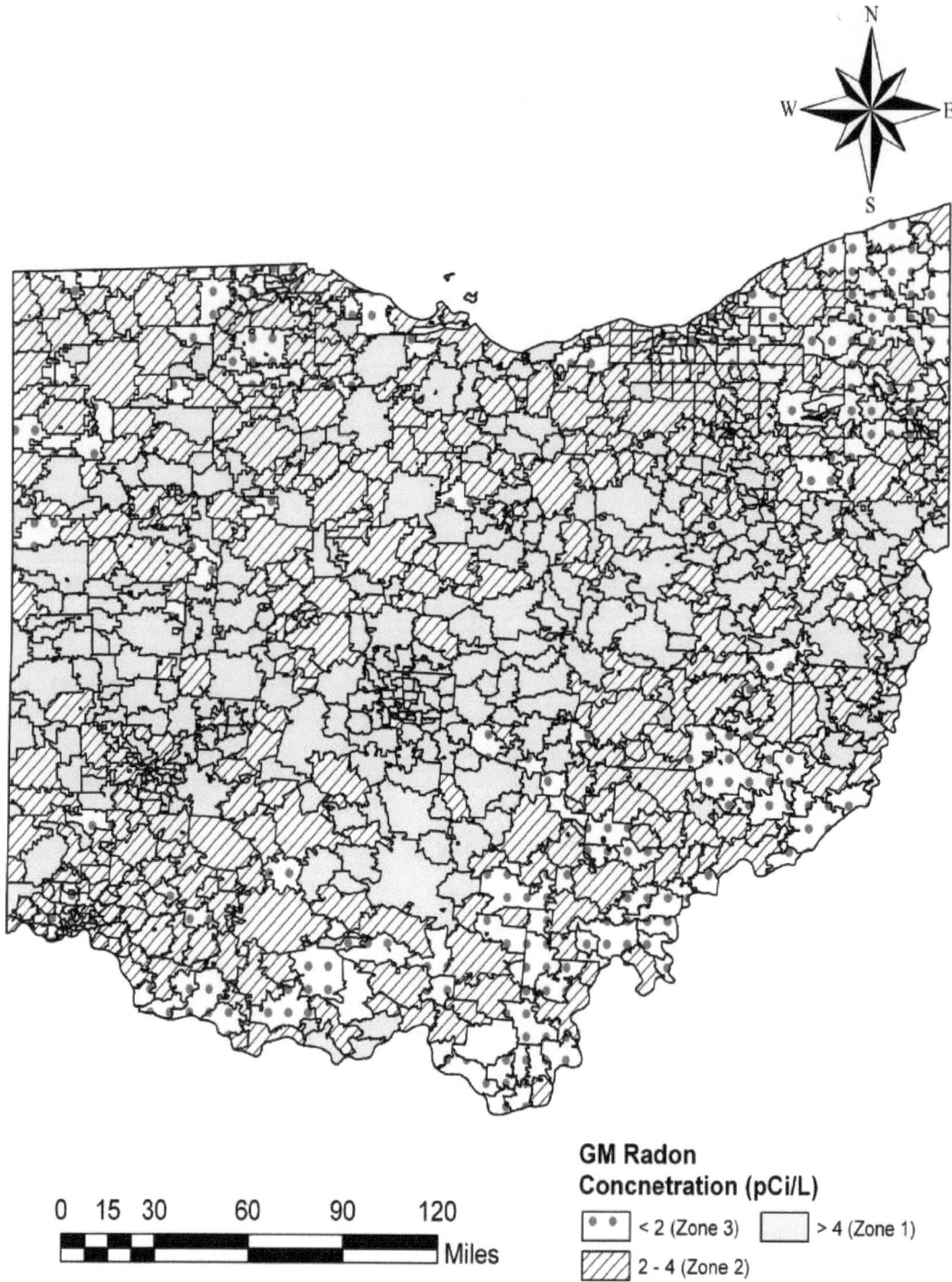

Figure 4b: GM ^{222}Rn concentrations in Ohio zip codes based on the USEPA classification.

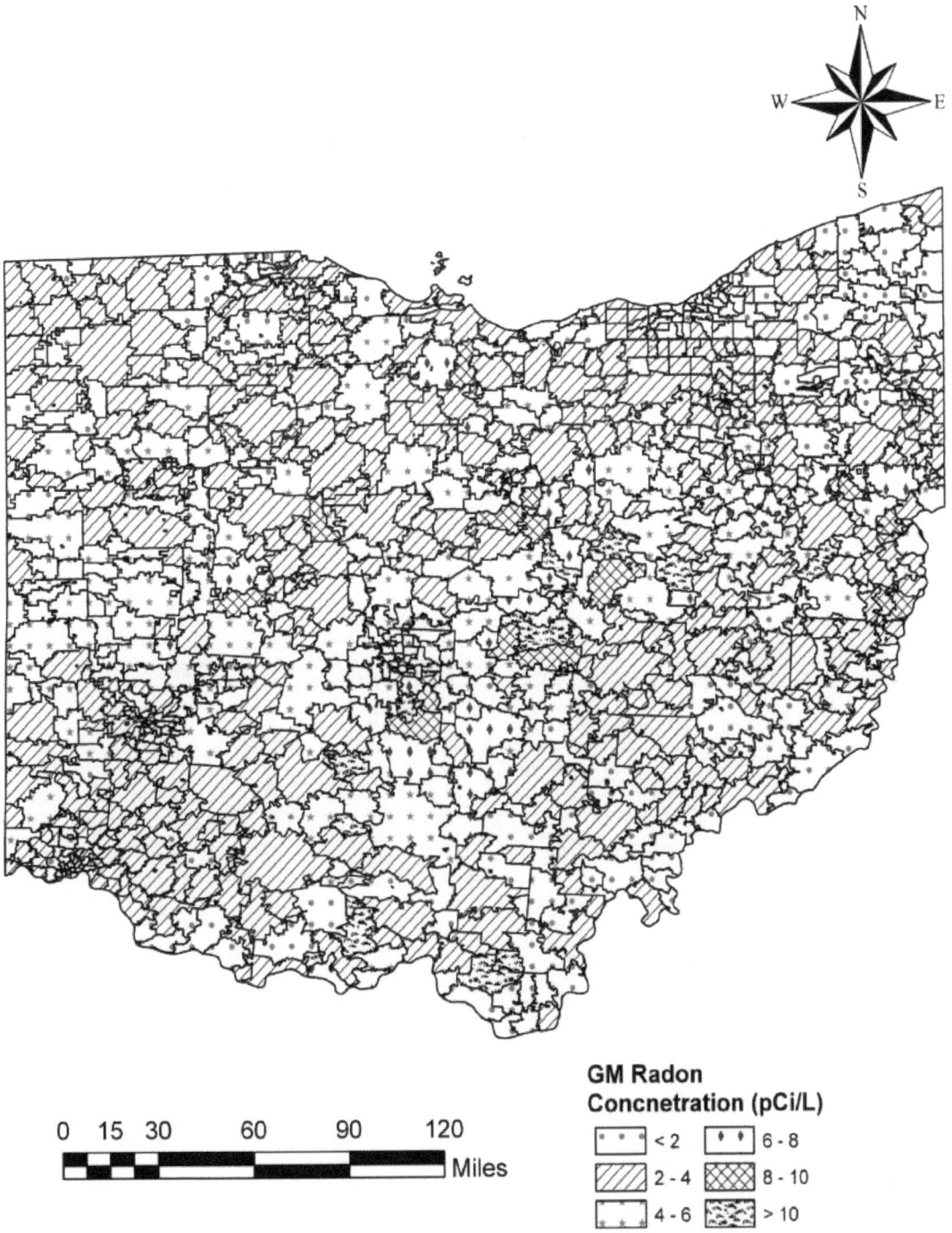

GM Radon Concnetration (pCi/L)

< 2	6 - 8
2 - 4	8 - 10
4 - 6	> 10

0   15   30   60   90   120

Miles

Figure 4c: GM [222]Rn concentrations in Ohio zip codes based on the 2 pCi/L breakdown classifications.

CHAPTER 3

Radon Statistics & Maps for Zip Code Areas in Counties of Ohio

Abstract: This chapter provides detailed information regarding the ^{222}Rn problem in each of the 88 counties in Ohio by using the GIS based ^{222}Rn GM concentration maps and computing different statistical metrics. The geospatial and statistical analyses are examined at a county level in great detail for all the available 88 counties in Ohio that assists in accurately assessing the radon problem in Ohio at a more microscopic level in reference to a specific location.

Keywords: Radon, radon in Ohio, ORIS, GIS, geospatial radon maps, radon statistical metrics, WHO-USEPA radon map for Ohio counties, USEPA radon map for Ohio counties, 2 pCi/L breakdown radon map for Ohio counties, number of radon homes tested statistics, minimum radon concentration statistics, maximum radon concentration statistics, arithmetic mean radon concentration statistics, geometric mean radon concentration statistics, standard deviation in radon measurements statistics, coefficient of variation in radon measurements statistics, median radon concentration statistics, first quartile radon concentration statistics, third quartile radon concentration statistics, radon assessment in individual Ohio counties.

3.1. A ZIP CODE BASED ^{222}RN ASSESSMENT FOR INDIVIDUAL OHIO COUNTIES

Tables **1** through **88** present a summary of the ^{222}Rn statistical metrics computed for different zip codes within an individual county, for all the 88 counties in Ohio. Figs. **1a** through **88a** illustrate the GIS based ^{222}Rn GM maps for the zip codes in individual Ohio counties based on the WHO-USEPA classification. Similarly, Figs. **1b** through **88b** and Figs. **1c** through **88c** illustrate the graphical representations of ^{222}Rn GM concentrations for the zip codes in individual Ohio counties based on the USEPA and the 2 pCi/L breakdown classifications, respectively.

Table 1: ^{222}Rn statistics for Adams county

Zip code	No.	Min.	Max.	AM	GM	SD	CV
45144	8	0.6	16.8	5.41	3.23	5.81	0.3
45616	2	1.6	4.7	3.15	2.74	2.19	0.11
45650	2	35.4	35.4	35.4	35.4	0	0
45660	65	0.1	42.5	5.09	1.71	8.41	0.43
45679	16	0.7	6.9	3.61	3.14	1.81	0.09
45684[*]	11	0.8	21.6	9.22	6.54	7.21	0.37
45693	10	0.1	4.8	1.62	0.96	1.47	0.08
45697	22	0.1	15.6	4.63	2.41	4.79	0.25
45105[**]	8	1.3	7.8	3.35	2.75	2.33	0.12

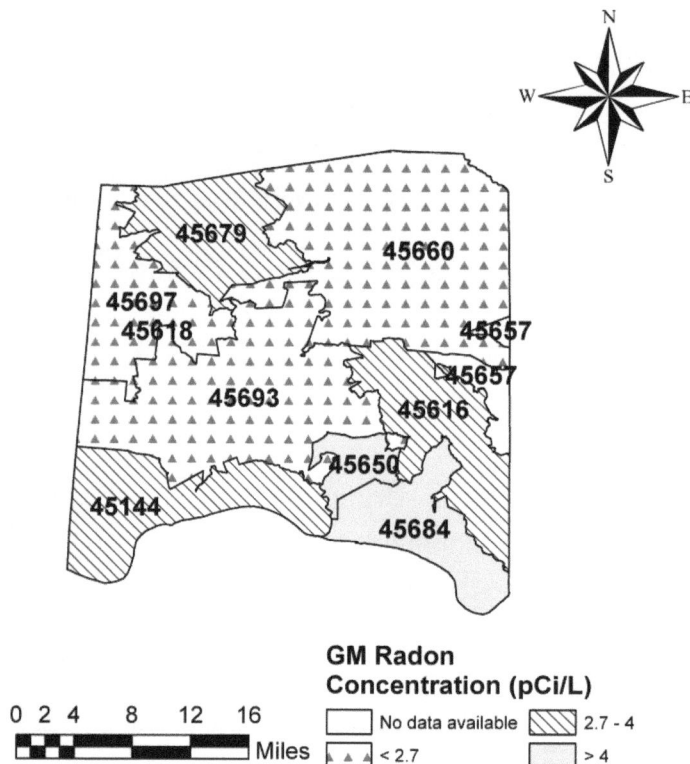

Figure 1a: GM ^{222}Rn concentrations in zip codes of Adams county based on the WHO-USEPA classification.

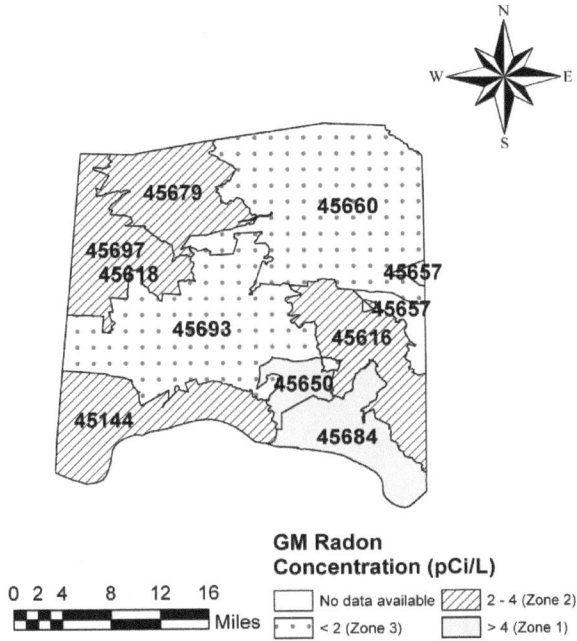

Figure 1b: GM ^{222}Rn concentrations in zip codes of Adams county based on the USEPA classification.

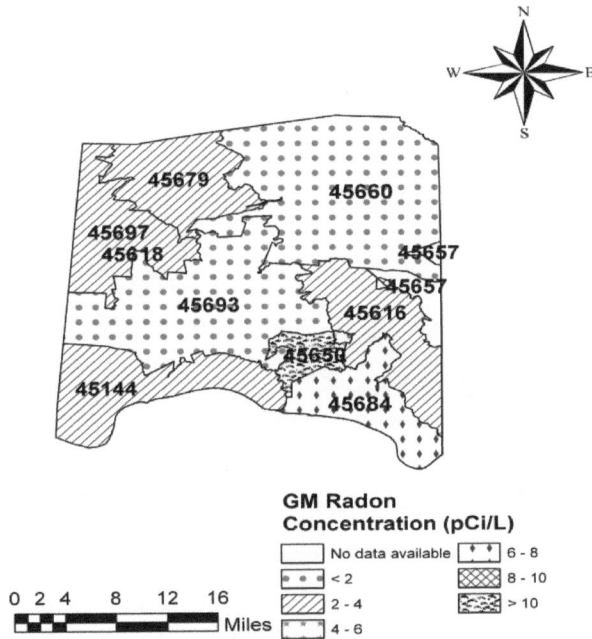

Figure 1c: GM ^{222}Rn concentrations in zip codes of Adams county based on the 2 pCi/L breakdown classification.

Table 2: ^{222}Rn statistics for Allen county

Zip code	No.	Min.	Max.	AM	GM	SD	CV
45801	51	0.1	30.8	7.27	3.74	8.27	0.42
45804	56	0.3	24.7	4.86	3.12	4.68	0.24
45805	238	0.1	95.8	5.27	3.05	8.3	0.42
45806	75	0.1	38.1	7.64	4.23	7.38	0.38
45807	96	0.1	26.7	5.79	3.77	4.94	0.25
45809	3	2	10.2	4.77	3.5	4.71	0.24
45810*	32	0.1	57.1	6.92	3.79	10.47	0.54
45817	38	0.5	62.6	7.51	4.02	12.79	0.65
45820	1	2	2	2	2	0	0
45830*	25	0.1	35.9	9.16	5.43	9.14	0.47
45833	81	0.2	47.5	7.93	4.78	7.51	0.38
45844*	36	1	28.7	7.38	4.99	6.97	0.36
45850	10	1.2	24.3	10.73	6.74	8.14	0.42
45854	1	2.2	2.2	2.2	2.2	0	0
45877*	20	0.2	28.1	9.7	6.49	7.83	0.4
45887	14	0.5	14.3	4.41	2.69	4.43	0.23
45896*	3	0.1	9.1	4.03	1.38	4.61	0.24
45802**	4	1.1	4.5	2.73	2.28	1.73	0.09

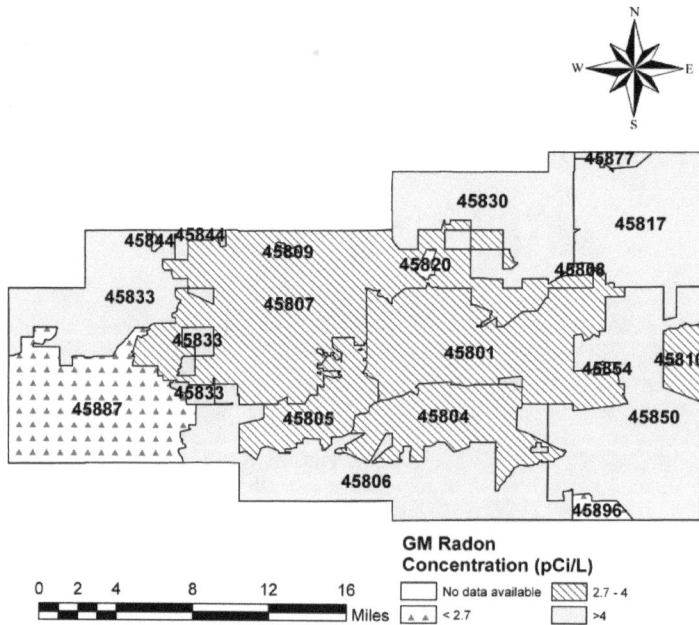

Figure 2a: GM ^{222}Rn concentrations in zip codes of Allen county based on the WHO-USEPA classification.

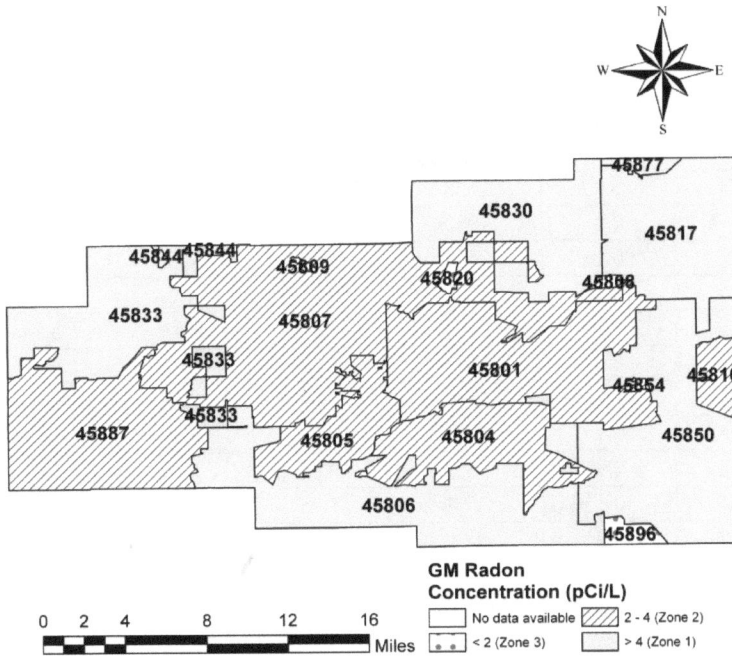

Figure 2b: GM ^{222}Rn concentrations in zip codes of Allen county based on the USEPA classification.

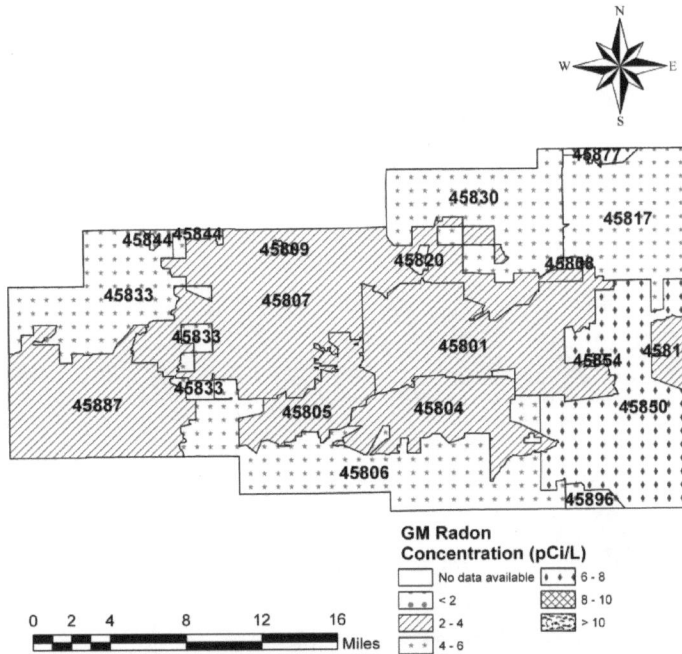

Figure 2c: GM ^{222}Rn concentrations in zip codes of Allen county based on the 2 pCi/L breakdown classification.

Table 3: ^{222}Rn statistics for Ashland county

Zip code	No.	Min.	Max.	AM	GM	SD	CV
44235*	21	0.8	44.2	6.23	3.89	9.32	0.48
44287*	39	0.1	17	4.78	2.85	4.49	0.23
44638*	7	0.9	11.8	4.79	3.75	3.56	0.18
44691*	867	0.1	498.4	11.58	4.47	29.06	1.49
44805	451	0.1	155.8	8.17	3.62	17.7	0.91
44837*	17	0.7	50.8	8.52	4.2	13.38	0.69
44838	6	1	23.7	6.15	3.23	8.74	0.45
44840	26	0.3	185.3	17.34	5.93	37.57	1.92
44842	52	0.1	29.9	6.49	3.53	6.88	0.35
44843*	35	0.1	179	24.3	8.47	42.04	2.15
44851*	32	0.9	13.7	5.01	4.14	3.28	0.17
44859	27	0.1	17.4	5.75	4.58	3.4	0.17
44864	42	0.9	115	15.46	7.68	23.56	1.21
44866	7	1.1	11.9	3.98	2.85	3.91	0.2
44878*	14	0.4	16.6	4.61	3.35	4.13	0.21
44880	13	0.6	17.7	5.42	3.18	5.48	0.28
44903*	494	0.1	150.4	8.02	4.43	12.43	0.64
44848**	2	6.8	10.9	8.85	8.61	2.9	0.15

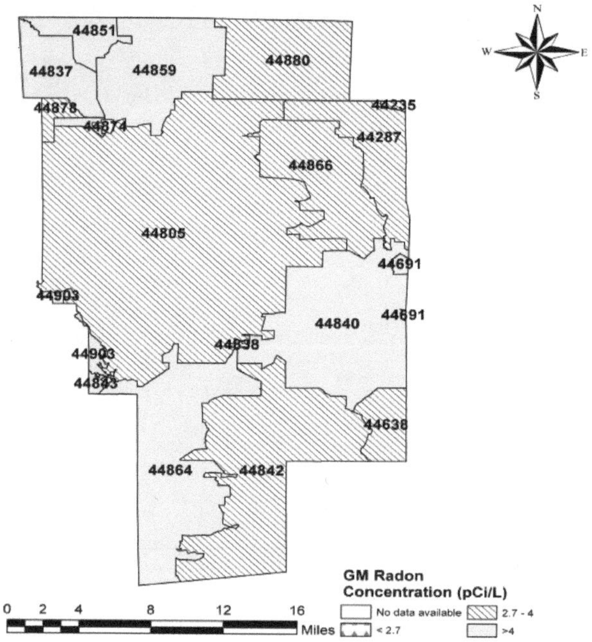

Figure 3a: GM ^{222}Rn concentrations in zip codes of Ashland county based on the WHO-USEPA classification.

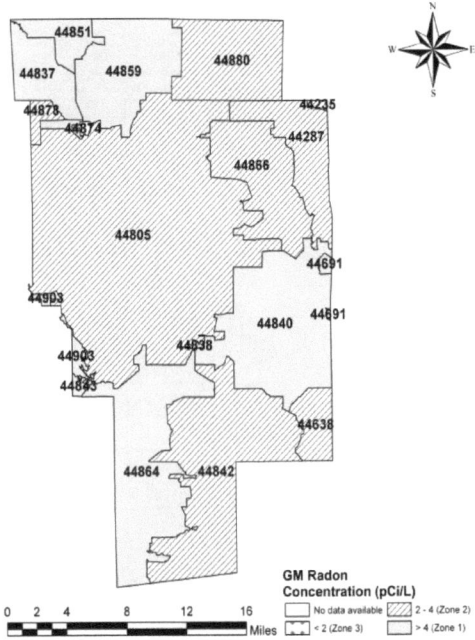

Figure 3b: GM ^{222}Rn concentrations in zip codes of Ashland county based on the USEPA classification.

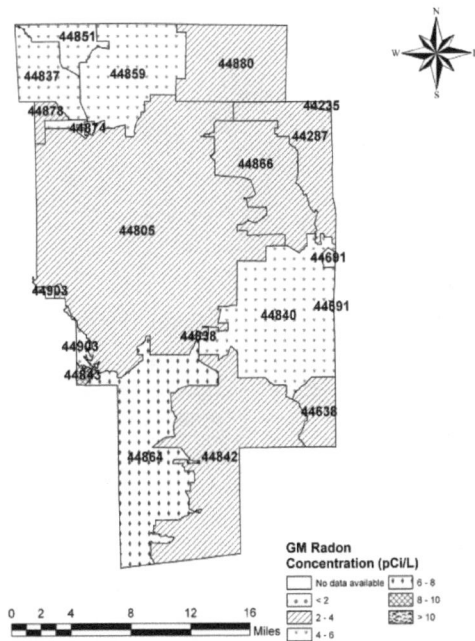

Figure 3c: GM ^{222}Rn concentrations in zip codes of Ashland county based on the 2 pCi/L breakdown classification.

Table 4: ^{222}Rn statistics for Ashtabula county

Zip code	No.	Min.	Max.	AM	GM	SD	CV
44003	32	0.2	4.7	1.27	1	0.94	0.05
44004	209	0.1	51.7	3.73	1.51	7.12	0.36
44010	25	0.4	5.3	2.05	1.54	1.49	0.08
44030	142	0.1	122.2	7.7	2.91	16.31	0.83
44032	11	0.3	3.9	1.3	0.95	1.14	0.06
44041	87	0.1	14	2.18	1.16	2.66	0.14
44047	92	0.1	210	3.67	1.11	21.78	1.12
44048	26	0.4	37.4	7.47	3.76	9.89	0.51
44057	152	0.1	37.8	3.97	1.86	6.77	0.35
44062	38	0.1	5.8	1.83	1.1	1.72	0.09
44064	12	0.4	12.5	3.98	2.69	3.77	0.19
44076	39	0.1	9.3	1.56	1.08	1.62	0.08
44082	19	0.3	8.4	2.12	1.42	2.07	0.11
44084	32	0.2	5.6	2.08	1.56	1.57	0.08
44085	52	0.2	14.1	2.63	1.66	2.88	0.15
44086	25	0.1	27.5	3.72	2.07	5.59	0.29
44093	13	0.3	8.5	2.23	1.61	2.1	0.11
44099	16	0.3	5.3	1.31	0.89	1.35	0.07
44428	21	0.3	10.7	2.73	1.74	2.79	0.14
44005[**]	1	1.5	1.5	1.5	1.5	0	0
44068[**]	24	0.1	51	12.02	4.17	14.82	0.76
44088[**]	3	12.7	28.3	22.1	20.87	8.28	0.42

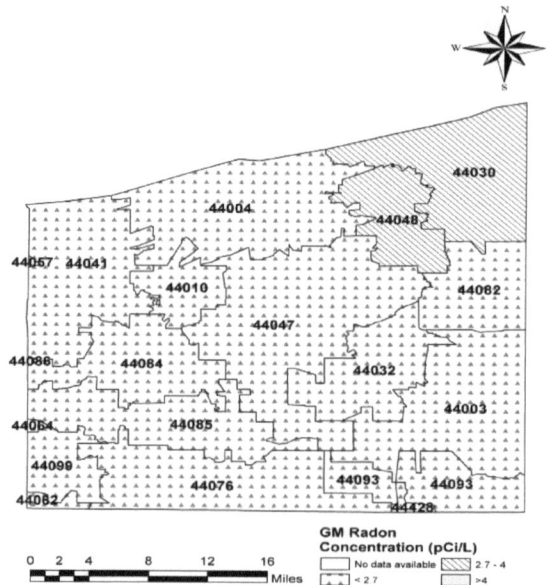

Figure 4a: GM ^{222}Rn concentrations in zip codes of Ashtabula county based on the WHO-USEPA classification.

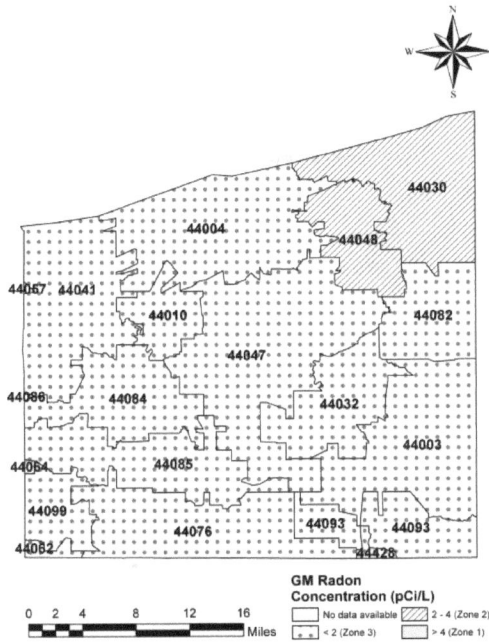

Figure 4b: GM ^{222}Rn concentrations in zip codes of Ashtabula county based on the USEPA classification.

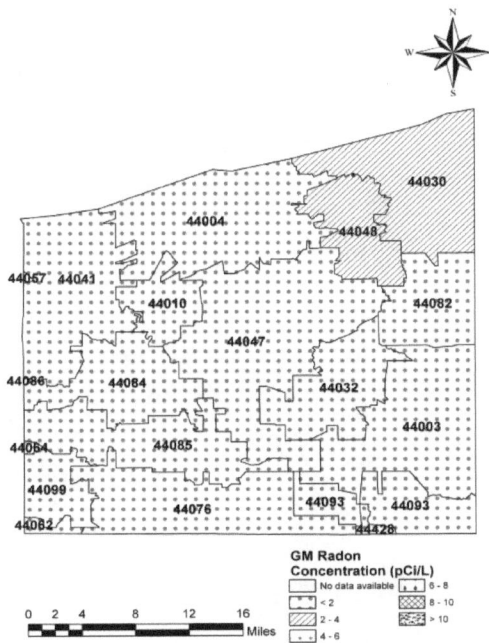

Figure 4c: GM ^{222}Rn concentrations in zip codes of Ashtabula county based on the 2 pCi/L breakdown classification.

Table 5: ^{222}Rn statistics for Athens county

Zip code	No.	Min.	Max.	AM	GM	SD	CV
43728	3	1.7	5	2.9	2.57	1.82	0.09
45701	719	0.1	82.1	5.24	3.68	5.32	0.27
45710	53	0.3	12.4	3.18	2.06	3.11	0.16
45711	12	0.4	4.5	1.96	1.36	1.67	0.09
45716	2	0.1	1.1	0.6	0.33	0.71	0.04
45719	6	1	8.4	3.82	2.63	3.38	0.17
45723	11	0.3	4.2	2.24	1.66	1.4	0.07
45732	17	0.1	7.9	3.21	2	2.69	0.14
45735	8	0.7	3.8	2.1	1.76	1.24	0.06
45740	4	1	9.5	4.85	3.7	3.5	0.18
45742	20	0.3	18	2.9	1.65	3.97	0.2
45761	12	0.1	15.5	2.92	1.12	4.33	0.22
45764	22	0.6	9.2	4.13	3.15	2.82	0.14
45766	7	0.8	3.6	2.1	1.75	1.29	0.07
45776	6	0.2	6.4	2.02	0.96	2.48	0.13
45778	7	0.7	9.4	3.5	2.52	3.08	0.16
45780	43	0.6	33	7.48	4.47	8.1	0.41
45717[**]	1	1.6	1.6	1.6	1.6	0	0

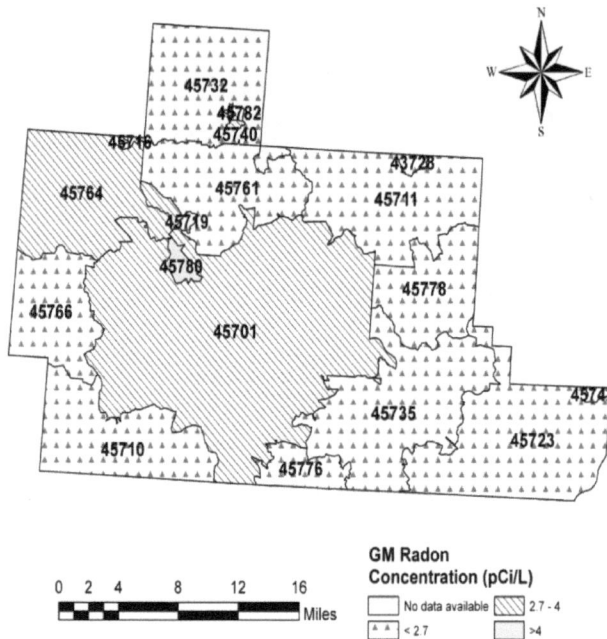

Figure 5a: GM ^{222}Rn concentrations in zip codes of Athens county based on the WHO-USEPA classification.

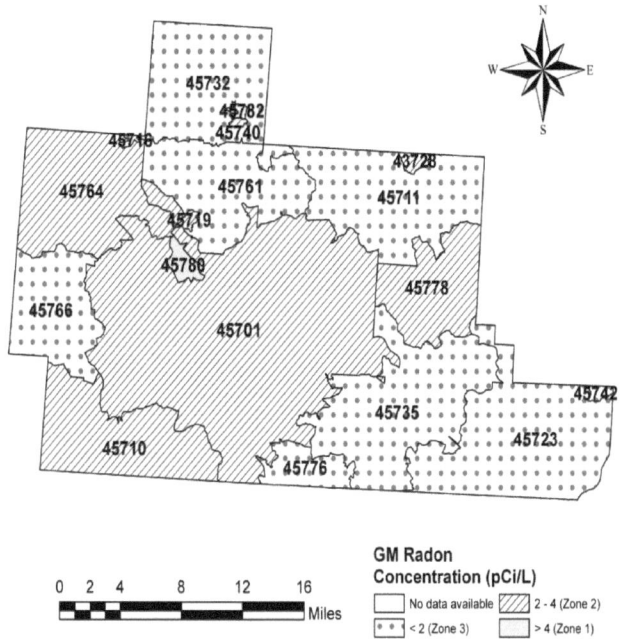

Figure 5b: GM ^{222}Rn concentrations in zip codes of Athens county based on the USEPA classification.

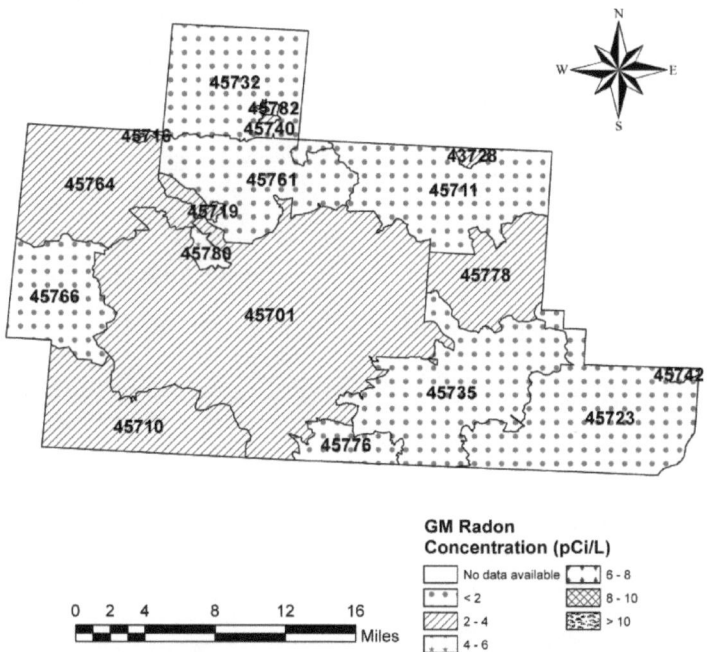

Figure 5c: GM ^{222}Rn concentrations in zip codes of Athens county based on the 2 pCi/L breakdown classification.

Table 6: ^{222}Rn statistics for Auglaize county

Zip code	No.	Min.	Max.	AM	GM	SD	CV
43331	19	0.5	8.7	2.06	1.47	2.08	0.11
45306	32	0.8	18.3	6.79	5.05	5.07	0.26
45334	19	0.7	17	3.7	2.54	4.05	0.21
45806*	75	0.1	38.1	7.64	4.23	7.38	0.38
45819	1	21.5	21.5	21.5	21.5	0	0
45845	60	0.5	48.6	8.24	5.98	8.11	0.42
45850*	10	1.2	24.3	10.73	6.74	8.14	0.42
45862	7	0.4	16.1	5.23	2.33	6.48	0.33
45865	85	0.1	64.5	8.56	4.35	10.44	0.53
45869	71	0.4	24.9	6.51	4.18	6.03	0.31
45871	33	0.1	38.9	9.63	5.23	9.82	0.5
45884	1	6	6	6	6	0	0
45885	75	0.1	31.1	6.12	3.35	6.65	0.34
45887*	14	0.5	14.3	4.41	2.69	4.43	0.23
45895	122	0.1	34.2	6.16	3.81	5.97	0.31
45896	3	0.1	9.1	4.03	1.38	4.61	0.24

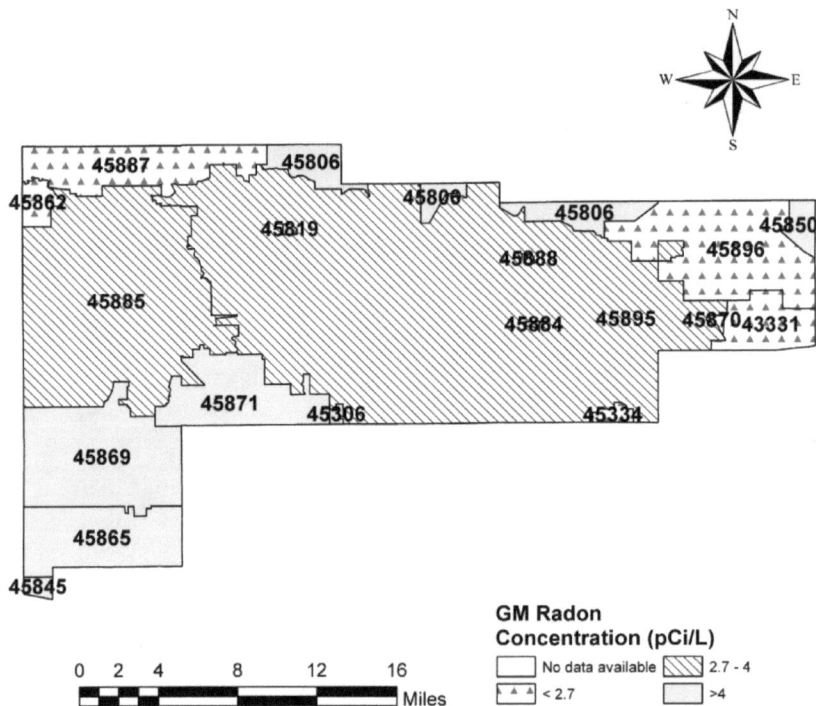

Figure 6a: GM ^{222}Rn concentrations in zip codes of Auglaize county based on the WHO-USEPA classification.

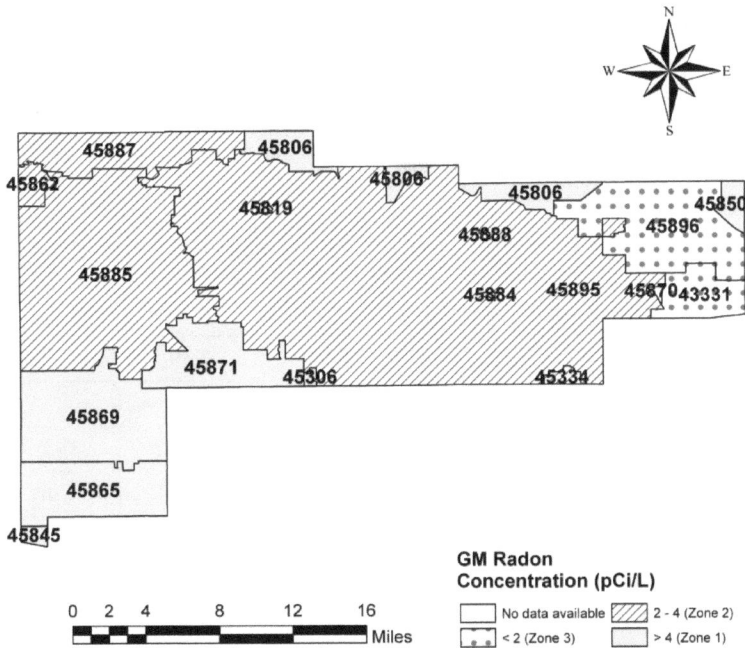

Figure 6b: GM ^{222}Rn concentrations in zip codes of Auglaize county based on the USEPA classification.

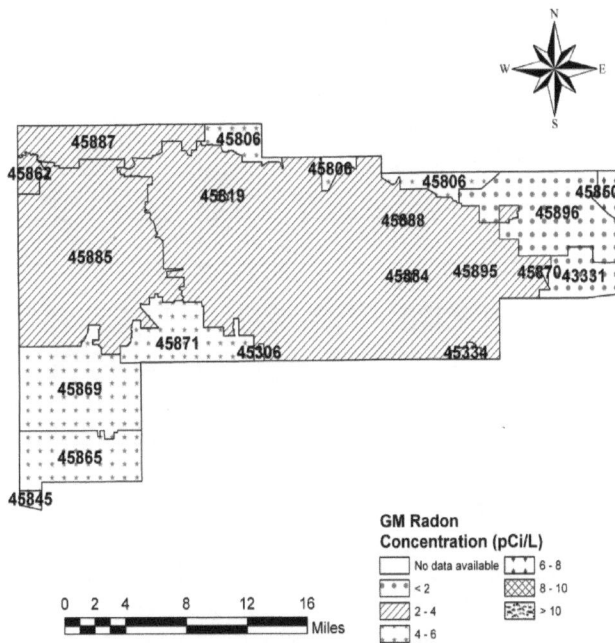

Figure 6c: GM ^{222}Rn concentrations in zip codes of Auglaize county based on the 2 pCi/L breakdown classification.

Table 7: ^{222}Rn statistics for Belmont county

Zip code	No.	Min.	Max.	AM	GM	SD	CV
43713	48	0.5	17.9	5.44	3.48	4.91	0.25
43716	39	0.6	51.2	8.27	4.27	10.97	0.56
43718	62	0.1	43.2	5.62	2.82	8.11	0.42
43719	9	1.3	14.3	5.62	4.28	4.7	0.24
43747	5	0.4	6.1	3.74	2.11	3.01	0.15
43759	5	2	2.6	2.4	2.39	0.24	0.01
43901	5	1.1	8.9	4.74	3.46	3.6	0.18
43902	2	0.8	3.1	1.95	1.57	1.63	0.08
43905	5	0.9	13.3	4.8	3.28	4.9	0.25
43906	57	0.8	91.3	8.52	5.06	13.51	0.69
43912	41	0.1	23.4	3.97	2.53	4.47	0.23
43917	26	2.3	65.3	11.88	7.91	13.93	0.71
43927	1	2.4	2.4	2.4	2.4	0	0
43933	13	0.5	54	17.93	6.04	21.79	1.12
43934	1	1.5	1.5	1.5	1.5	0	0
43935	64	0.1	13.4	3.28	2.39	2.57	0.13
43940	4	1	20.5	11.43	6.92	9.34	0.48
43942	18	0.5	15.3	5.57	3.06	5.32	0.27
43943	36	0.2	180	29.74	9.69	40.64	2.08
43947	56	0.1	288.6	10.4	3.4	38.47	1.97
43950	338	0.1	119.4	7.15	3.26	16.31	0.84
43967	2	0.9	0.9	0.9	0.9	0	0
43971	3	0.8	6.7	3.37	2.41	3.02	0.15
43972	1	2.2	2.2	2.2	2.2	0	0
43977	15	0.9	26.9	5.99	3.35	7.28	0.37
43983	6	0.6	10.4	4.22	2.94	3.66	0.19
43985	2	0.5	5.9	3.2	1.72	3.82	0.2
43909[**]	2	2.3	2.3	2.3	2.3	0	0
43916[**]	8	2.7	108.5	60.89	25.91	49.2	2.52
43937[**]	4	3.1	24.3	9.95	7.31	9.69	0.5

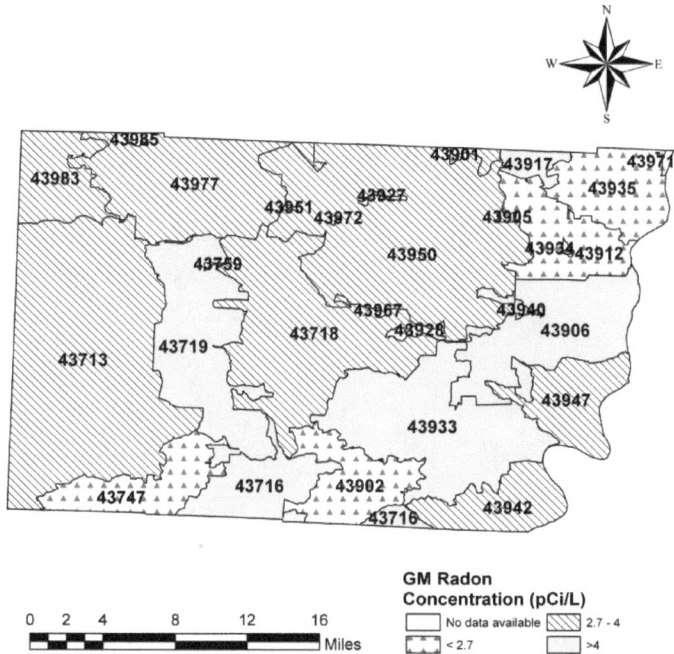

Figure 7a: GM ^{222}Rn concentrations in zip codes of Belmont county based on the WHO-USEPA classification.

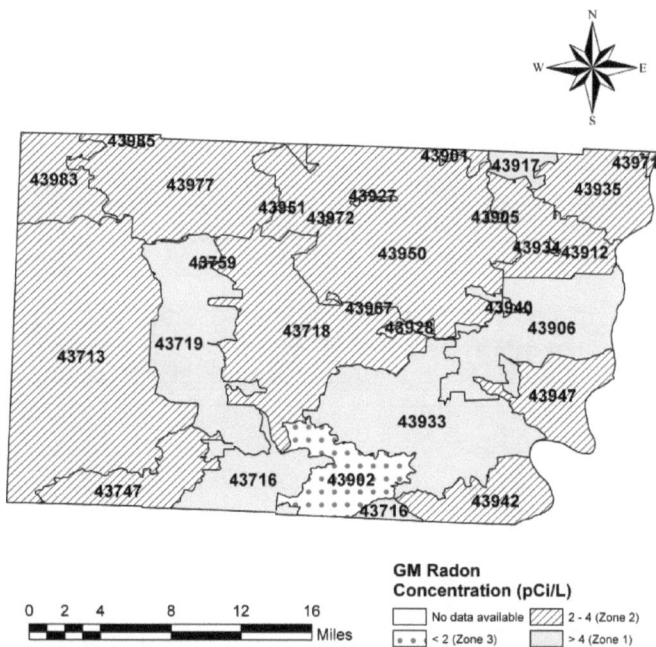

Figure 7b: GM ^{222}Rn concentrations in zip codes of Belmont county based on the USEPA classification.

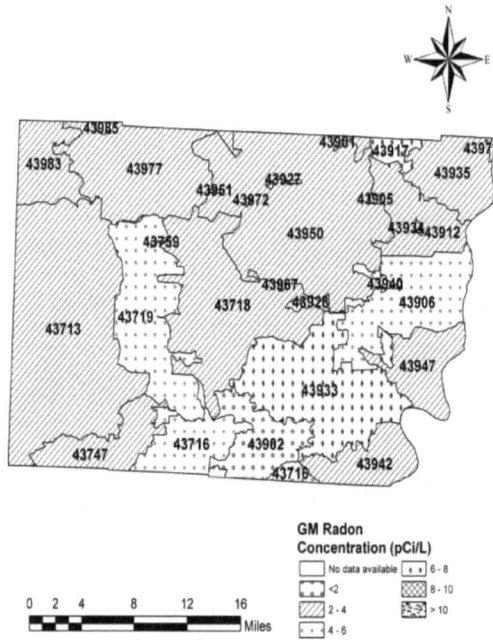

Figure 7c: GM ^{222}Rn concentrations in zip codes of Belmont county based on the 2 pCi/L breakdown classification.

Table 8: ^{222}Rn statistics for Brown county

Zip code	No.	Min.	Max.	AM	GM	SD	CV
45101	16	0.9	9.8	3.88	2.92	3.01	0.15
45106	59	0.1	13.6	3.39	2.08	2.99	0.15
45107	66	0.1	9.8	3.17	2.23	2.39	0.12
45118	15	0.1	8	2.93	1.86	2.39	0.12
45120	20	0.5	18.2	3.22	1.96	4.18	0.21
45121	42	0.2	20.1	3.13	1.5	4.16	0.21
45130	3	0.3	1.6	0.77	0.58	0.72	0.04
45131	2	1.1	1.1	1.1	1.1	0	0
45133	126	0.1	31.1	4.14	2.49	4.51	0.23
45142	25	0.5	9.4	4.15	3.05	2.78	0.14
45144*	8	0.6	16.8	5.41	3.23	5.81	0.3
45148	7	0.8	7.3	4.5	3.59	2.64	0.13
45154	19	0.5	18.3	3.42	1.88	4.35	0.22
45167	10	0.6	8.2	3.09	2.1	2.8	0.14
45168	2	1.8	2.5	2.15	2.12	0.49	0.03
45171	27	0.4	9.8	3.01	2.05	2.64	0.14
45176	19	0.5	11.1	4.02	2.86	3.17	0.16
45697*	22	0.1	15.6	4.63	2.41	4.79	0.25

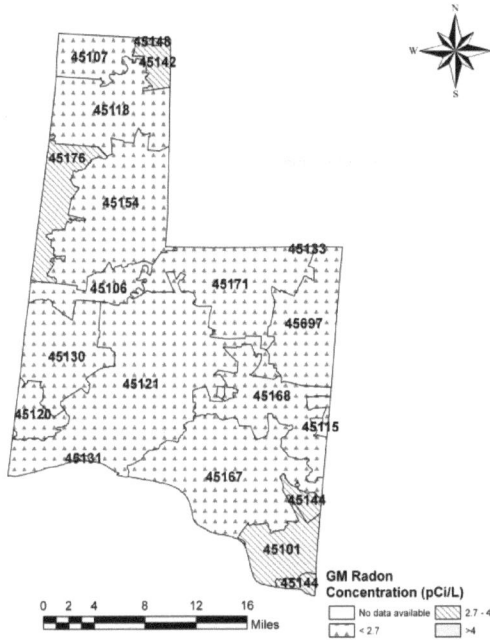

Figure 8a: GM ^{222}Rn concentrations in zip codes of Brown county based on the WHO-USEPA classification.

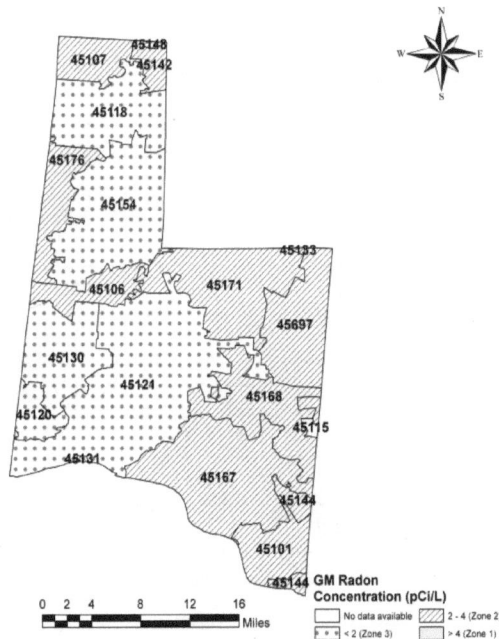

Figure 8b: GM ^{222}Rn concentrations in zip codes of Brown county based on the USEPA classification.

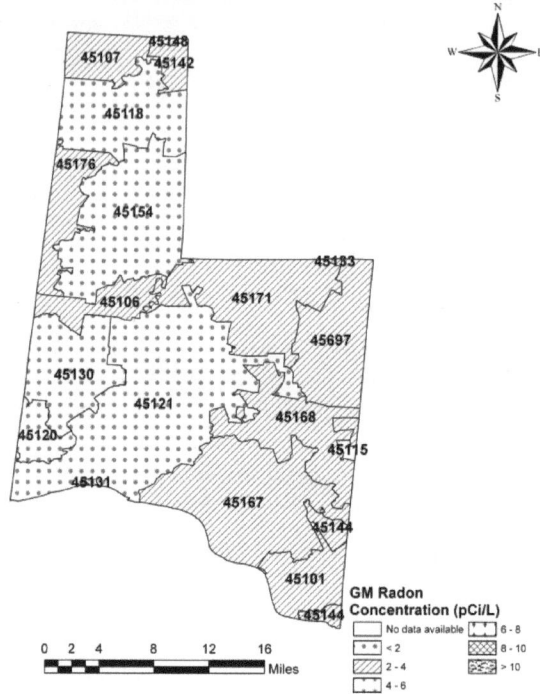

Figure 8c: GM ^{222}Rn concentrations in zip codes of Brown county based on the 2 pCi/L breakdown classification.

Table 9: ^{222}Rn statistics for Butler county

Zip code	No.	Min.	Max.	AM	GM	SD	CV
45003*	10	0.9	14	6.01	4.4	4.43	0.23
45011	1111	0.1	85.9	4.9	3.18	5.37	0.27
45013	562	0.1	470	49.54	5.07	114.91	5.88
45014	776	0.1	61.1	6.45	3.44	7.75	0.4
45015	72	0.1	49.7	5.6	3.05	7.34	0.38
45042	640	0.1	143.8	4.51	1.96	7.76	0.4
45044	963	0.1	50.2	4.83	3.14	4.94	0.25
45050	153	0.1	30.6	4.75	2.78	5.09	0.26
45053	34	0.1	11	3.93	2.77	2.69	0.14
45056	366	0.1	61.1	4.93	2.95	5.99	0.31
45062	11	2.6	15.2	8.1	6.99	4.36	0.22

Zip code	No.	Min.	Max.	AM	GM	SD	CV
45064	18	0.5	30.9	6.14	2.77	9.39	0.48
45067	126	0.1	48.9	9.89	5.71	10.16	0.52
45069	1942	0.1	69.8	4.52	2.92	4.71	0.24
45241	1001	0.1	42.3	4.07	2.48	4.41	0.23
45246	233	0.1	23.5	3.89	2.5	3.85	0.2
45327	233	0.1	86.1	9.23	5.43	11.52	0.59
45012[**]	1	0.9	0.9	0.9	0.9	0	0
45018[**]	5	1.3	19.7	6.66	3.93	7.72	0.4
45061[**]	5	2.7	19.4	7.1	5.41	6.92	0.35
45063[**]	3	2.9	5.7	3.97	3.79	1.51	0.08
45071[**]	4	1	1.8	1.38	1.34	0.35	0.02

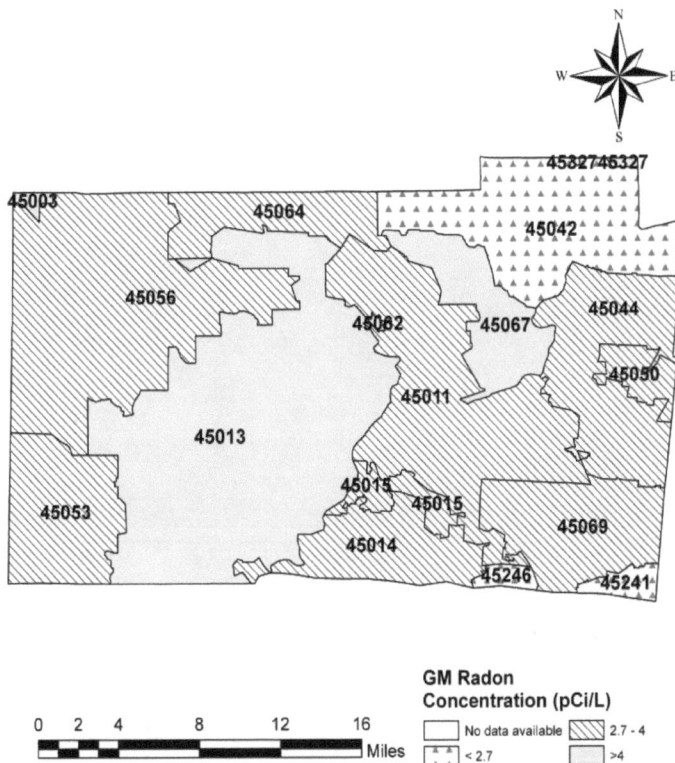

Figure 9a: GM ^{222}Rn concentrations in zip codes of Butler county based on the WHO-USEPA classification.

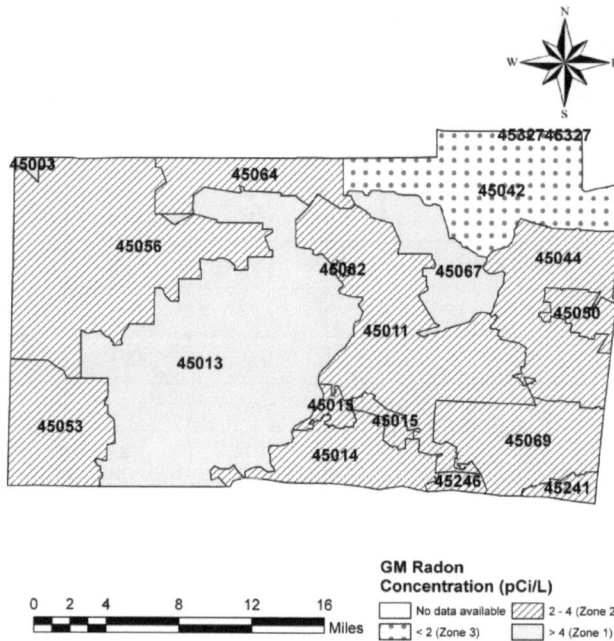

Figure 9b: GM ^{222}Rn concentrations in zip codes of Butler county based on the USEPA classification.

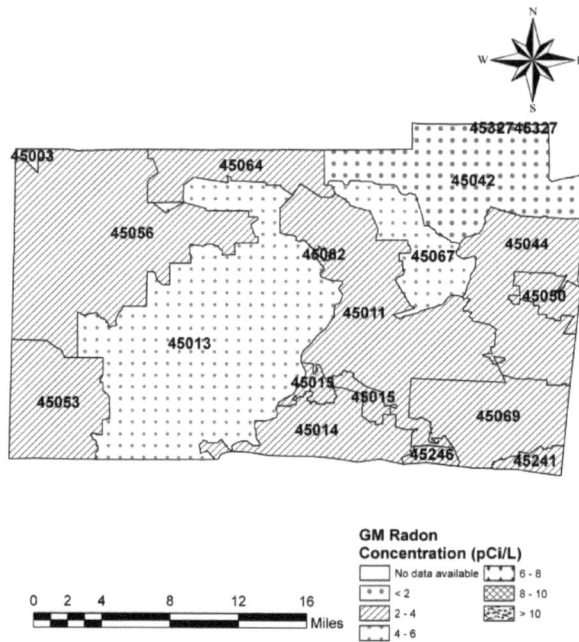

Figure 9c: GM ^{222}Rn concentrations in zip codes of Butler county based on the 2 pCi/L breakdown classification.

Table 10: ^{222}Rn statistics for Carroll county

Zip Code	No.	Min.	Max.	AM	GM	SD	CV
43903	19	0.1	23	3.58	1.85	5.22	0.27
43908*	3	1.9	3.5	2.47	2.37	0.9	0.05
43945*	29	0.6	177.8	20.52	5.57	41.09	2.1
43986*	20	0.7	23.9	7.56	3.75	8.25	0.42
43988*	28	0.8	124.5	43.55	21.87	40.86	2.09
44427*	19	0.5	219.1	31.13	8.1	57.5	2.94
44607	5	0.3	18.2	5.42	2.46	7.33	0.38
44615	258	0.1	220.2	13.5	3.95	30.9	1.58
44620	29	0.9	15.2	4.84	3.65	3.82	0.2
44621*	25	0.1	6.7	2.72	2.1	1.57	0.08
44625*	21	0.6	48.3	7.29	3.7	11.59	0.59
44643*	64	0.2	95.7	14.5	6.82	21.45	1.1
44644	103	0.1	69.5	10.72	4.73	14.08	0.72
44651	9	0.4	9.5	3.98	2.59	3.25	0.17
44656*	36	0.2	47.5	11.71	5.44	12.81	0.66
44657*	111	0.1	580	17.61	6.09	56.16	2.88
44675	12	0.5	55.6	10.6	4.68	16.79	0.86
44688*	34	0.6	55.7	6.47	3.95	9.53	0.49
44695*	18	0.1	33	9.53	5.77	8.54	0.44
44730*	47	0.1	63.6	6.88	3.51	11.13	0.57
44639**	3	7.9	9.5	8.47	8.44	0.9	0.05

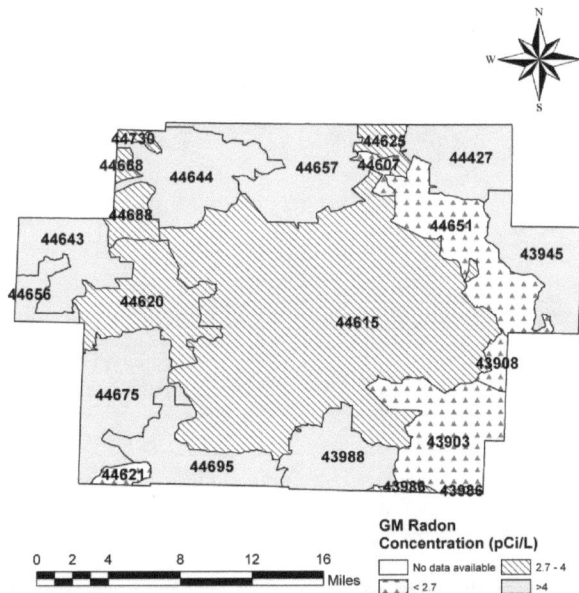

Figure 10a: GM ^{222}Rn concentrations in zip codes of Carroll county based on the WHO-USEPA classification.

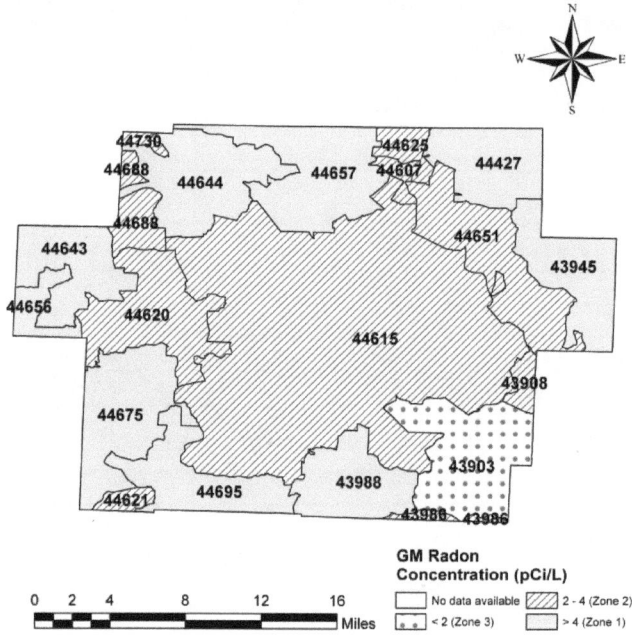

Figure 10b: GM ^{222}Rn concentrations in zip codes of Carroll county based on the USEPA classification.

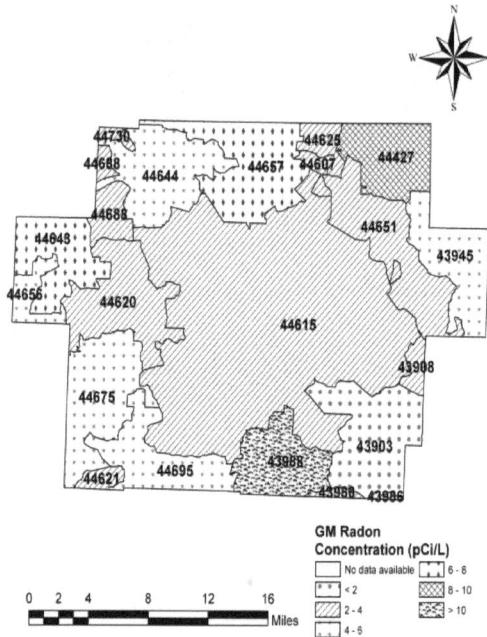

Figure 10c: GM ^{222}Rn concentrations in zip codes of Carroll county based on the 2 pCi/L breakdown classification.

Table 11: ^{222}Rn statistics for Champaign county

Zip code	No.	Min.	Max.	AM	GM	SD	CV
43009	63	0.6	106.5	14.13	6.68	21.32	1.09
43044	110	0.1	41.5	7.51	4.32	7.59	0.39
43045[*]	87	0.1	26.6	6.05	4	5.17	0.26
43060[*]	51	0.2	62.7	8.24	4.51	11.03	0.56
43070	34	0.1	4.7	1.26	0.83	1.1	0.06
43072	217	0.1	46	4.33	2.01	6.95	0.36
43078	719	0.1	265.3	9.3	5.34	14.18	0.73
43084	12	0.1	18.6	5.54	2.53	5.78	0.3
43318[*]	59	0.8	150.3	9.64	5.33	20.02	1.03
43343[*]	18	0.1	38.8	14.47	6.01	14.88	0.76
43357[*]	207	0.1	75	16	8.17	17.07	0.87
45312[*]	43	0.1	59.2	9.64	6.01	10.29	0.53
45317	16	0.2	19.6	8.84	6.2	5.7	0.29
45344[*]	440	0.1	150.1	11.3	5.75	16.58	0.85
45365[*]	567	0.1	111.1	7.67	4.63	8.64	0.44
45389	12	1.4	30.8	11.13	7.62	8.57	0.44
45502[*]	652	0.1	860	10.34	4.52	41.04	2.1
43047[**]	1	15.3	15.3	15.3	15.3	0	0

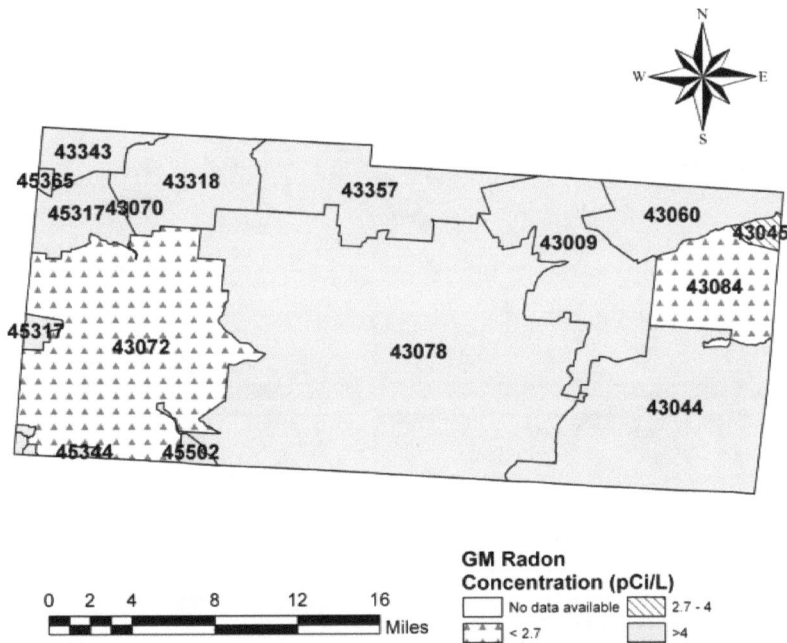

Figure 11a: GM ^{222}Rn concentrations in zip codes of Champaign county based on the WHO-USEPA classification.

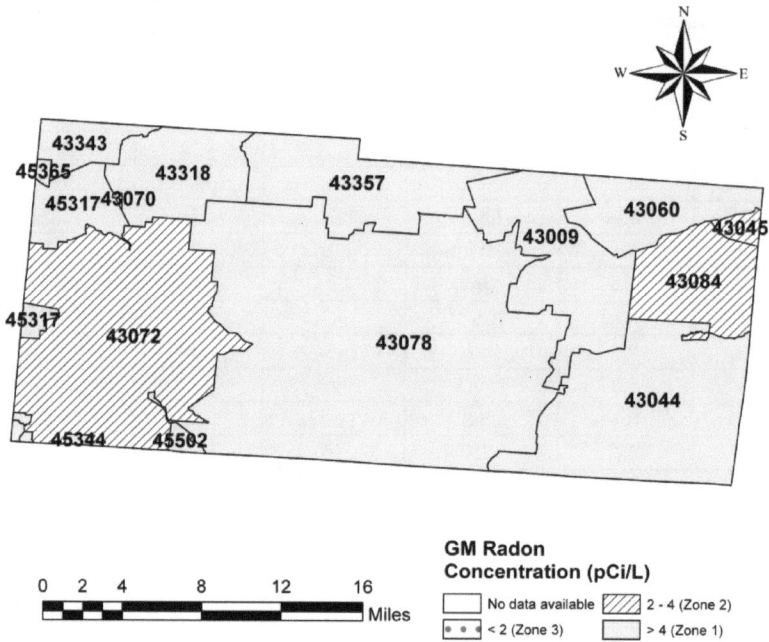

Figure 11b: GM ^{222}Rn concentrations in zip codes of Champaign county based on the USEPA classification.

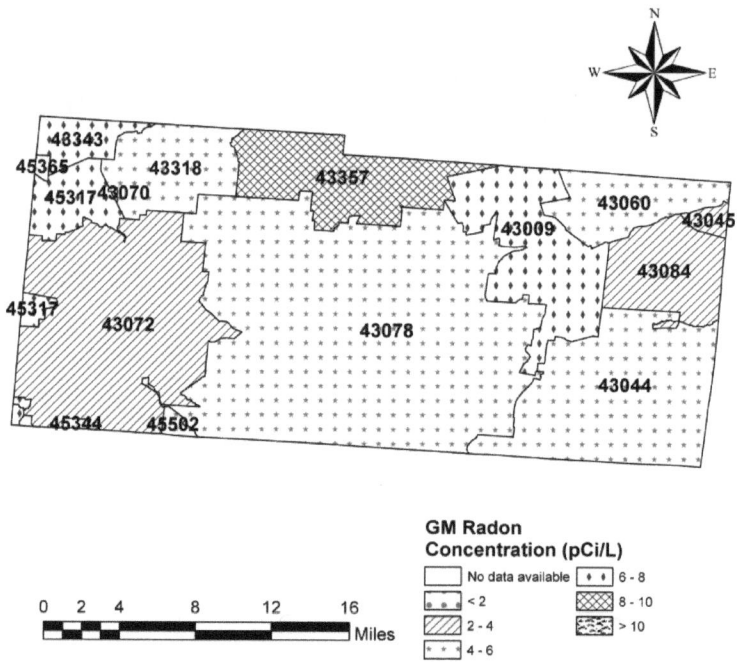

Figure 11c: GM ^{222}Rn concentrations in zip codes of Champaign county based on the 2 pCi/L breakdown classification.

Table 12: ^{222}Rn statistics for Clark county

Zip Code	No.	Min.	Max.	AM	GM	SD	CV
43010	1	3.2	3.2	3.2	3.2	0	0
43044[*]	110	0.1	41.5	7.51	4.32	7.59	0.39
43078[*]	719	0.1	265.3	9.3	5.34	14.18	0.73
43140[*]	397	0.1	82.1	7.87	4.88	9.39	0.48
43153[*]	8	0.9	10.9	5.1	4.1	3.25	0.17
45314[*]	176	0.1	97.7	5.4	2.71	10.73	0.55
45316[*]	7	0.2	9.3	3.06	1.87	3.05	0.16
45319	19	0.2	26.6	5.6	3.48	6.03	0.31
45323	287	0.1	67.1	9.81	5.7	10.81	0.55
45324[*]	1357	0.1	163	7.91	4.13	10.98	0.56
45341	151	0.1	750.1	27.41	11.31	85.32	4.37
45344	440	0.1	150.1	11.3	5.75	16.58	0.85
45349	12	0.8	6.6	3.33	2.75	1.88	0.1
45368	85	0.2	35.8	5.33	3.18	5.63	0.29
45369	88	0.1	21.3	5.09	3.26	4.74	0.24
45372	21	0.2	37.5	8.49	5.52	8.28	0.42
45387[*]	653	0.1	111.9	6.83	4.02	9.2	0.47
45502	652	0.1	860	10.34	4.52	41.04	2.1
45503	857	0.1	106	7.33	4.49	8.07	0.41
45504	567	0.1	100	6.05	3.71	7.88	0.4
45505	379	0.1	87.7	6.89	4.25	8.63	0.44
45506	246	0.2	47.1	6.71	4.01	7.37	0.38
45501[**]	37	0.1	88.1	21.56	7.82	29.14	1.49

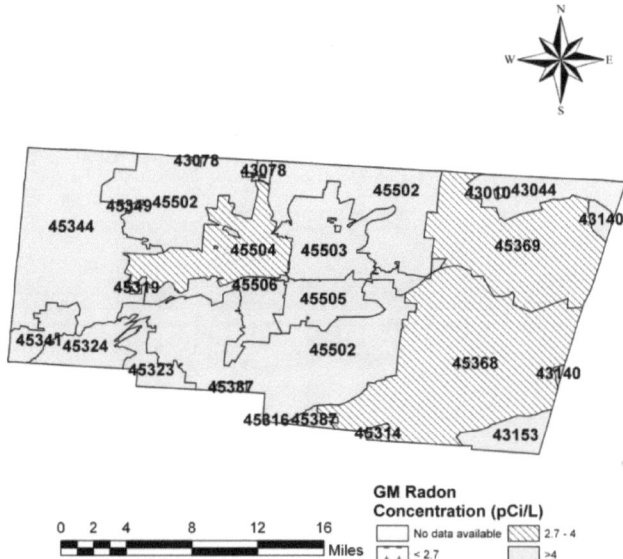

Figure 12a: GM ^{222}Rn concentrations in zip codes of Clark county based on the WHO-USEPA classification.

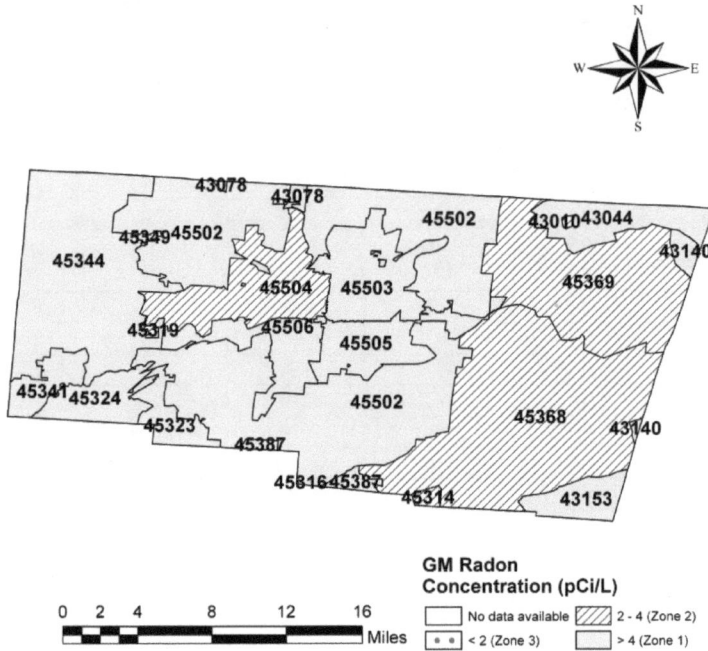

Figure 12b: GM ^{222}Rn concentrations in zip codes of Clark county based on the USEPA classification.

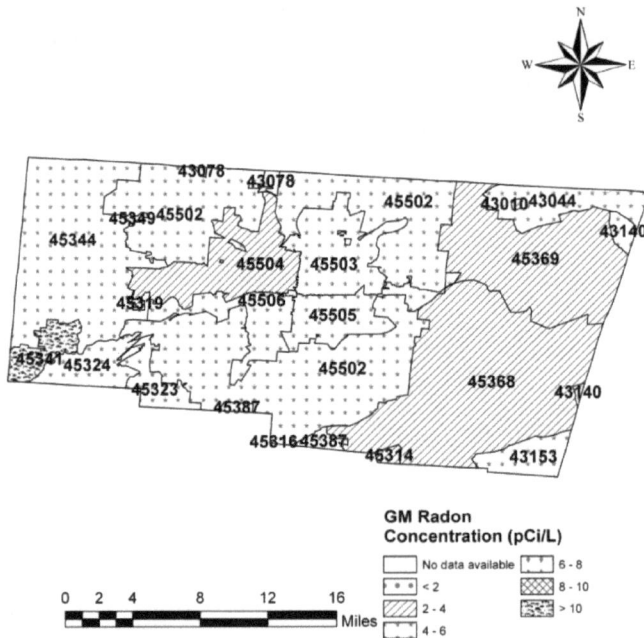

Figure 12c: GM ^{222}Rn concentrations in zip codes of Clark county based on the 2 pCi/L breakdown classification.

Table 13: ^{222}Rn statistics for Clermont county

Zip code	No.	Min.	Max.	AM	GM	SD	CV
45102	190	0.1	26.3	4.42	2.7	4.16	0.21
45103	325	0.1	37.2	4.72	3.06	4.61	0.24
45106	59	0.1	13.6	3.39	2.08	2.99	0.15
45107*	66	0.1	9.8	3.17	2.23	2.39	0.12
45118*	15	0.1	8	2.93	1.86	2.39	0.12
45120	20	0.5	18.2	3.22	1.96	4.18	0.21
45121*	42	0.2	20.1	3.13	1.5	4.16	0.21
45122	59	0.3	15.2	3.77	2.83	3.03	0.16
45130*	3	0.3	1.6	0.77	0.58	0.72	0.04
45140	1852	0.1	470.8	4.95	2.99	12.09	0.62
45147	4	2.3	4.9	3.28	3.11	1.24	0.06
45150	644	0.1	83.5	5.14	3.3	6.53	0.33
45153	9	0.6	2.9	1.09	0.91	0.81	0.04
45157	80	0.1	25.2	3.85	2.54	4.23	0.22
45160	3	1	1.4	1.2	1.19	0.2	0.01
45162*	35	0.2	6.4	2.05	1.51	1.61	0.08
45176	19	0.5	11.1	4.02	2.86	3.17	0.16
45244	860	0.1	40.5	4.16	2.84	4.19	0.21
45245	338	0.1	23.7	5.03	3.3	4.96	0.25
45255*	557	0.1	42.4	4.19	2.72	4.05	0.21

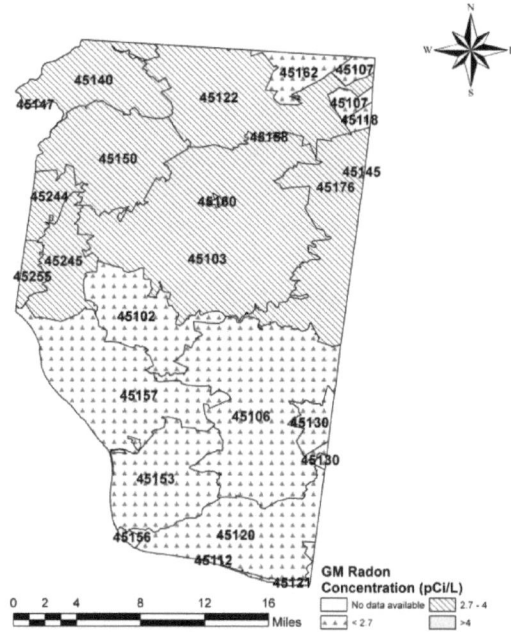

Figure 13a: GM ^{222}Rn concentrations in zip codes of Clermont county based on the WHO-USEPA classification.

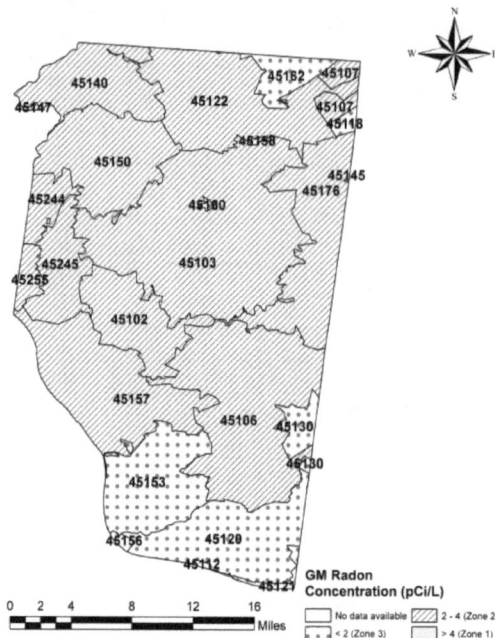

Figure 13b: GM ^{222}Rn concentrations in zip codes of Clermont county based on the USEPA classification.

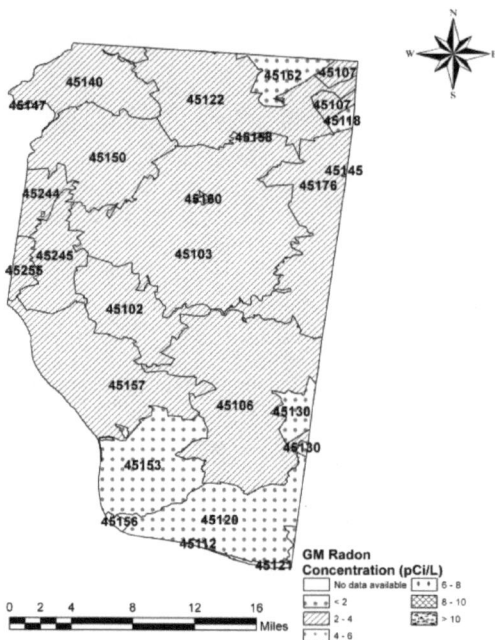

Figure 13c: GM ^{222}Rn concentrations in zip codes of Clermont county based on the 2 pCi/L breakdown classification.

Table 14: ^{222}Rn statistics for Clinton county

Zip code	No.	Min.	Max.	AM	GM	SD	CV
45068[*]	335	0.1	44.9	5.63	3.58	6	0.31
45107	66	0.1	9.8	3.17	2.23	2.39	0.12
45113	44	0.6	54.2	5.94	3.36	8.98	0.46
45135[*]	10	0.5	15.4	3.54	1.94	4.58	0.23
45142[*]	25	0.5	9.4	4.15	3.05	2.78	0.14
45146	10	0.2	40.3	7.98	3.49	11.83	0.61
45148	7	0.8	7.3	4.5	3.59	2.64	0.13
45159	37	0.1	30.9	6.73	2.94	7.69	0.39
45166	1	2	2	2	2	0	0
45169	23	0.4	12	4.24	3.11	3.46	0.18
45177	317	0.1	55.6	4.8	2.85	6.47	0.33
45335[*]	195	0.1	39.8	4.88	2.68	6.69	0.34
45138[**]	9	1.9	42.1	11.28	6.18	14.07	0.72

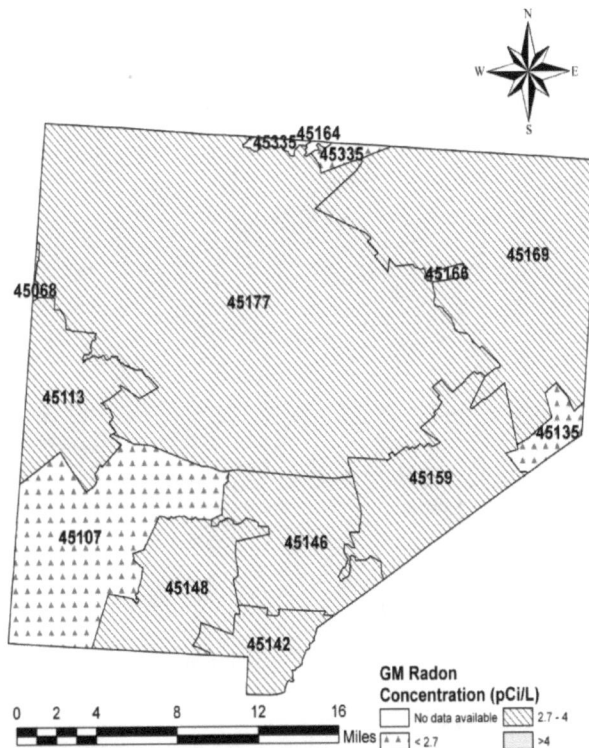

Figure 14a: GM ^{222}Rn concentrations in zip codes of Clinton county based on the WHO-USEPA classification.

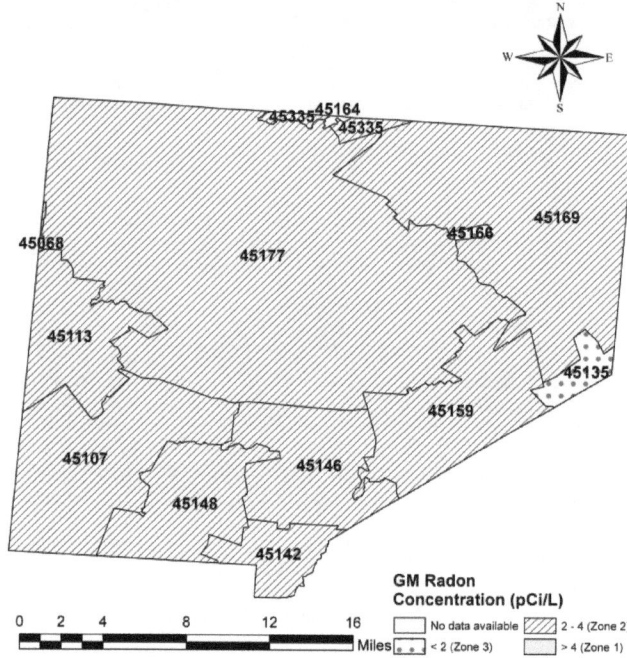

Figure 14b: GM ^{222}Rn concentrations in zip codes of Clinton county based on the USEPA classification.

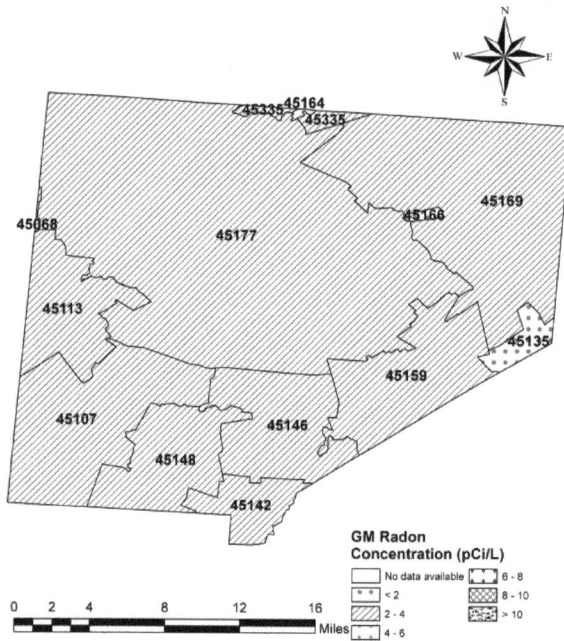

Figure 14c: GM ^{222}Rn concentrations in zip codes of Clinton county based on the 2 pCi/L breakdown classification.

Table 15: ^{222}Rn statistics for Columbiana county

Zip code	No.	Min.	Max.	AM	GM	SD	CV
43920	286	0.1	83.4	12.14	5.57	15.93	0.82
43930*	4	3.5	39	15.05	9.08	16.74	0.86
43945	29	0.6	177.8	20.52	5.57	41.09	2.1
43962	2	3.7	3.7	3.7	3.7	0	0
43968	43	0.1	14.6	3.66	2.39	3.41	0.17
44408	157	0.1	234.2	6.31	3.07	19.41	0.99
44413	60	0.1	21.3	3.15	2.26	3.22	0.16
44423	26	0.1	41	12.72	6.13	14.05	0.72
44427	19	0.5	219.1	31.13	8.1	57.5	2.94
44431	56	0.2	15.3	4	3.13	2.94	0.15
44432	123	0.1	136.5	17.72	6.29	27.84	1.43
44441	46	0.2	41.6	14.62	8.44	11.77	0.6
44443*	31	0.1	207	27.88	6.65	50.99	2.61
44445	27	0.2	39	5.23	2.84	8.09	0.41
44454*	10	1.1	26.8	6.03	3.82	7.6	0.39
44455	11	0.6	41.1	13.93	5.79	15.42	0.79
44460	226	0.1	76.7	6.56	2.95	12.28	0.63
44490	2	3.5	4	3.75	3.74	0.35	0.02
44493	3	2.3	6.2	4	3.68	2	0.1
44601*	214	0.1	16.3	2.62	1.89	2.21	0.11
44609*	17	0.3	5.3	1.43	1.06	1.31	0.07
44625	21	0.6	48.3	7.29	3.7	11.59	0.59
44634	17	0.2	10.2	2.59	1.78	2.56	0.13
44657*	111	0.1	580	17.61	6.09	56.16	2.88
44415**	3	0.9	9.2	3.93	2.41	4.58	0.23
44665**	2	2	2.6	2.3	2.28	0.42	0.02

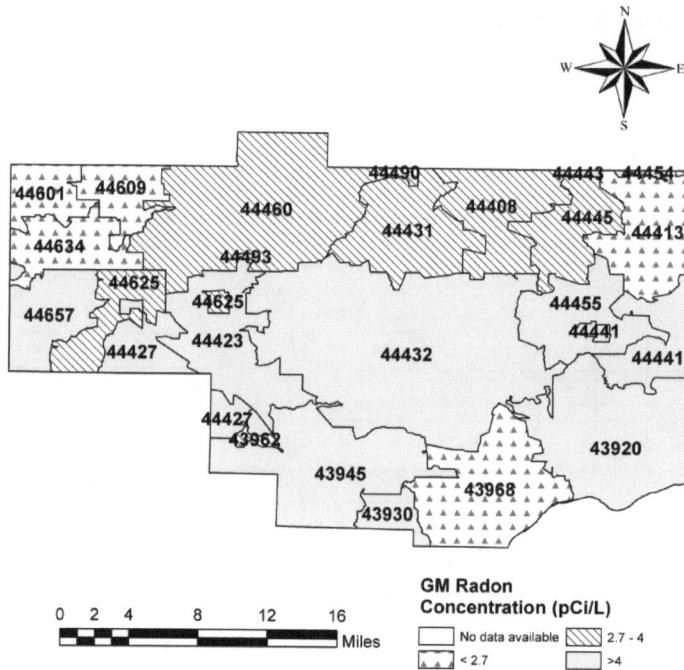

Figure 15a: GM ^{222}Rn concentrations in zip codes of Columbiana county based on the WHO-USEPA classification.

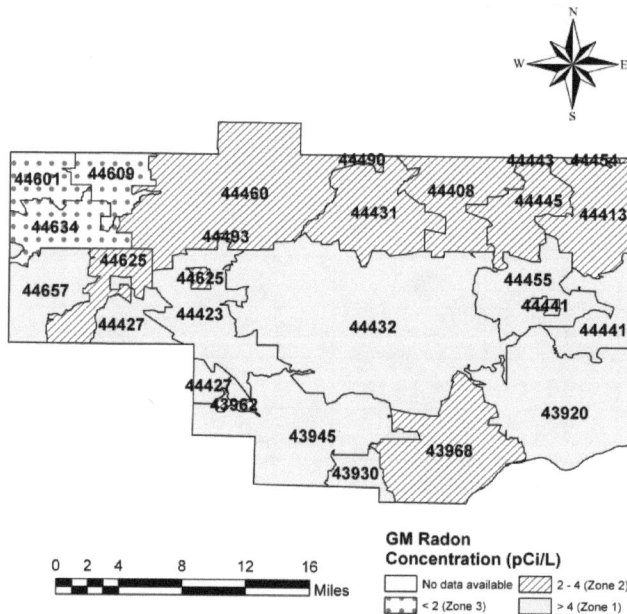

Figure 15b: GM ^{222}Rn concentrations in zip codes of Columbiana county based on the USEPA classification.

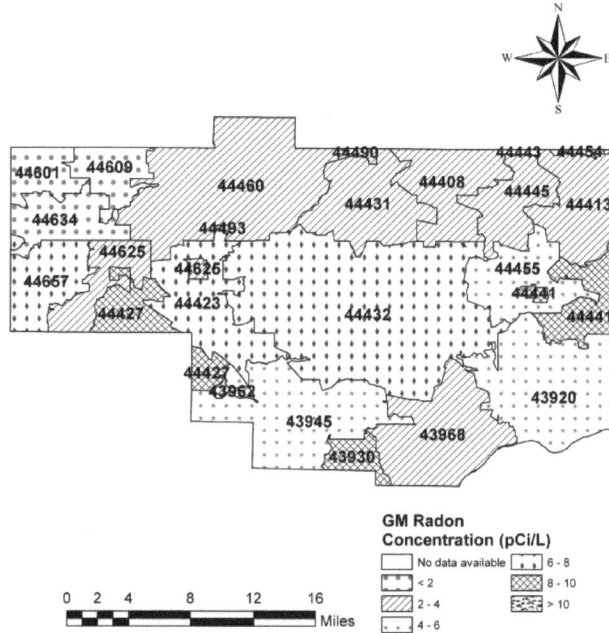

Figure 15c: GM ^{222}Rn concentrations in zip codes of Columbiana county based on the 2 pCi/L breakdown classification.

Table 16: ^{222}Rn statistics for Coshocton county

Zip code	No.	Min.	Max.	AM	GM	SD	CV
43006*	4	1	15.9	10.35	6.83	7.12	0.36
43749*	12	0.1	115.1	12.85	2.11	32.85	1.68
43804*	2	0.8	2.5	1.65	1.41	1.2	0.06
43805	1	2.3	2.3	2.3	2.3	0	0
43811	12	0.4	51.7	15.86	7.25	18.56	0.95
43812	180	0.1	95.6	11.78	5.17	18.64	0.95
43821*	30	0.2	14.6	4.16	2.9	3.15	0.16
43822*	84	0.2	131.1	11.75	5.7	18.43	0.94
43824	27	0.7	62.5	25.7	11.75	24.91	1.28
43832*	26	0.2	20.4	6.77	3.22	7.41	0.38
43840	9	1.1	10.5	5.31	3.97	3.71	0.19
43843	11	1.6	60.5	10.39	4.74	17.77	0.91
43844	17	0.6	117.7	25.59	9.82	35.85	1.84
43845	29	0.2	40.8	10.46	6.68	8.42	0.43
44637*	19	0.6	140.3	27.02	9.77	43.48	2.23
44654*	77	0.1	43.7	7.24	3.74	9.54	0.49
43828**	1	8.8	8.8	8.8	8.8	0	0

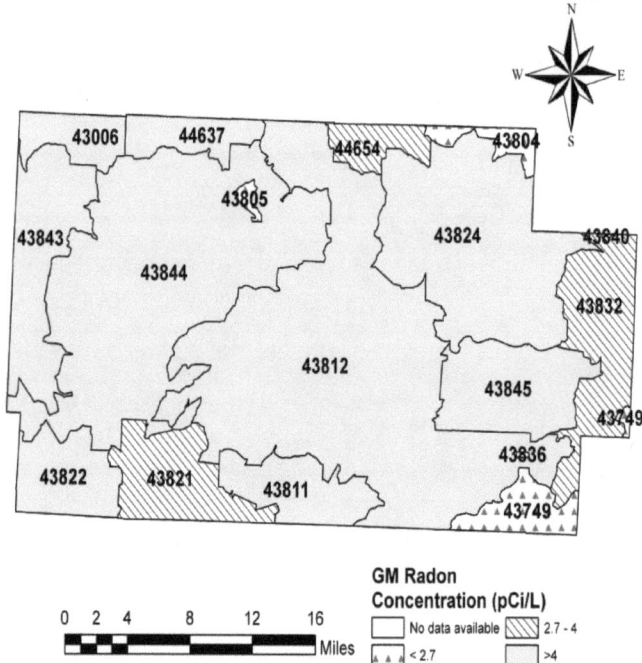

Figure 16a: GM ^{222}Rn concentrations in zip codes of Coshocton county based on the WHO-USEPA classification.

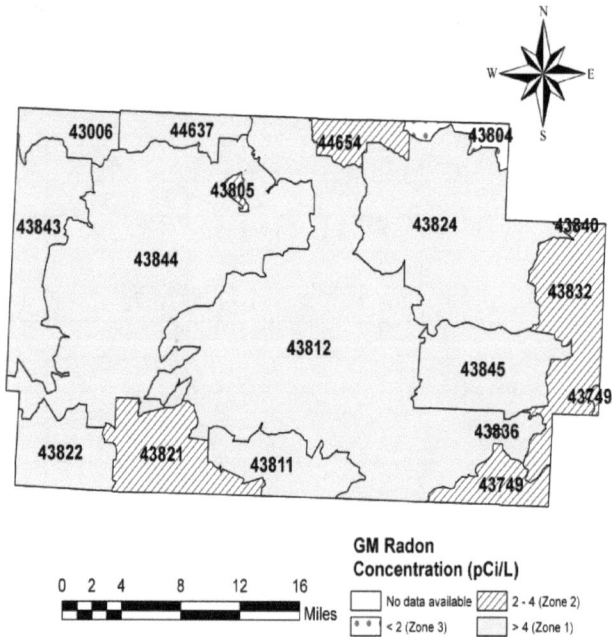

Figure 16b: GM ^{222}Rn concentrations in zip codes of Coshocton county based on the USEPA classification.

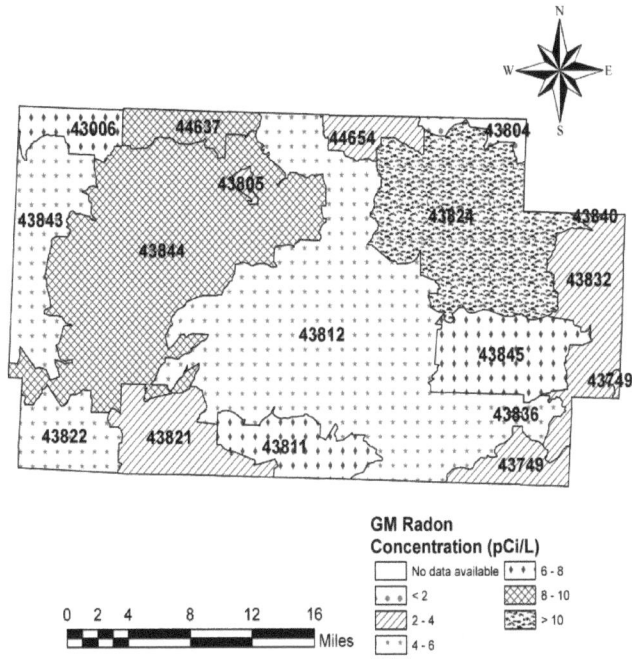

Figure 16c: GM ^{222}Rn concentrations in zip codes of Coshocton county based on the 2 pCi/L breakdown classification.

Table 17: ^{222}Rn statistics for Crawford county

Zip code	No.	Min.	Max.	AM	GM	SD	CV
43302*	911	0.1	67.1	6.25	3.47	7.2	0.37
43314*	83	0.1	26.3	4.65	3.38	4.12	0.21
44818*	21	0.1	16.4	4.3	2.72	3.99	0.2
44820	223	0.1	163	6.61	4.11	11.93	0.61
44827	181	0.1	56.1	3.18	1.25	5.67	0.29
44833	493	0.1	140.2	7.75	4.41	12.72	0.65
44849*	52	0.2	20.9	4.15	2.88	4.36	0.22
44854	15	0.2	7.1	2.96	2.21	1.78	0.09
44856	3	1.7	5.6	3	2.53	2.25	0.12
44865*	15	2.3	38.1	10.89	7.47	11.62	0.6
44875*	169	0.1	32.6	5.68	4.02	5.21	0.27
44881	1	3.9	3.9	3.9	3.9	0	0
44882*	32	0.5	9.1	4.19	2.91	3	0.15
44887	7	1.8	7.8	4.16	3.57	2.57	0.13
44860**	3	3.4	7.3	4.7	4.39	2.25	0.12

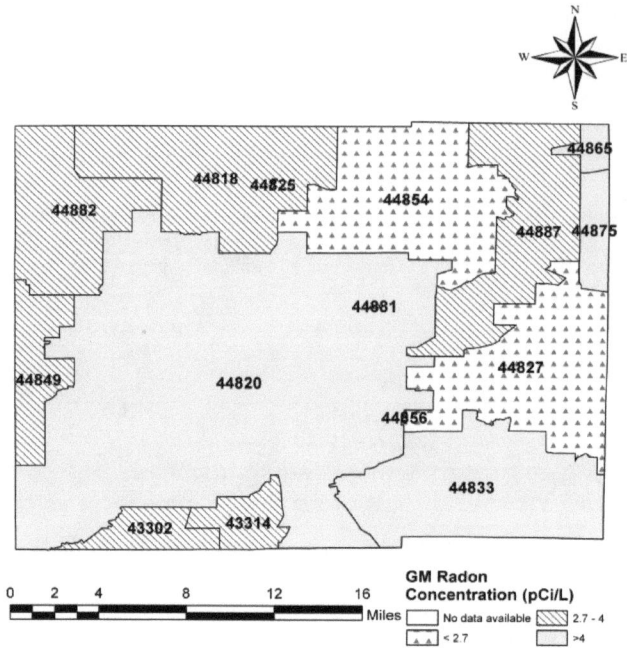

Figure 17a: GM ^{222}Rn concentrations in zip codes of Crawford county based on the WHO-USEPA classification.

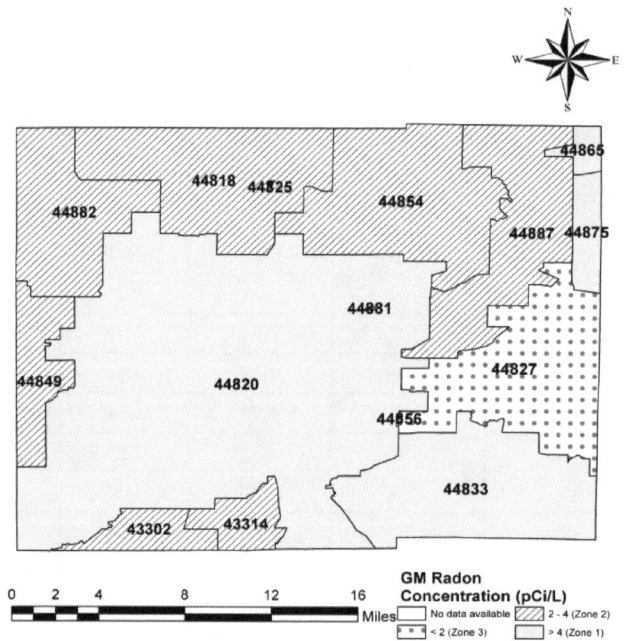

Figure 17b: GM ^{222}Rn concentrations in zip codes of Crawford county based on the USEPA classification.

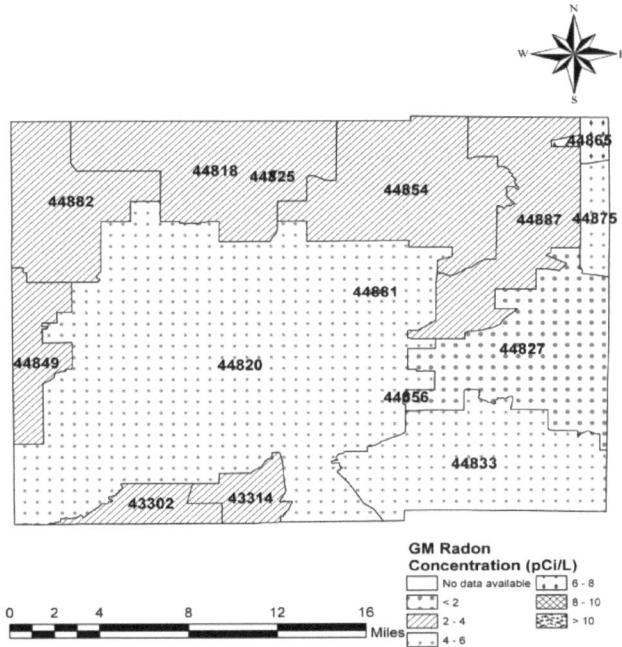

Figure 17c: GM ^{222}Rn concentrations in zip codes of Crawford county based on the 2 pCi/L breakdown classification.

Table 18: ^{222}Rn statistics for Cuyahoga county

Zip code	No.	Min.	Max.	AM	GM	SD	CV	Zip code	No.	Min.	Max.	AM	GM	SD	CV
44017	297	0.1	30.3	3.79	2.43	3.66	0.19	44125	138	0.1	16	2.2	1.35	2.95	0.15
44022	698	0.1	733.8	3.66	1.95	27.79	1.42	44126	303	0.1	89.9	3.35	1.94	7.25	0.37
44040	133	0.3	22.4	2.95	2.09	3.33	0.17	44127	16	0.6	7.4	2.41	1.93	1.81	0.09
44070	465	0.1	95.8	4.85	2.86	7.29	0.37	44128	83	0.1	14.5	1.17	0.76	1.77	0.09
44101	15	0.2	4.1	1.43	0.95	1.22	0.06	44129	224	0.1	29.3	2.07	1.24	3.16	0.16
44102	107	0.1	13.3	2.14	1.26	2.34	0.12	44130	457	0.1	105	2.61	1.49	6.01	0.31
44103	36	0.1	10.3	1.66	0.87	2.41	0.12	44131	288	0.1	79.2	3.28	2.17	5.85	0.3
44104	23	0.2	10.1	2.08	1.08	2.75	0.14	44132	63	0.1	14.8	1.3	0.79	2.13	0.11
44105	85	0.1	13.7	2.31	1.34	2.33	0.12	44133	345	0.1	15.2	2.87	2.09	2.34	0.12
44106	223	0.1	24.9	1.96	1.3	2.33	0.12	44134	263	0.1	23.1	1.85	1.27	2.05	0.11
44107	397	0.1	17.5	1.63	1	2.04	0.1	44135	120	0.1	14.9	2.33	1.46	2.66	0.14
44108	50	0.4	8.5	2.64	1.83	2.34	0.12	44136	560	0.1	656.2	5.12	2.16	38.94	1.99
44109	134	0.1	14.4	2.29	1.32	2.56	0.13	44137	92	0.1	6.9	1.32	0.96	1.19	0.06
44110	30	0.1	7.9	1.32	0.88	1.49	0.08	44138	328	0.1	68.4	4.01	2.56	4.85	0.25
44111	217	0.1	30	2.32	1.36	3.09	0.16	44139	718	0.1	156.4	4.12	1.87	14.7	0.75
44112	76	0.1	10	1.26	0.71	1.49	0.08	44140	859	0.1	37	5.06	3.34	4.73	0.24

Zip code	No.	Min.	Max.	AM	GM	SD	CV	Zip code	No.	Min.	Max.	AM	GM	SD	CV
44113	115	0.1	51.8	3.3	1.66	5.5	0.28	44141	524	0.1	72.9	3.53	2.23	4.67	0.24
44114	33	0.1	20	2.03	0.79	3.73	0.19	44142	108	0.1	6.5	1.52	1.01	1.43	0.07
44115	10	0.5	14.3	2.82	1.61	4.18	0.21	44143	391	0.1	23.3	3.04	2.22	2.61	0.13
44116	444	0.1	112.4	2.68	1.71	6.1	0.31	44144	117	0.1	8.7	1.7	1.09	1.79	0.09
44117	58	0.1	6.3	1.26	0.87	1.23	0.06	44145	1063	0.1	68.7	4.69	2.99	5.36	0.27
44118	844	0.1	161.8	1.78	0.94	6	0.31	44146	169	0.1	23.7	2.07	1.31	2.6	0.13
44119	82	0.1	17.1	1.53	0.98	2.11	0.11	44147	331	0.1	19.6	2.72	2.04	2.24	0.11
44120	369	0.1	10.5	1.64	1.12	1.55	0.08	44149	194	0.1	12.7	3.09	2.42	2.32	0.12
44121	372	0.1	114.6	1.87	1.04	6.13	0.31	44190[**]	1	1.1	1.1	1.1	1.1	0	0
44122	802	0.1	114.6	1.81	1.24	4.28	0.22	44194[**]	1	6.1	6.1	6.1	6.1	0	0
44123	83	0.1	6.3	1.31	0.79	1.42	0.07	44195[**]	1	5	5	5	5	0	0
44124	678	0.1	19	2.13	1.54	2.18	0.11	44199[**]	7	1.8	4.6	2.93	2.71	1.24	0.06

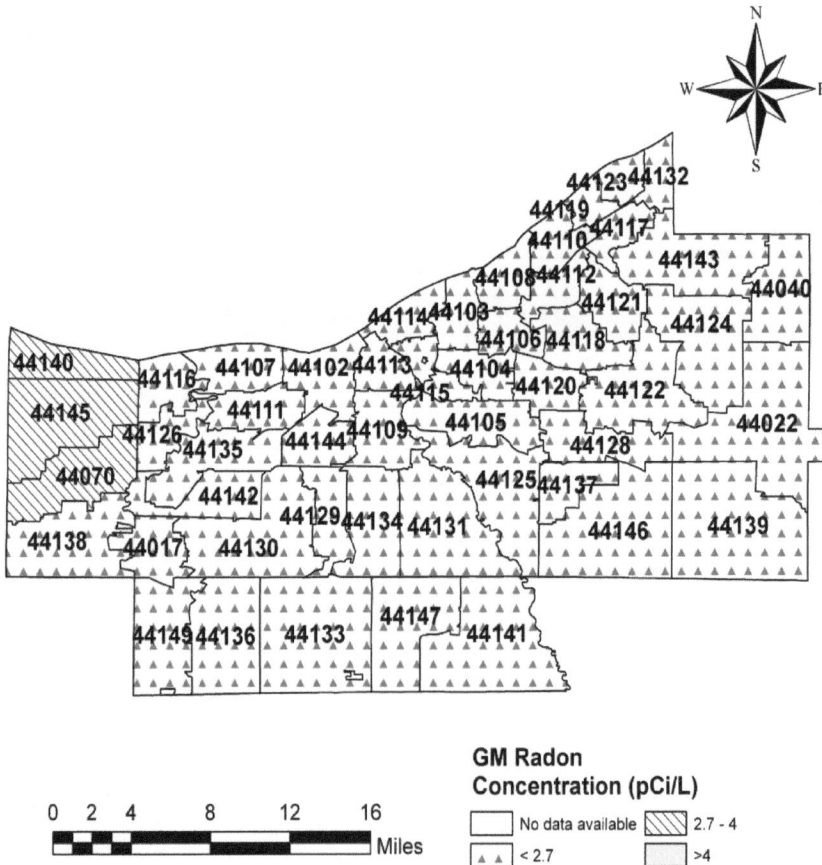

Figure 18a: GM ^{222}Rn concentrations in zip codes of Cuyahoga county based on the WHO-USEPA classification.

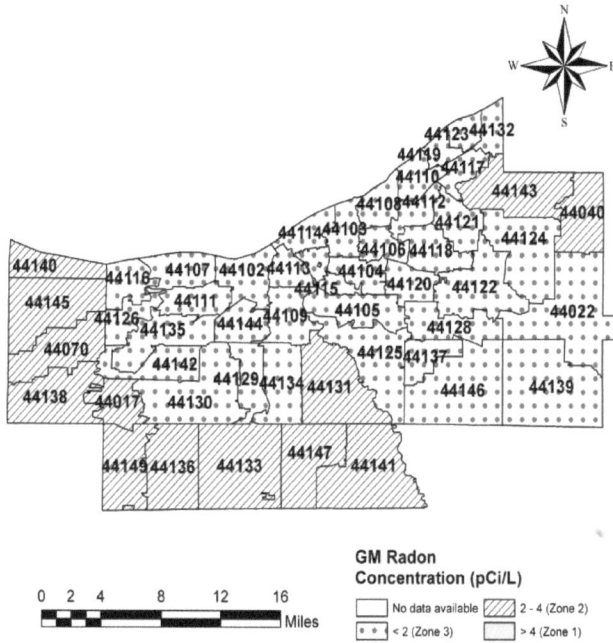

Figure 18b: GM ^{222}Rn concentrations in zip codes of Cuyahoga county based on the USEPA classification.

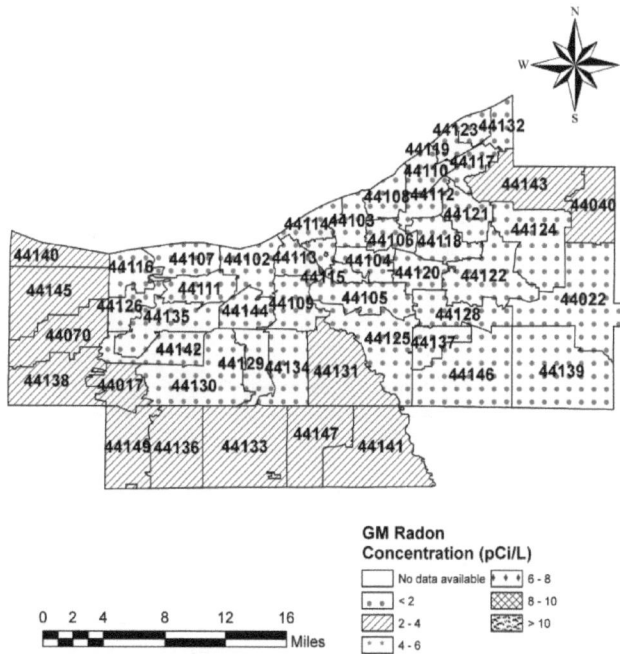

Figure 18c: GM ^{222}Rn concentrations in zip codes of Cuyahoga county based on the 2 pCi/L breakdown classification.

Table 19: ^{222}Rn statistics for Darke county

Zip code	No.	Min.	Max.	AM	GM	SD	CV
45303	15	2	125	15.37	5.8	32.18	1.65
45304	136	0.1	49	7.09	3.95	7.55	0.39
45308	66	0.1	71	8.43	4.69	12.25	0.63
45309*	428	0.1	267.8	7.08	3.65	14.87	0.76
45310*	8	0.7	13	5.96	4.16	4.33	0.22
45318*	88	0.1	252	11.42	5.54	27.38	1.4
45328	3	0.8	2.7	1.5	1.29	1.04	0.05
45331	334	0.1	84.5	8.88	4.28	11.47	0.59
45332	5	1.1	9	5.68	4.29	3.7	0.19
45337*	49	0.2	18.1	6.04	4.54	4.44	0.23
45346	41	0.4	62.3	10.07	5.54	12.21	0.63
45347*	41	0.6	44.4	8.94	5.67	8.97	0.46
45348	11	1.2	33.9	8.8	3.87	12.7	0.65
45350	3	0.8	2.2	1.53	1.41	0.7	0.04
45351	10	0.8	38.2	10.3	5.5	11.32	0.58
45352	1	10.8	10.8	10.8	10.8	0	0
45358	3	0.9	2.4	1.7	1.57	0.75	0.04
45362	9	0.5	47.7	13.72	5.64	19.07	0.98
45380	149	0.1	41.6	7.38	4.65	7.48	0.38
45382*	18	0.1	45.1	10.11	4.19	11.91	0.61
45388	13	1.7	16.2	5.85	4.27	5	0.26
45390	21	1.3	16.8	6.75	5.38	4.59	0.24
45846*	42	0.1	38.5	5.88	3.37	7.2	0.37

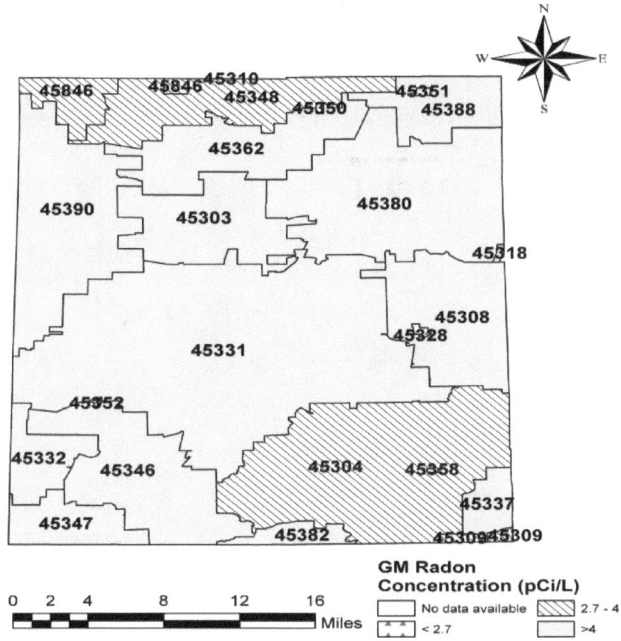

Figure 19a: GM ^{222}Rn concentrations in zip codes of Darke county based on the WHO-USEPA classification.

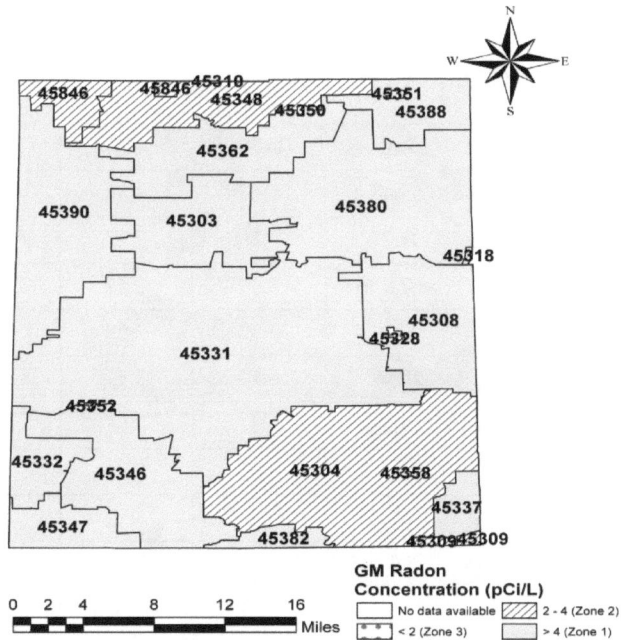

Figure 19b: GM ^{222}Rn concentrations in zip codes of Darke county based on the USEPA classification.

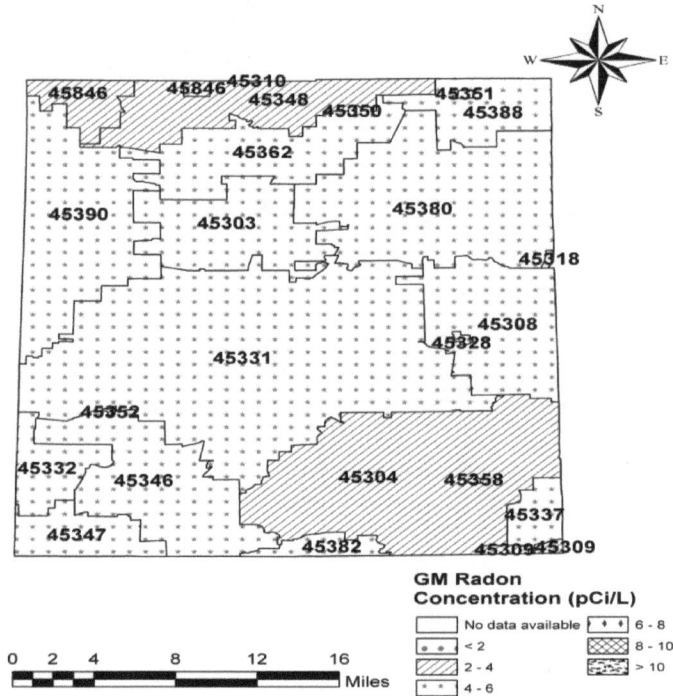

Figure 19c: GM ^{222}Rn concentrations in zip codes of Darke county based on the 2 pCi/L breakdown classification.

Table 20: ^{222}Rn statistics for Defiance county

Zip code	No.	Min.	Max.	AM	GM	SD	CV
43506[*]	94	0.1	53.4	6.67	3.41	8.72	0.45
43512	211	0.1	21.2	3.26	2.12	3.24	0.17
43517[*]	11	0.4	67.8	8.99	2.51	19.89	1.02
43526	12	0.1	5.6	3.32	2.52	1.62	0.08
43527[*]	12	0.6	29.4	6.14	3.11	8.26	0.42
43545[*]	79	0.1	19.5	3.81	2.58	3.24	0.17
43548[*]	10	1.5	22.9	8.33	6.26	6.81	0.35
43549	2	4.4	4.6	4.5	4.5	0.14	0.01
43556	9	0.4	5.3	2.5	1.77	1.89	0.1
45813[*]	15	1	7.6	3.17	2.64	1.99	0.1
45821[*]	1	3	3	3	3	0	0
45831[*]	12	0.6	12.1	3.78	2.7	3.21	0.16

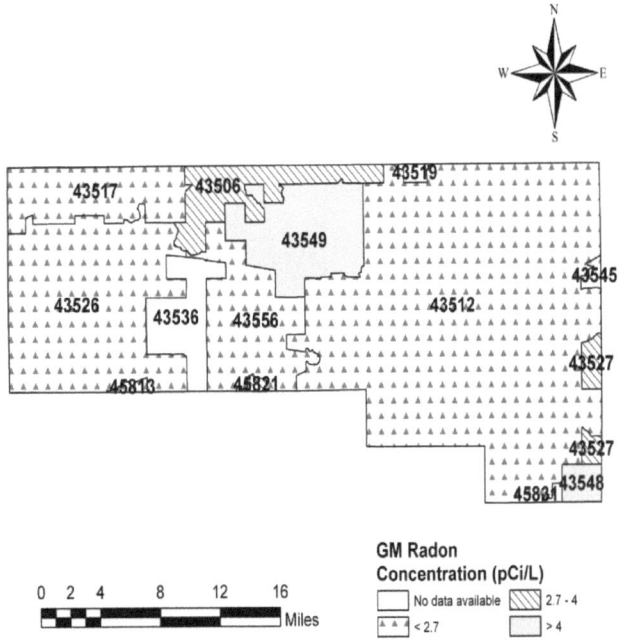

Figure 20a: GM [222]Rn concentrations in zip codes of Defiance county based on the WHO-USEPA classification.

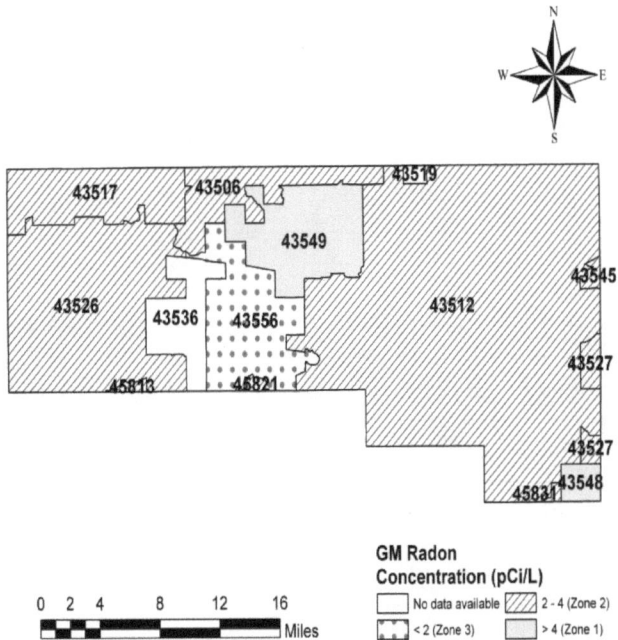

Figure 20b: GM [222]Rn concentrations in zip codes of Defiance county based on the USEPA classification.

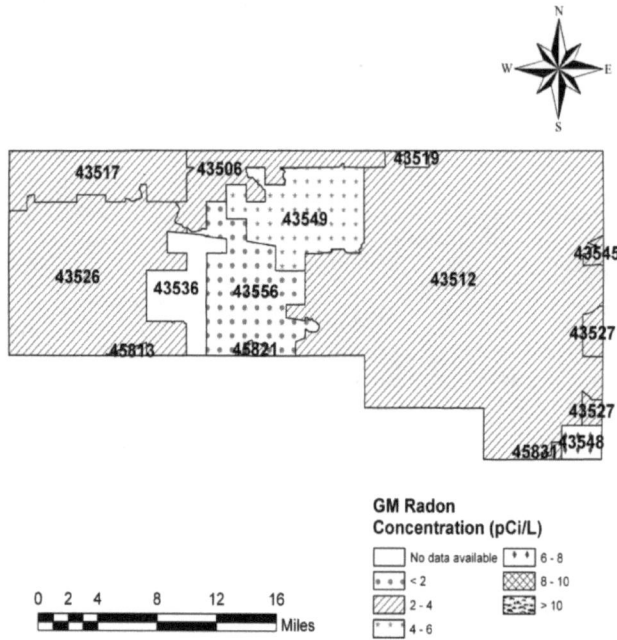

Figure 20c: GM ^{222}Rn concentrations in zip codes of Defiance county based on the 2 pCi/L breakdown classification.

Table 21: ^{222}Rn statistics for Delaware county

Zip code	No.	Min.	Max.	AM	GM	SD	CV
43003	44	0.1	50.1	6.1	2.83	9.28	0.48
43011*	154	0.1	61.7	8.61	5.53	9.87	0.51
43015	1518	0.1	116.5	7.57	4.65	9.4	0.48
43016*	1225	0.1	137.2	7.83	4.57	8.55	0.44
43017*	2855	0.1	193.7	10.78	5.93	13.25	0.68
43021	591	0.1	31.3	6.47	4.34	5.38	0.28
43031*	398	0.1	49.9	6.72	4.21	7.01	0.36
43032	9	0.1	9	3.49	1.95	2.64	0.13
43035	719	0.1	101.9	6.59	4.16	7.24	0.37
43040*	1063	0.1	63.5	5.28	3.28	6.13	0.31
43061	133	0.1	29.8	6.1	3.97	5.38	0.28
43065	1938	0.1	735	7.62	4.65	18.64	0.95
43066	75	0.1	80.7	6.83	2.45	13.38	0.68
43074	547	0.1	75.7	6.52	3.49	8.28	0.42
43082	1068	0.1	58.8	7.37	4.58	6.9	0.35
43240	41	0.1	16.1	5.13	3.5	3.96	0.2
43334*	94	0.1	62.9	9.01	4.68	12.88	0.66
43342*	58	0.6	21.3	9.15	6.79	5.92	0.3
43344*	153	0.1	61.8	5.54	3.14	7.7	0.39
43356*	33	0.7	28.8	6.16	3.94	6.15	0.32

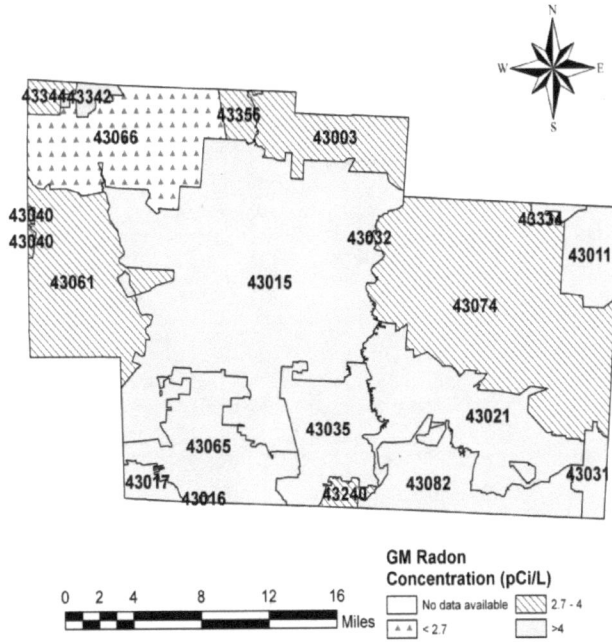

Figure 21a: GM ^{222}Rn concentrations in zip codes of Delaware county based on the WHO-USEPA classification.

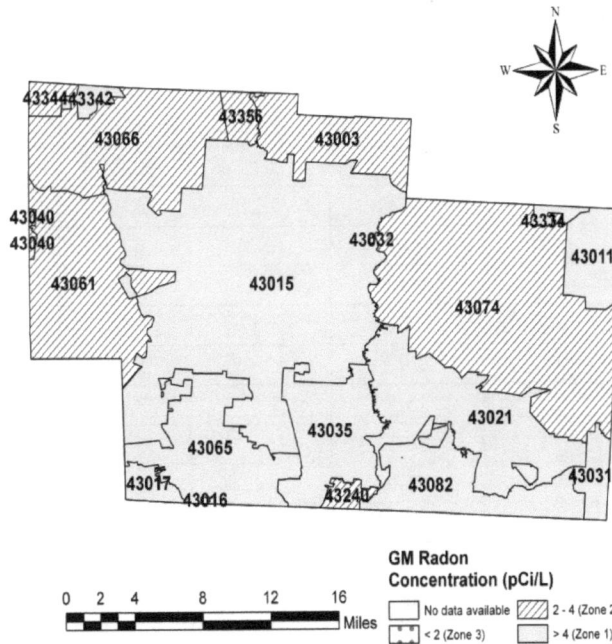

Figure 21b: GM ^{222}Rn concentrations in zip codes of Delaware county based on the USEPA classification.

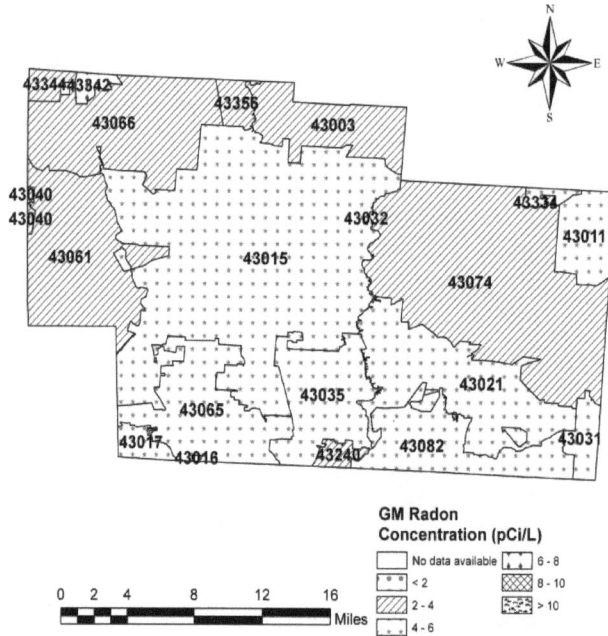

Figure 21c: GM ^{222}Rn concentrations in zip codes of Delaware county based on the 2 pCi/L breakdown classification.

Table 22: ^{222}Rn statistics for Erie county

Zip code	No.	Min.	Max.	AM	GM	SD	CV
43438	7	1.1	3.9	2.16	1.92	1.16	0.06
43464*	7	0.2	9	2.14	0.98	3.17	0.16
44089	270	0.1	89.6	7.21	4.03	10.7	0.55
44811*	262	0.1	159	15.21	6.34	24.33	1.25
44814	77	0.1	247.5	13.11	5.86	28.79	1.47
44816	6	0.5	11	3.87	2.71	3.64	0.19
44824	129	0.1	279.2	13.43	5.23	36.71	1.88
44826*	15	0.6	15.8	5.37	3.42	5.12	0.26
44839	464	0.1	70.1	6.4	3.72	7.78	0.4
44846	101	0.1	81.6	9.22	4.5	13.68	0.7
44847*	132	0.1	149.7	23.18	9.05	35.72	1.83
44857*	335	0.1	104.1	6.41	3.23	10.19	0.52
44870	674	0.1	134.9	4.89	2.89	7.31	0.37
44889*	44	0.1	24.8	4.74	2.88	4.77	0.24

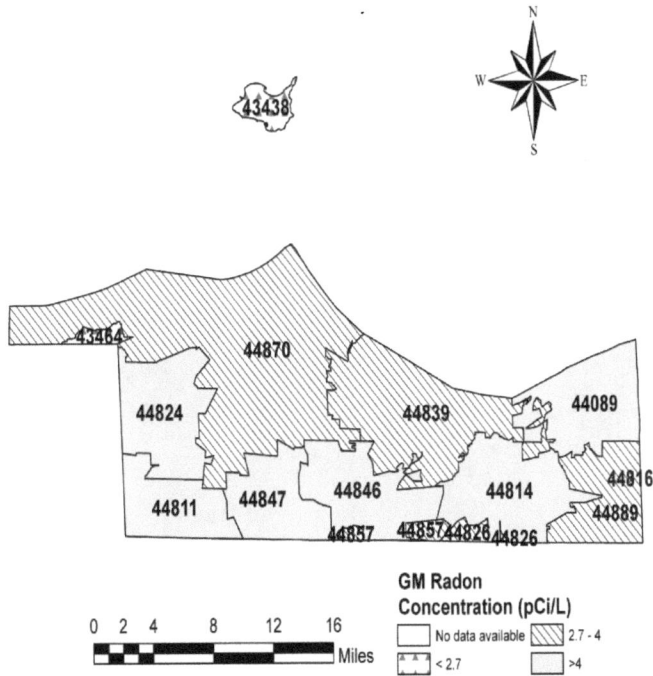

Figure 22a: GM ^{222}Rn concentrations in zip codes of Erie county based on the WHO-USEPA classification.

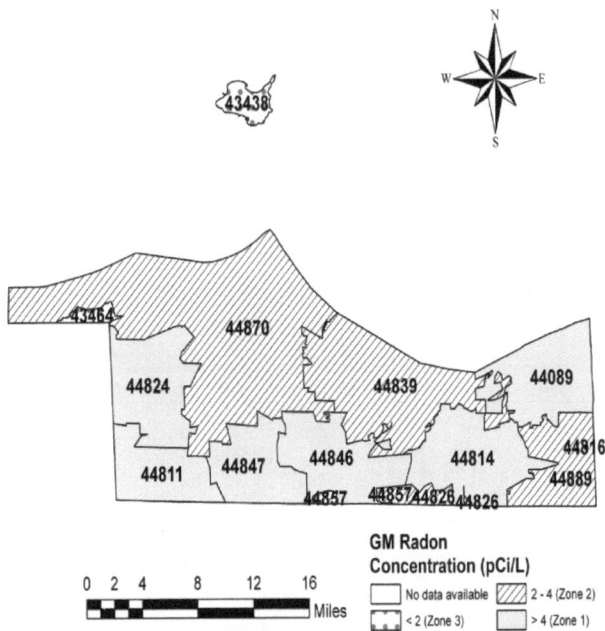

Figure 22b: GM ^{222}Rn concentrations in zip codes of Erie county based on the USEPA classification.

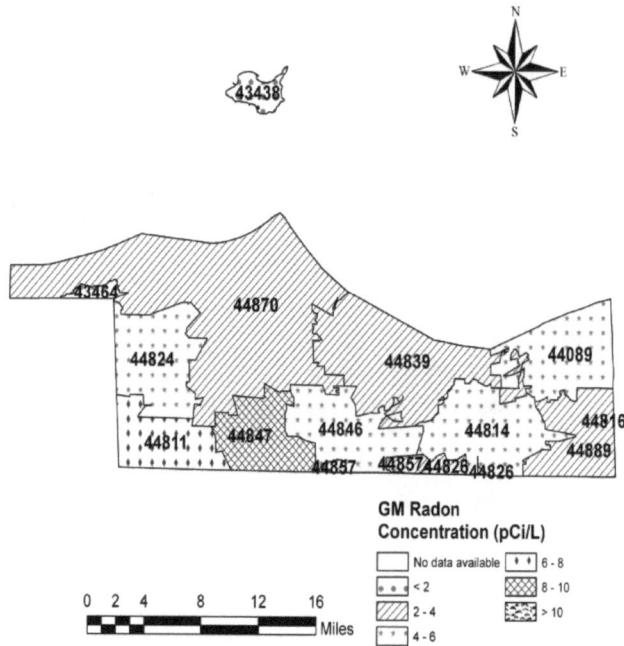

Figure 22c: GM ^{222}Rn concentrations in zip codes of Erie county based on the 2 pCi/L breakdown classification.

Table 23: ^{222}Rn statistics for Fairfield county

Zip code	No.	Min.	Max.	AM	GM	SD	CV
43046	53	0.9	29	5.24	3.6	5.27	0.27
43062*	910	0.1	470	9.39	5	21.67	1.11
43068*	951	0.1	180.5	8.6	5.03	12.12	0.62
43076*	168	0.2	274.5	13.41	4.4	36.3	1.86
43102	18	0.2	33.4	8.13	3.79	9.88	0.51
43105	261	0.1	44.5	4.63	1.8	7.08	0.36
43107	32	0.8	260	16.24	5.71	46.2	2.37
43110*	438	0.1	112	10.56	6.45	11.49	0.59
43112	130	0.1	238.5	12.25	6.33	22.96	1.18
43113*	342	0.2	86.6	11.96	7.89	12.53	0.64
43130	700	0.1	340.5	10.97	5.97	21.42	1.1
43136	9	3.8	23.8	14.68	13.15	6	0.31
43147	1316	0.1	110.1	9.34	5.41	9.88	0.51
43148	24	0.1	45.7	8.63	5.28	9.74	0.5
43150	6	0.3	5.9	2.45	1.49	2.39	0.12
43154	20	0.3	19.1	7.31	4.86	5.79	0.3
43155	21	0.8	40.2	9.25	4.28	12.24	0.63

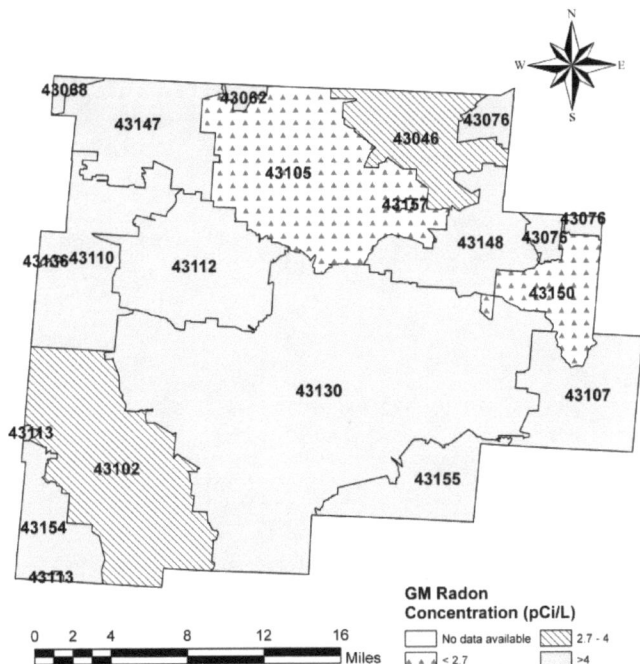

Figure 23a: GM ^{222}Rn concentrations in zip codes of Fairfield county based on the WHO-USEPA classification.

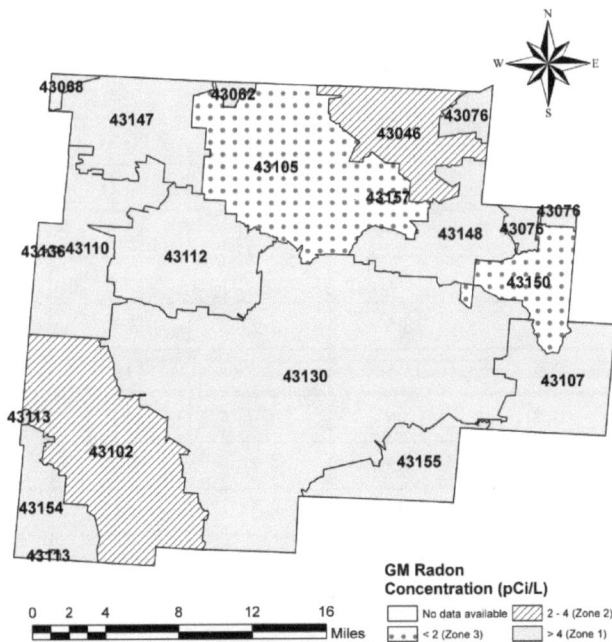

Figure 23b: GM ^{222}Rn concentrations in zip codes of Fairfield county based on the USEPA classification.

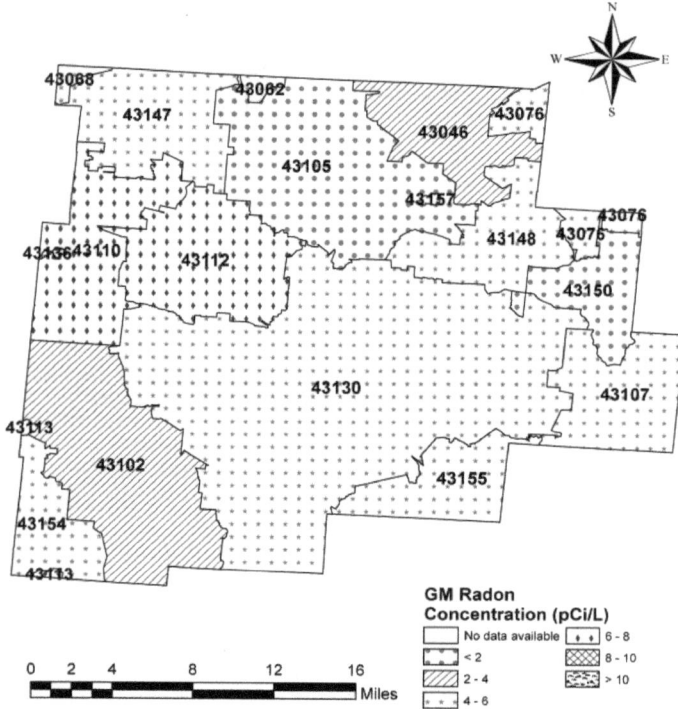

Figure 23c: GM ^{222}Rn concentrations in zip codes of Fairfield county based on the 2 pCi/L breakdown classification.

Table 24: ^{222}Rn statistics for Fayette county

Zip code	No.	Min.	Max.	AM	GM	SD	CV
43106	7	0.1	36.7	11.6	4.23	13.47	0.69
43128	17	0.5	300	21.38	3.5	71.87	3.68
43142	1	9.6	9.6	9.6	9.6	0	0
43143*	58	0.1	28.5	5.21	3.37	4.88	0.25
43145*	4	2.9	41.1	16.88	11.32	16.68	0.85
43153*	8	0.9	10.9	5.1	4.1	3.25	0.17
43160	115	0.2	38.3	4.71	2.89	5.09	0.26
45123*	20	0.2	43	8.27	4.41	10.5	0.54
45135*	10	0.5	15.4	3.54	1.94	4.58	0.23
45169*	23	0.4	12	4.24	3.11	3.46	0.18
45335*	195	0.1	39.8	4.88	2.68	6.69	0.34

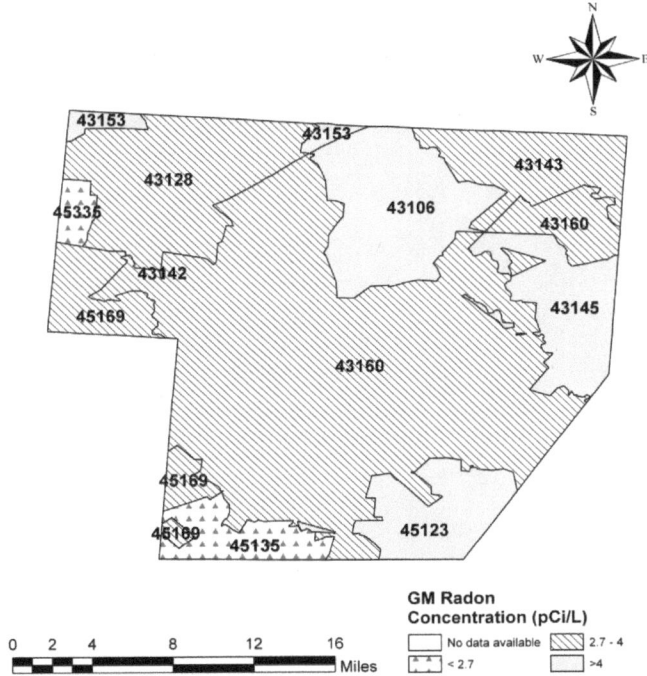

Figure 24a: GM ^{222}Rn concentrations in zip codes of Fayette county based on the WHO-USEPA classification.

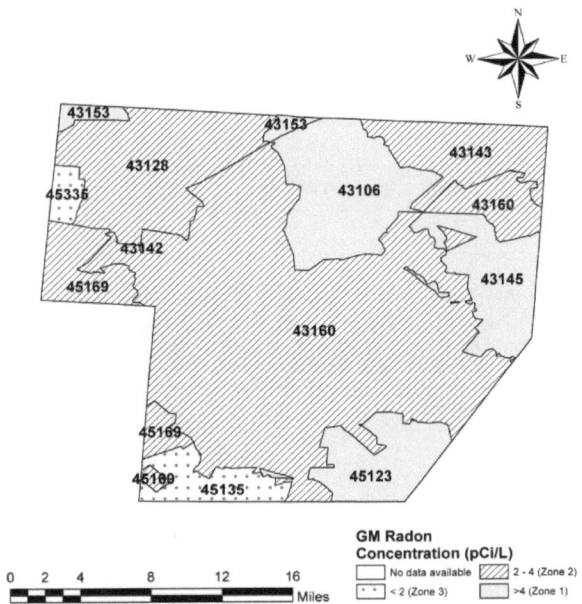

Figure 24b: GM ^{222}Rn concentrations in zip codes of Fayette county based on the USEPA classification.

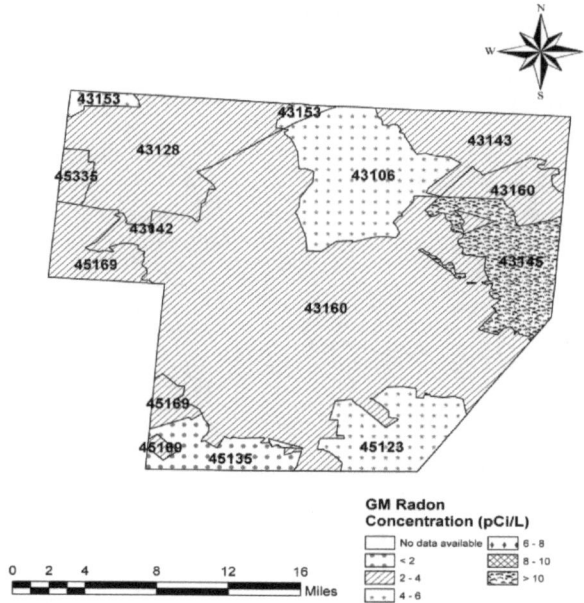

Figure 24c: GM ^{222}Rn concentrations in zip codes of Fayette county based on the 2 pCi/L breakdown classification.

Table 25: ^{222}Rn statistics for Franklin county

Zip code	No.	Min.	Max.	AM	GM	SD	CV	Zip code	No.	Min.	Max.	AM	GM	SD	CV
43002	11	2.7	40.1	12.21	7.82	12.41	0.64	43211	90	0.1	25.9	5.17	3.39	4.54	0.23
43004	469	0.1	126.5	11.5	6.1	16.93	0.87	43212	589	0.1	178.2	6.17	4.09	10.38	0.53
43016	1225	0.1	137.2	7.83	4.57	8.55	0.44	43213	591	0.1	79.6	9.77	5.2	11.63	0.6
43017	2855	0.1	193.7	10.78	5.93	13.25	0.68	43214	1307	0.1	939	10.85	6.36	28.21	1.44
43026	1847	0.1	151	9.33	5.69	9.99	0.51	43215	318	0.1	174.8	8.95	4.33	16.92	0.87
43054	1061	0.1	60.1	6.34	4	6.7	0.34	43217	16	0.9	16	4.66	3.46	3.93	0.2
43064*	329	0.5	70.4	8.28	5.12	10.09	0.52	43219	129	0.1	112.2	10.8	5.67	17.05	0.87
43065*	1938	0.1	735	7.62	4.65	18.64	0.95	43220	1513	0.1	68.1	9.4	6.53	8.07	0.41
43068	951	0.1	180.5	8.6	5.03	12.12	0.62	43221	1877	0.1	165.9	9.36	6.3	10.31	0.53
43081	2239	0.1	114.8	8.13	4.92	8.99	0.46	43222	17	0.4	16.4	6.49	4.54	4.97	0.25
43085	2323	0.1	369.9	12.36	7.48	15.83	0.81	43223	157	0.1	25.4	5.34	3.58	4.61	0.24
43109	4	3.4	20.5	9.33	7.48	7.61	0.39	43224	356	0.1	333.9	8.16	5.34	18.27	0.94
43110	438	0.1	112	10.56	6.45	11.49	0.59	43227	309	0.1	76.7	9.32	6.14	8.73	0.45
43119	393	0.1	63.9	8.33	5.44	7.39	0.38	43228	716	0.1	73.5	8.17	4.48	9.03	0.46

Zip code	No.	Min.	Max.	AM	GM	SD	CV	Zip code	No.	Min.	Max.	AM	GM	SD	CV
43123	920	0.1	91.1	8.02	5.57	7.61	0.39	43229	854	0.1	102.8	7.75	4.95	7.69	0.39
43125	253	0.1	94.8	13.06	8.2	13.75	0.7	43230	1508	0.1	417.7	11.09	5.9	17.28	0.88
43126	6	3.1	18.5	10.08	8.36	6.17	0.32	43231	278	0.1	52	7.92	4.74	8.78	0.45
43137	21	0.9	100.5	24.05	13.41	27.44	1.41	43232	579	0.1	142.8	16.55	8.08	17.83	0.91
43140*	397	0.1	82.1	7.87	4.88	9.39	0.48	43235	1657	0.1	720	10.69	5.97	21.06	1.08
43146*	143	0.5	43.1	9.05	6.17	8.39	0.43	43086**	9	1	5.7	2.82	2.39	1.61	0.08
43147*	1316	0.1	110.1	9.34	5.41	9.88	0.51	43216**	25	1.1	232	42.25	19.5	57.33	2.94
43201	357	0.1	73.6	8.37	4.67	9.78	0.5	43218**	4	3.8	10.1	6.93	6.5	2.7	0.14
43202	445	0.1	152.9	7.21	4.66	9.91	0.51	43226**	6	2.1	14.7	8.22	6.4	5.4	0.28
43203	48	0.1	47.4	5.91	1.45	9.84	0.5	43234**	4	2.4	18.1	9.28	7.4	6.51	0.33
43204	395	0.1	114.6	6.91	4.1	9.37	0.48	43236**	3	3.9	18.8	9.6	7.65	8.04	0.41
43205	59	0.4	50.4	9.5	5.41	11.53	0.59	43266**	99	0.1	35	3.79	1.23	8.56	0.44
43206	359	0.2	35.3	6.43	4.62	5.81	0.3	43270**	1	3.7	3.7	3.7	3.7	0	0
43207	255	0.2	125.3	11.24	6.62	15.06	0.77	43271**	1	5.2	5.2	5.2	5.2	0	0
43209	1302	0.1	65.1	7.07	4.66	6.92	0.35	43279**	1	10.4	10.4	10.4	10.4	0	0
43210	59	0.1	30.7	5.73	2.51	6.52	0.33								

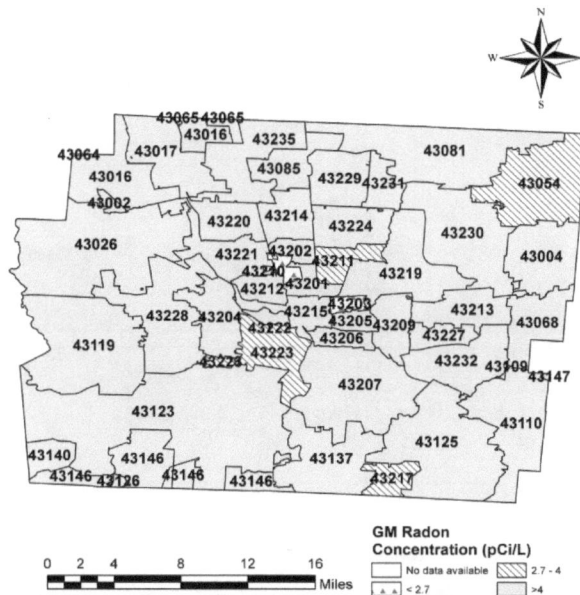

Figure 25a: GM ^{222}Rn concentrations in zip codes of Franklin county based on the WHO-USEPA classification.

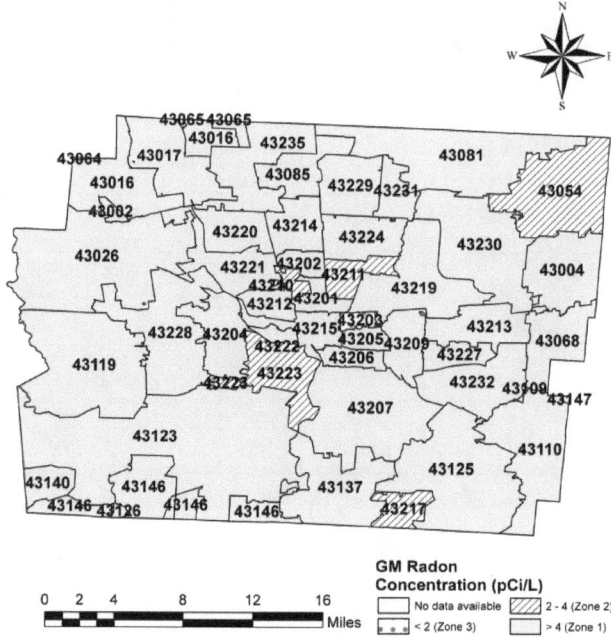

Figure 25b: GM ^{222}Rn concentrations in zip codes of Franklin county based on the USEPA classification.

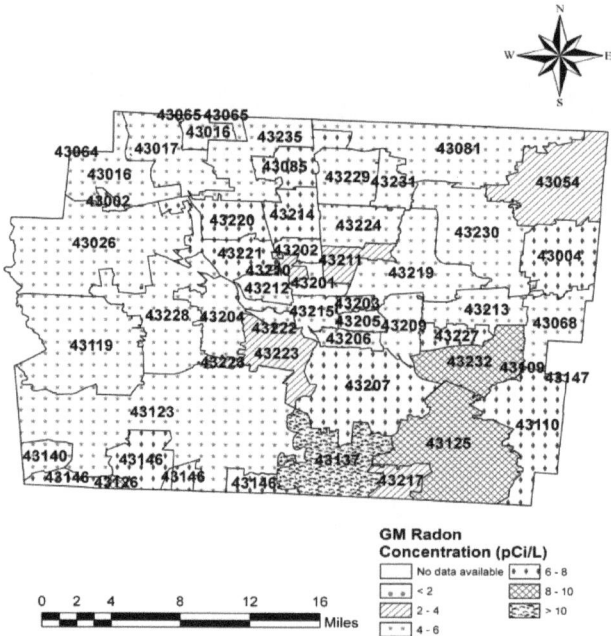

Figure 25c: GM ^{222}Rn concentrations in zip codes of Franklin county based on the 2 pCi/L breakdown classification.

Table 26: ^{222}Rn statistics for Fulton county

Zip code	No.	Min.	Max.	AM	GM	SD	CV
43502	91	0.2	19.4	5.06	3.4	4.41	0.23
43515	54	0.1	20.2	4.36	2.75	4.25	0.22
43521	13	0.6	12.9	4.51	3.4	3.32	0.17
43533	9	1.2	10	4.74	3.58	3.43	0.18
43540	9	0.1	5.2	2	1.23	1.64	0.08
43553	5	1.5	21.9	8	4.77	8.67	0.44
43558	85	0.1	20.4	2.75	1.61	3.26	0.17
43567	59	0.1	16.2	3.79	2.2	3.56	0.18
43570*	16	0.7	22.9	6.53	3.81	6.7	0.34

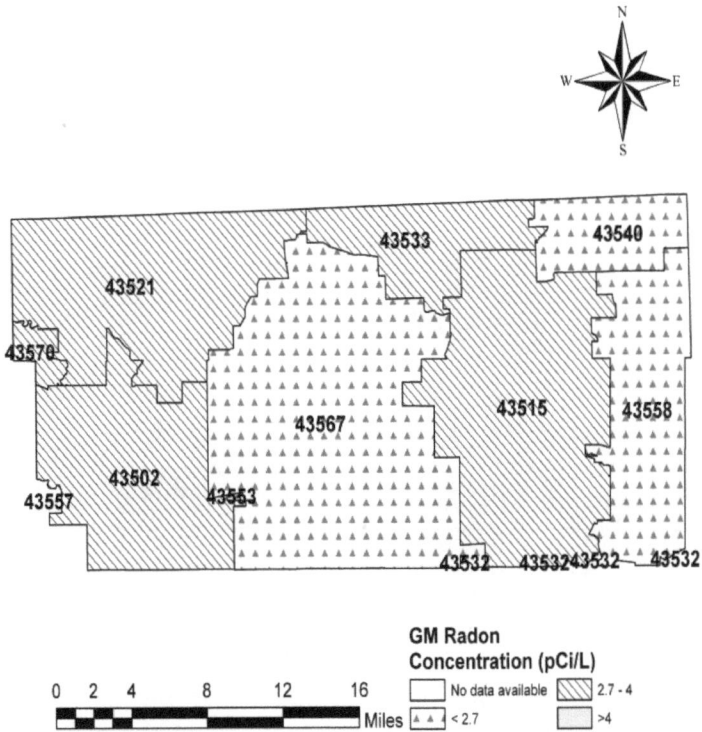

Figure 26a: GM ^{222}Rn concentrations in zip codes of Fulton county based on the WHO-USEPA classification.

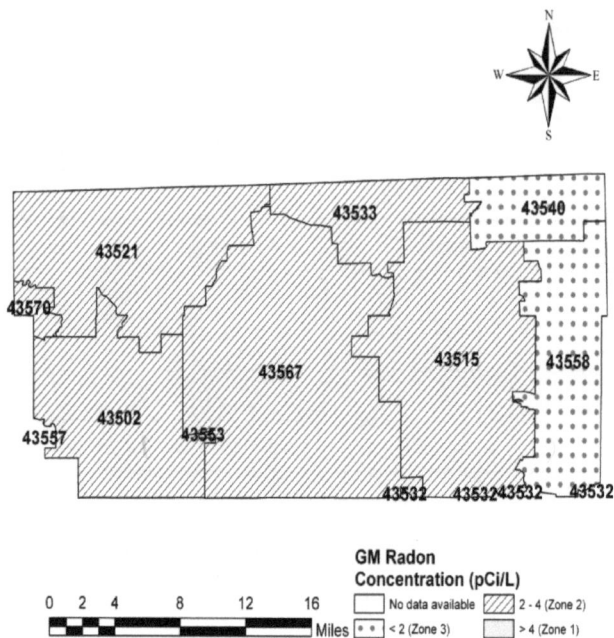

Figure 26b: GM ^{222}Rn concentrations in zip codes of Fulton county based on the USEPA classification.

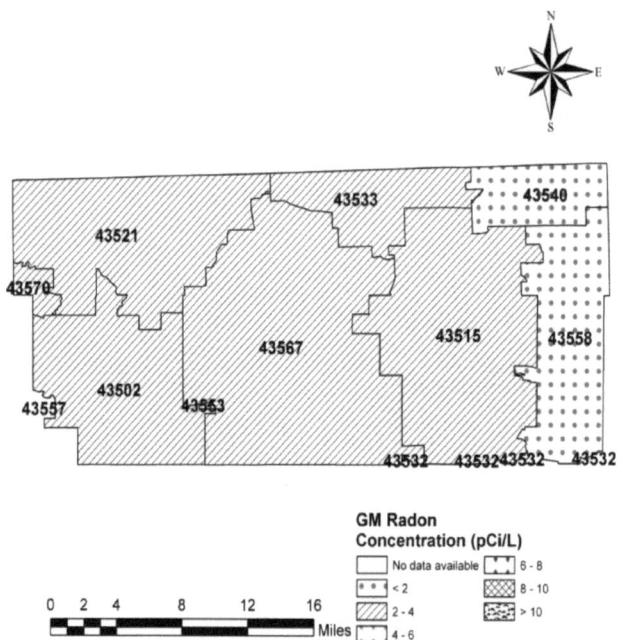

Figure 26c: GM ^{222}Rn concentrations in zip codes of Fulton county based on the 2 pCi/L breakdown classification.

Table 27: ^{222}Rn statistics for Gallia county

Zip code	No.	Min.	Max.	AM	GM	SD	CV
45614	21	0.3	10.6	2.17	1.52	2.37	0.12
45620	5	1.6	4.1	2.78	2.65	0.92	0.05
45623	5	0.6	4.9	1.8	1.32	1.78	0.09
45631	103	0.2	9.2	2.78	2.11	1.99	0.1
45656*	14	0.6	12.4	3.62	2.3	3.89	0.2
45658	6	0.3	7	2.52	1.56	2.5	0.13
45674	6	0.7	5.2	2.17	1.7	1.68	0.09
45678*	8	0.1	1	0.75	0.59	0.37	0.02
45685	2	1.7	1.7	1.7	1.7	0	0
45686	9	0.5	3.7	1.59	1.41	0.88	0.05

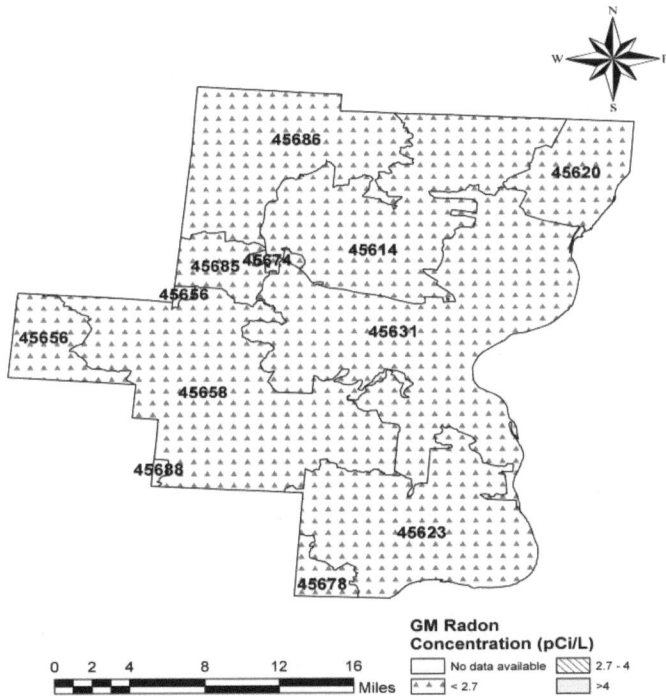

Figure 27a: GM ^{222}Rn concentrations in zip codes of Gallia county based on the WHO-USEPA classification.

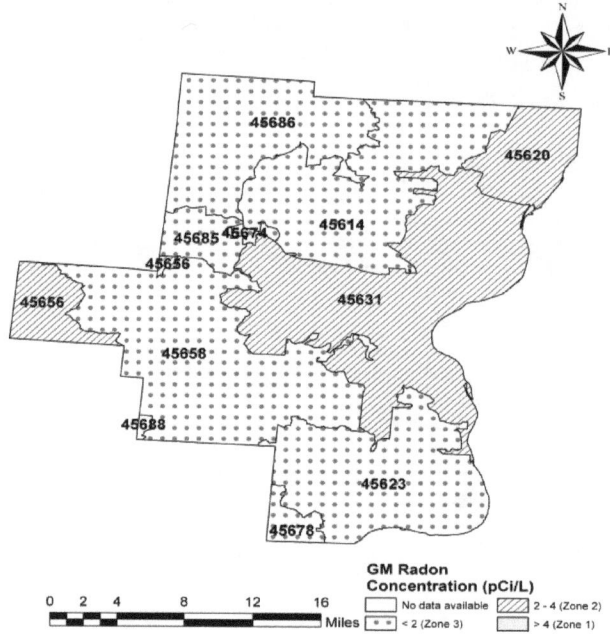

Figure 27b: GM ^{222}Rn concentrations in zip codes of Gallia county based on the USEPA classification.

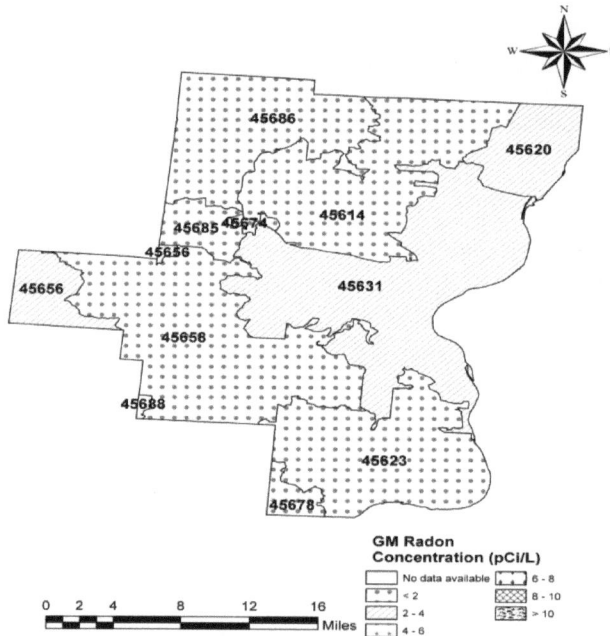

Figure 27c: GM ^{222}Rn concentrations in zip codes of Gallia county based on the 2 pCi/L breakdown classification.

Table 28: ^{222}Rn statistics for Geauga county

Zip code	No.	Min.	Max.	AM	GM	SD	CV
44021	82	0.1	13.3	2.71	1.94	2.3	0.12
44022*	698	0.1	733.8	3.66	1.95	27.79	1.42
44023	441	0.1	46.6	3.21	2.32	3.33	0.17
44024	319	0.1	19.6	2.83	2.07	2.51	0.13
44026	194	0.1	22.8	2.23	1.61	2.19	0.11
44046	18	0.8	260	18.49	4.27	60.33	3.09
44057*	152	0.1	37.8	3.97	1.86	6.77	0.35
44060*	899	0.1	49.9	3.78	2.06	5.86	0.3
44062	38	0.1	5.8	1.83	1.1	1.72	0.09
44064	12	0.4	12.5	3.98	2.69	3.77	0.19
44065	41	0.1	16.5	3.73	2.12	4.18	0.21
44072	123	0.1	20.2	2.39	1.54	2.7	0.14
44080	4	1.6	12.5	5.85	4.01	5.25	0.27
44086	25	0.1	27.5	3.72	2.07	5.59	0.29
44099*	16	0.3	5.3	1.31	0.89	1.35	0.07
44202*	465	0.1	22.4	2.43	1.65	2.35	0.12
44231*	63	0.4	14.1	3.46	2.64	2.65	0.14
44234*	36	0.5	29.8	3.48	2.2	5.37	0.27
44491*	14	0.1	5.1	1.59	1.16	1.27	0.07
44033**	2	0.8	2.7	1.75	1.47	1.34	0.07
44073**	1	1.2	1.2	1.2	1.2	0	0

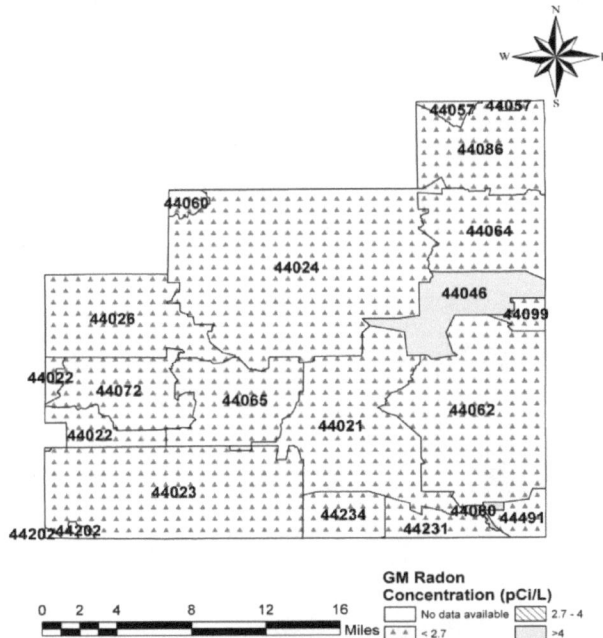

Figure 28a: GM ^{222}Rn concentrations in zip codes of Geauga county based on the WHO-USEPA classification.

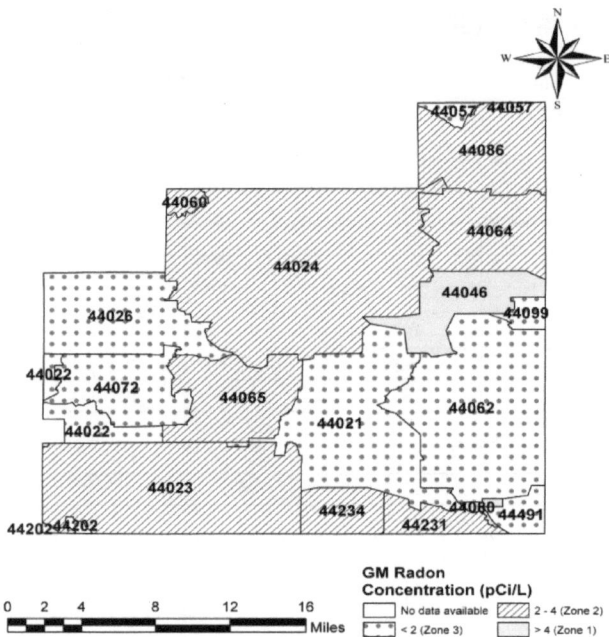

Figure 28b: GM ^{222}Rn concentrations in zip codes of Geauga county based on the USEPA classification.

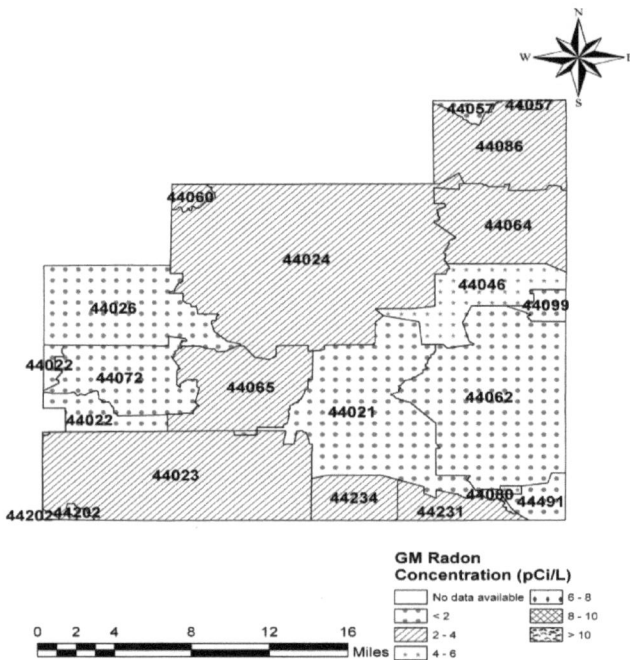

Figure 28c: GM ^{222}Rn concentrations in zip codes of Geauga county based on the 2 pCi/L breakdown classification.

Table 29: ^{222}Rn statistics for Greene county

Zip code	No.	Min.	Max.	AM	GM	SD	CV
43153*	8	0.9	10.9	5.1	4.1	3.25	0.17
45068*	335	0.1	44.9	5.63	3.58	6	0.31
45169*	23	0.4	12	4.24	3.11	3.46	0.18
45301	6	1.8	17.9	13.45	10.69	6.71	0.34
45305	841	0.1	79.5	6.79	4.44	6.93	0.35
45307	8	0.8	8.8	3.26	2.65	2.42	0.12
45314	176	0.1	97.7	5.4	2.71	10.73	0.55
45316	7	0.2	9.3	3.06	1.87	3.05	0.16
45324	1357	0.1	163	7.91	4.13	10.98	0.56
45335	195	0.1	39.8	4.88	2.68	6.69	0.34
45368*	85	0.2	35.8	5.33	3.18	5.63	0.29
45370	346	0.1	129	8.17	4.28	13.24	0.68
45384	35	0.1	26.9	4.7	2.42	6.88	0.35
45385	2910	0.1	142	8.21	4.78	9.9	0.51
45387	653	0.1	111.9	6.83	4.02	9.2	0.47
45424*	1187	0.1	83.2	5.16	2.89	6.83	0.35
45430	652	0.1	75.9	7.29	4.28	8.39	0.43
45431	975	0.1	117	5.65	3.5	6.94	0.36
45432	864	0.1	117	5.85	3.7	6.68	0.34
45433	31	0.1	268.8	44.31	3.62	83.01	4.25
45434	1100	0.1	820	7.47	4.23	25.62	1.31
45440*	1520	0.1	359	5.35	3.2	12.18	0.62
45458*	1266	0.1	35.5	3.82	2.27	3.94	0.2
45459*	2159	0.1	153	5.56	3.22	7.93	0.41
45435**	23	0.1	12.9	1.77	0.44	3.36	0.17

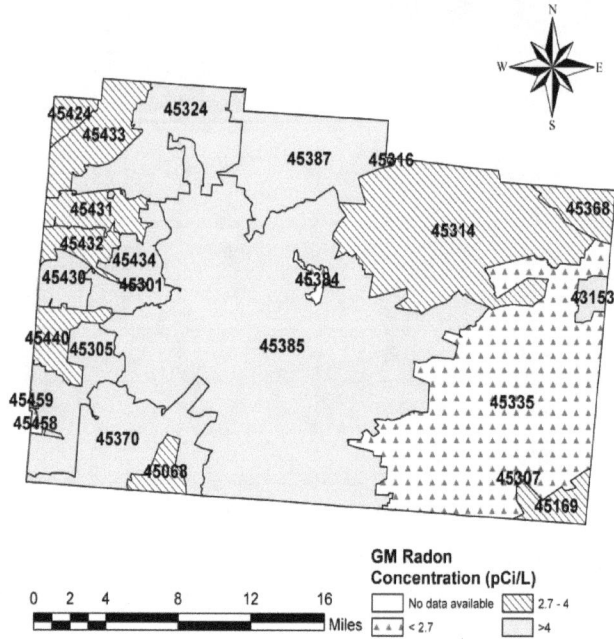

Figure 29a: GM ^{222}Rn concentrations in zip codes of Greene county based on the WHO-USEPA classification.

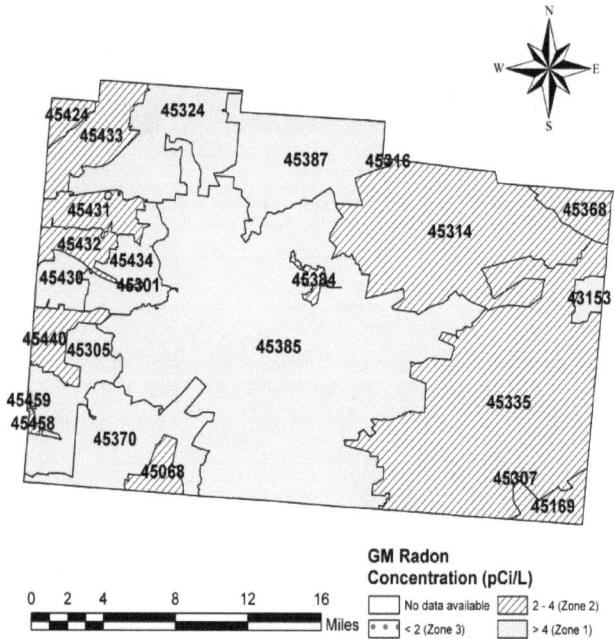

Figure 29b: GM ^{222}Rn concentrations in zip codes of Greene county based on the USEPA classification.

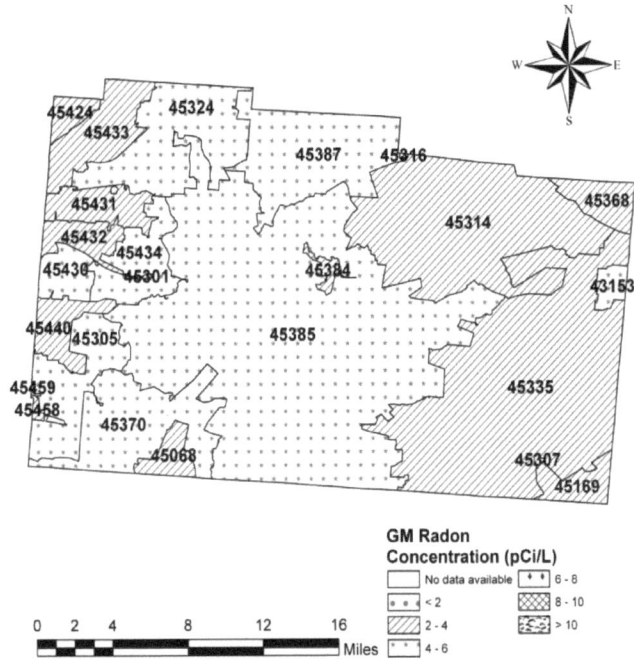

Figure 29c: GM ^{222}Rn concentrations in zip codes of Greene county based on the 2 pCi/L breakdown classification.

Table 30: ^{222}Rn statistics for Guernsey county

Zip code	No.	Min.	Max.	AM	GM	SD	CV
43723	24	0.1	28.1	5.68	2.73	7.63	0.39
43725	187	0.1	108.1	6.04	3.16	10.31	0.53
43732	13	0.1	8.6	2.93	1.66	3.04	0.16
43749	12	0.1	115.1	12.85	2.11	32.85	1.68
43750	2	3.8	9.1	6.45	5.88	3.75	0.19
43755	6	0.1	5.7	2.32	1.17	2.19	0.11
43762*	112	0.1	129	6.7	3.21	13.62	0.7
43768	4	1.9	2.9	2.2	2.17	0.48	0.02
43772	13	0.1	8.4	2.91	1.37	3.02	0.15
43773	5	0.6	9.2	3.28	2.21	3.42	0.18
43778	3	1.7	5.9	3.7	3.27	2.11	0.11
43780	9	0.5	4.6	1.58	1.04	1.73	0.09
43832*	26	0.2	20.4	6.77	3.22	7.41	0.38
43837*	16	0.8	40.3	9.74	5.88	10.43	0.53
43973*	2	0.6	1.4	1	0.92	0.57	0.03
43983*	6	0.6	10.4	4.22	2.94	3.66	0.19
44699*	7	0.7	28.7	10.8	5.1	12.18	0.62

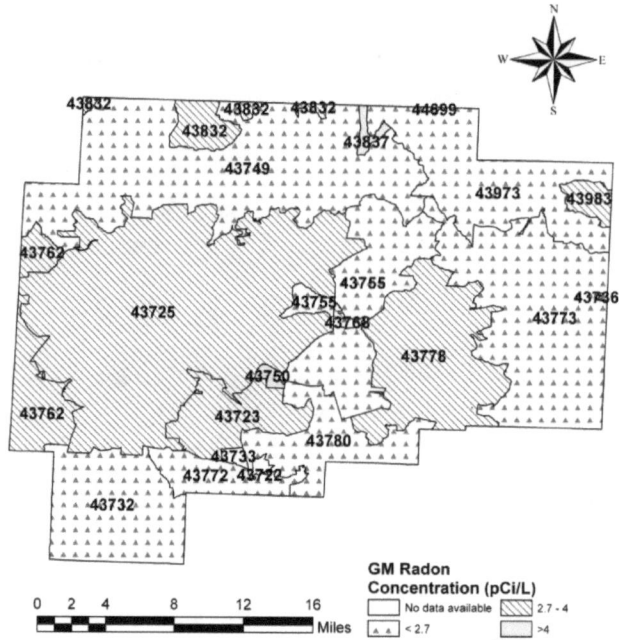

Figure 30a: GM ^{222}Rn concentrations in zip codes of Guernsey county based on the WHO-USEPA classification.

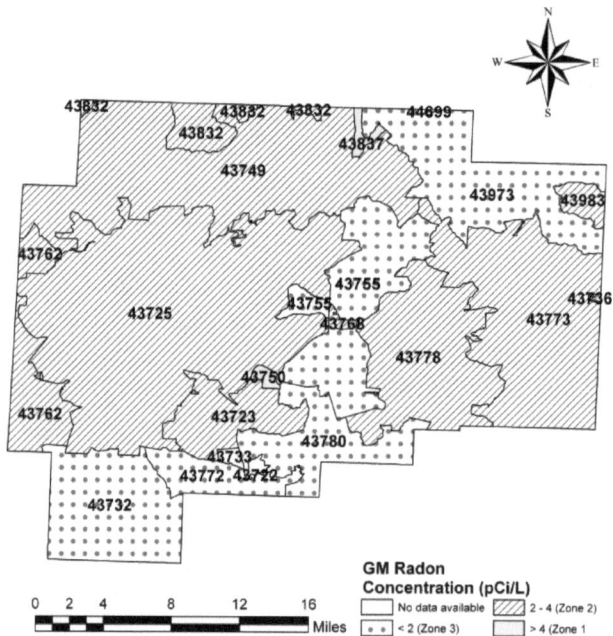

Figure 30b: GM ^{222}Rn concentrations in zip codes of Guernsey county based on the USEPA classification.

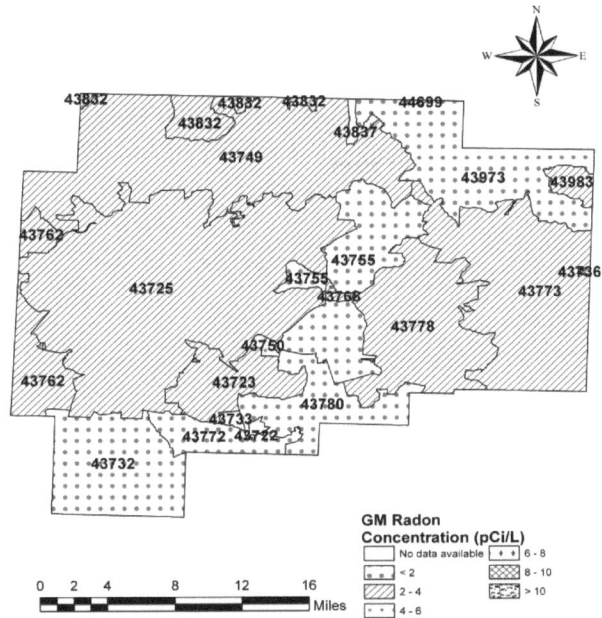

Figure 30c: GM ^{222}Rn concentrations in zip codes of Guernsey county based on the 2 pCi/L breakdown classification.

Table 31: ^{222}Rn statistics for Hamilton county

Zip code	No.	Min.	Max.	AM	GM	SD	CV	Zip code	No.	Min.	Max.	AM	GM	SD	CV
45001	5	1.6	3.6	2.74	2.63	0.84	0.04	45229	64	0.3	19.2	2.43	1.71	2.7	0.14
45002	125	0.1	28	3.78	2.66	3.56	0.18	45230	686	0.1	27	3.11	2.25	2.78	0.14
45013*	562	0.1	470	49.54	5.07	114.91	5.88	45231	607	0.1	40.1	2.99	1.94	3.3	0.17
45030	261	0.1	59	7.21	4.35	7.4	0.38	45232	35	0.3	23.4	3.86	2.23	5.25	0.27
45033	4	1	6	3.38	2.78	2.14	0.11	45233	202	0.2	26.7	3.66	2.63	3.34	0.17
45041	3	1.2	3.2	2.1	1.94	1.01	0.05	45236	653	0.1	87	2.76	1.77	4.32	0.22
45051	1	7.4	7.4	7.4	7.4	0	0	45237	151	0.1	19.4	3	2.09	2.66	0.14
45052	36	0.4	21.8	3.87	2.74	3.84	0.2	45238	374	0.1	42.4	2.68	1.6	3.99	0.2
45111	7	2.6	15.6	9.06	7.79	4.55	0.23	45239	333	0.1	23.5	2.67	1.6	3.16	0.16
45140*	1852	0.1	470.8	4.95	2.99	12.09	0.62	45240	285	0.1	18.6	2.57	1.68	2.69	0.14
45147*	4	2.3	4.9	3.28	3.11	1.24	0.06	45241	1001	0.1	42.3	4.07	2.48	4.41	0.23
45150*	644	0.1	83.5	5.14	3.3	6.53	0.33	45242	1039	0.1	33.5	3.91	2.59	4.1	0.21
45174	212	0.1	74.6	6.27	3.98	7.14	0.37	45243	820	0.1	26.5	2.99	2.21	2.67	0.14
45202	105	0.1	27	2.29	1.23	3.79	0.19	45244*	860	0.1	40.5	4.16	2.84	4.19	0.21
45203	24	0.1	5.3	1.6	1.22	1.3	0.07	45246	233	0.1	23.5	3.89	2.5	3.85	0.2
45204	15	0.1	8.6	2.27	1.41	2.37	0.12	45247	371	0.1	15.9	3.08	2.12	2.67	0.14
45205	76	0.1	14.1	2.24	1.45	2.25	0.12	45248	263	0.1	316.65	4.28	2	19.69	1.01

Zip code	No.	Min.	Max.	AM	GM	SD	CV	Zip code	No.	Min.	Max.	AM	GM	SD	CV
45206	107	0.1	15.2	2.57	1.69	2.58	0.13	45249	621	0.1	34.3	4.22	2.75	4.22	0.22
45207	152	0.1	4.9	0.99	0.71	0.97	0.05	45251	188	0.1	45.5	3.21	2.12	4.35	0.22
45208	868	0.1	35.4	2.79	2.06	2.64	0.14	45252	73	0.1	22.5	3.76	2.71	3.38	0.17
45209	264	0.1	19.5	2.52	1.75	2.53	0.13	45255	557	0.1	42.4	4.19	2.72	4.05	0.21
45211	315	0.1	25.2	2.69	1.69	3.17	0.16	45201**	10	0.1	6.3	2.25	1.49	1.79	0.09
45212	196	0.1	26.3	2.22	1.46	2.63	0.13	45221**	7	0.2	13.6	2.9	1.09	4.8	0.25
45213	203	0.1	49.5	3.35	2.29	4.4	0.23	45234**	6	0.5	26.4	6.18	2.83	9.95	0.51
45214	27	0.1	6.5	2.7	1.88	1.95	0.1	45235**	3	0.1	5.7	2.97	1.21	2.8	0.14
45215	634	0.1	20.6	3	1.96	2.87	0.15	45250**	1	3.3	3.3	3.3	3.3	0	0
45216	47	0.1	16.8	3.89	2.46	3.79	0.19	45253**	1	4	4	4	4	0	0
45217	55	0.5	30.5	2.99	2.02	4.25	0.22	45254**	2	2.3	4.3	3.3	3.14	1.41	0.07
45218	56	0.1	17.5	2.92	1.74	3.46	0.18	45258**	1	1.3	1.3	1.3	1.3	0	0
45219	71	0.1	18.7	2.23	1.15	2.9	0.15	45262**	1	3	3	3	3	0	0
45220	167	0.1	58.5	3.24	1.85	6.57	0.34	45263**	1	20.5	20.5	20.5	20.5	0	0
45223	99	0.1	17.5	3.19	2.11	2.93	0.15	45264**	1	2.8	2.8	2.8	2.8	0	0
45224	233	0.1	32.4	3.42	2.04	4.06	0.21	45271**	2	0.5	1.7	1.1	0.92	0.85	0.04
45225	12	0.3	5.9	1.83	1.13	1.96	0.1	45274**	2	1.5	3.1	2.3	2.16	1.13	0.06
45226	185	0.2	17.8	2.03	1.46	2.15	0.11	45280**	2	4.3	5.8	5.05	4.99	1.06	0.05
45227	348	0.1	64.8	4.29	2.69	5.67	0.29								

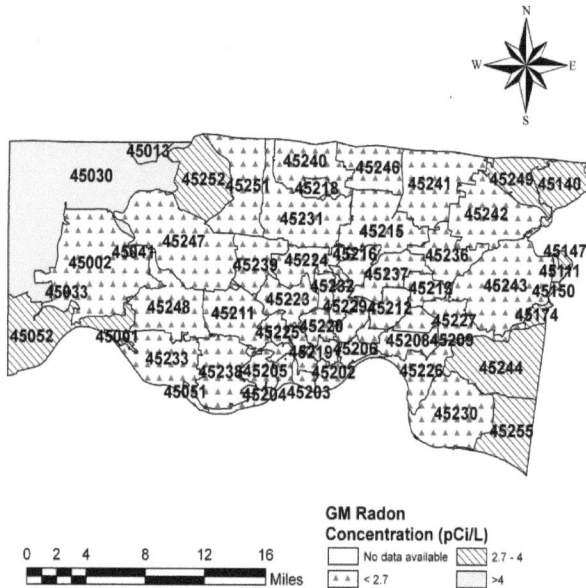

Figure 31a: GM ^{222}Rn concentrations in zip codes of Hamilton county based on the WHO-USEPA classification.

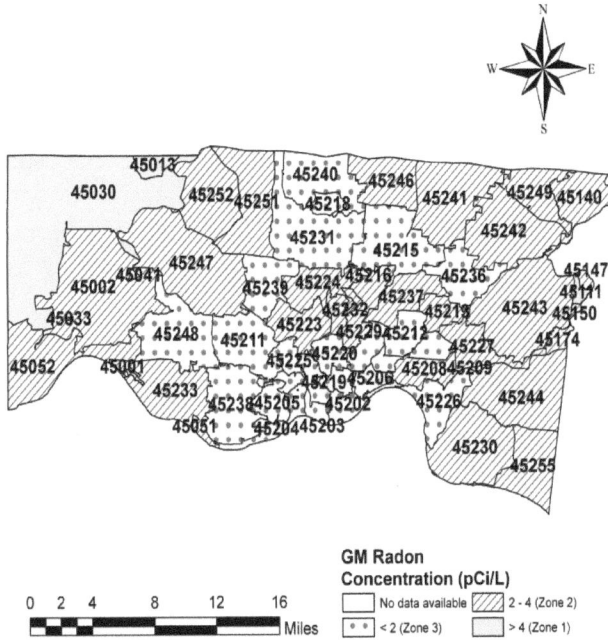

Figure 31b: GM ^{222}Rn concentrations in zip codes of Hamilton county based on the USEPA classification.

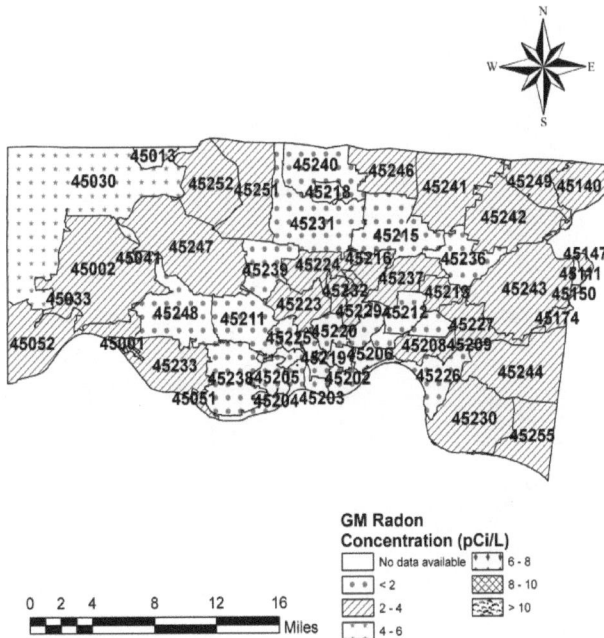

Figure 31c: GM ^{222}Rn concentrations in zip codes of Hamilton county based on the 2 pCi/L breakdown classification.

Table 32: ^{222}Rn statistics for Hancock county

Zip code	No.	Min.	Max.	AM	GM	SD	CV
43316*	63	0.6	42.1	4.7	3.02	6.45	0.33
43516*	13	0.7	6.1	2.35	2.04	1.39	0.07
44802*	19	0.1	120	11.21	4.05	26.65	1.36
44804	15	0.8	19.9	8.59	6.18	6.45	0.33
44817*	30	0.1	18.4	4.31	2.59	3.91	0.2
44830*	272	0.1	86.2	6.32	3.21	8.51	0.44
45810*	32	0.1	57.1	6.92	3.79	10.47	0.54
45814	18	0.5	10.5	3.24	2.07	3.18	0.16
45817*	38	0.5	62.6	7.51	4.02	12.79	0.65
45840	1041	0.1	73.9	5.12	3.06	5.93	0.3
45841	8	1.1	14.4	5.98	4.56	4.2	0.21
45843*	23	0.1	30.2	6.92	4.35	7.51	0.38
45856*	50	0.5	52.5	6.57	4.03	8.93	0.46
45858	40	0.1	62.6	6.09	2.97	10.32	0.53
45867	7	0.2	80.2	14.73	3.12	29.17	1.49
45868	11	0.8	21	6.83	3.97	7.37	0.38
45872*	28	0.1	14.8	3.42	2.11	3.65	0.19
45877*	20	0.2	28.1	9.7	6.49	7.83	0.4
45881	7	1.8	130	24.51	9.12	46.63	2.39
45889	25	0.1	33.8	7.34	3.72	7.46	0.38
45890	7	2.3	34.7	11.3	8.09	10.96	0.56
45839**	11	0.1	10.6	2.35	1.28	2.96	0.15

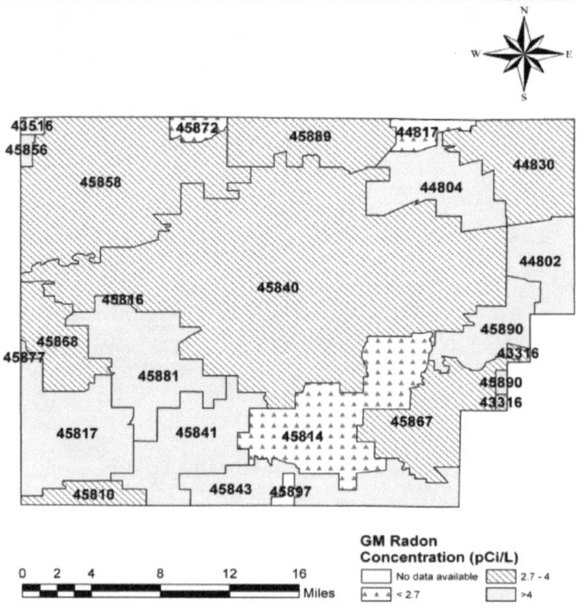

Figure 32a: GM ^{222}Rn concentrations in zip codes of Hancock county based on the WHO-USEPA classification.

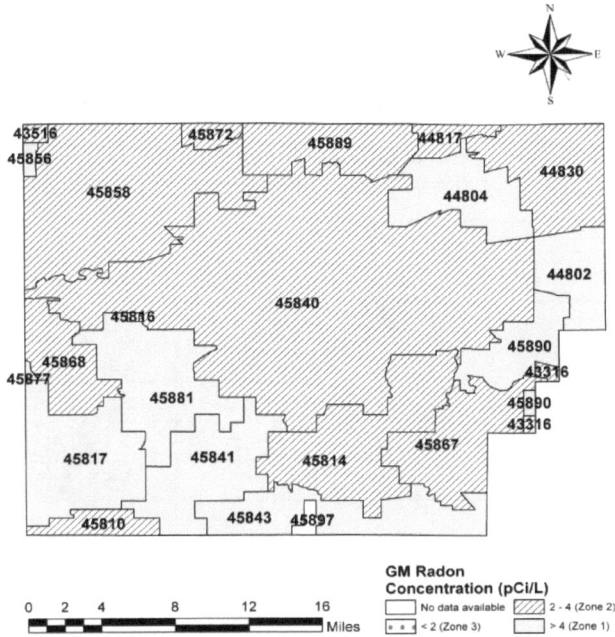

Figure 32b: GM ^{222}Rn concentrations in zip codes of Hancock county based on the USEPA classification.

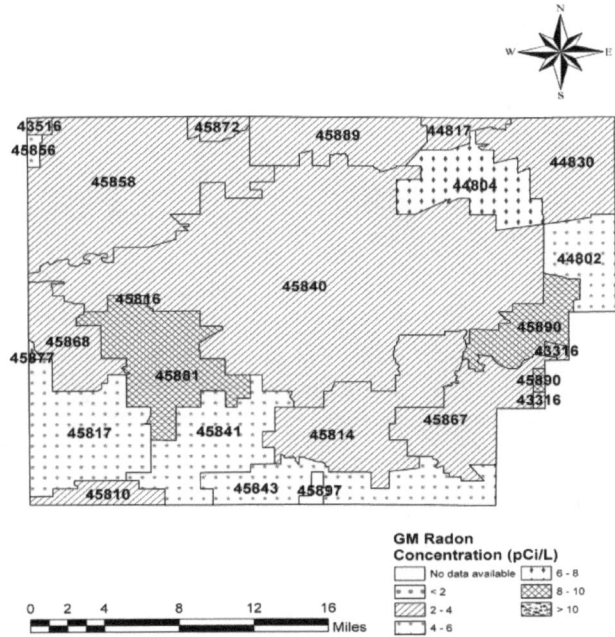

Figure 32c: GM ^{222}Rn concentrations in zip codes of Hancock county based on the 2 pCi/L breakdown classification.

Table 33: ^{222}Rn statistics for Hardin county

Zip code	No.	Min.	Max.	AM	GM	SD	CV
43310*	32	0.5	29.8	7.51	5.19	6.84	0.35
43326	101	0.1	16.1	3.71	2.4	3.51	0.18
43331*	19	0.5	8.7	2.06	1.47	2.08	0.11
43332*	18	2	30.3	12.28	8.78	8.91	0.46
43340	9	1.1	13.8	3.52	2.41	4.02	0.21
43345	6	1.2	6.4	3.08	2.37	2.39	0.12
43347*	7	0.7	4.9	2.6	2.07	1.71	0.09
45810	32	0.1	57.1	6.92	3.79	10.47	0.54
45812	10	1.5	8.3	3.5	3.14	1.92	0.1
45836	4	0.5	6.4	2.73	1.54	2.83	0.14
45843	23	0.1	30.2	6.92	4.35	7.51	0.38
45850*	10	1.2	24.3	10.73	6.74	8.14	0.42
45896*	3	0.1	9.1	4.03	1.38	4.61	0.24

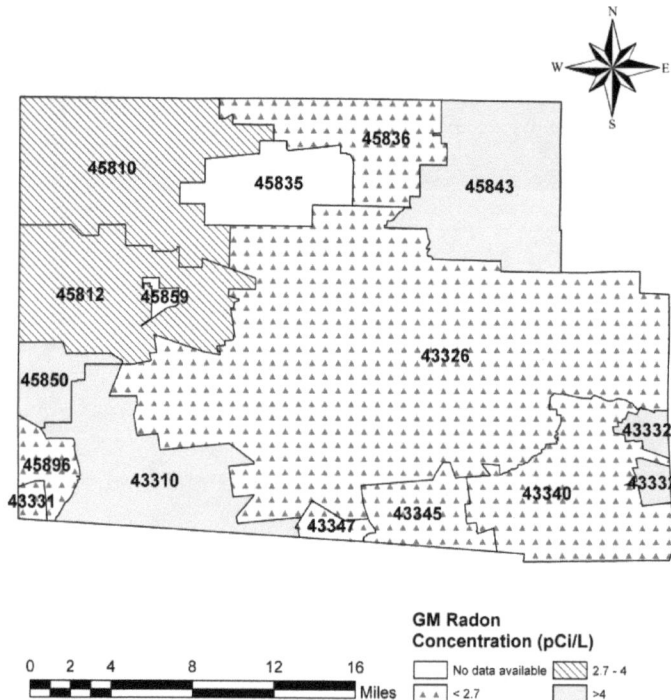

Figure 33a: GM ^{222}Rn concentrations in zip codes of Hardin county based on the WHO-USEPA classification.

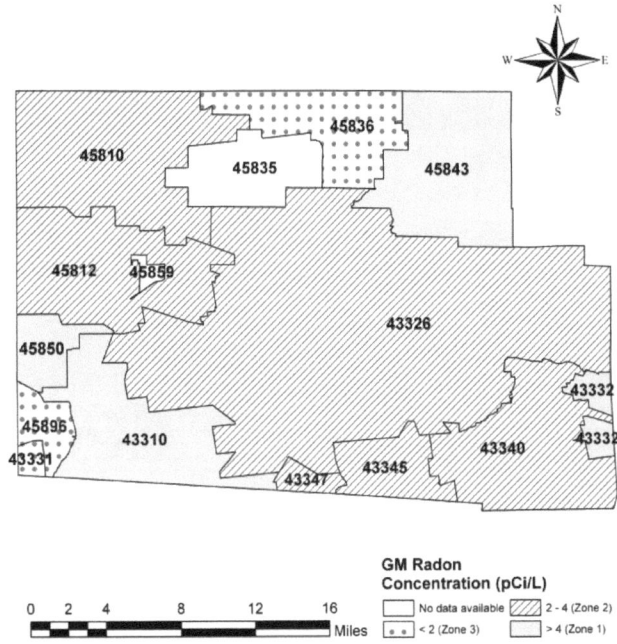

Figure 33b: GM ^{222}Rn concentrations in zip codes of Hardin county based on the USEPA classification.

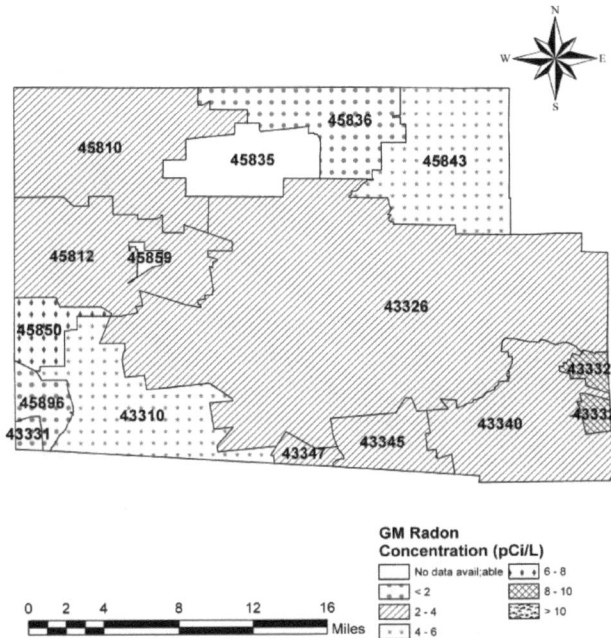

Figure 33c: GM ^{222}Rn concentrations in zip codes of Hardin county based on the 2 pCi/L breakdown classification.

Table 34: ^{222}Rn statistics for Harrison county

Zip code	No.	Min.	Max.	AM	GM	SD	CV
43901[*]	5	1.1	8.9	4.74	3.46	3.6	0.18
43907	43	1.2	42.6	9.52	5.37	11.78	0.6
43910[*]	23	0.6	44.7	9.69	3.73	14.14	0.72
43950[*]	338	0.1	119.4	7.15	3.26	16.31	0.84
43973	2	0.6	1.4	1	0.92	0.57	0.03
43974	3	0.8	2.1	1.47	1.36	0.65	0.03
43976	10	1.1	20.7	4.67	3.02	5.94	0.3
43977[*]	15	0.9	26.9	5.99	3.35	7.28	0.37
43983[*]	6	0.6	10.4	4.22	2.94	3.66	0.19
43986	20	0.7	23.9	7.56	3.75	8.25	0.42
43988	28	0.8	124.5	43.55	21.87	40.86	2.09
44621[*]	25	0.1	6.7	2.72	2.1	1.57	0.08
44683[*]	38	0.2	46.6	8.02	3.45	11.44	0.59
44695	18	0.1	33	9.53	5.77	8.54	0.44
44699	7	0.7	28.7	10.8	5.1	12.18	0.62

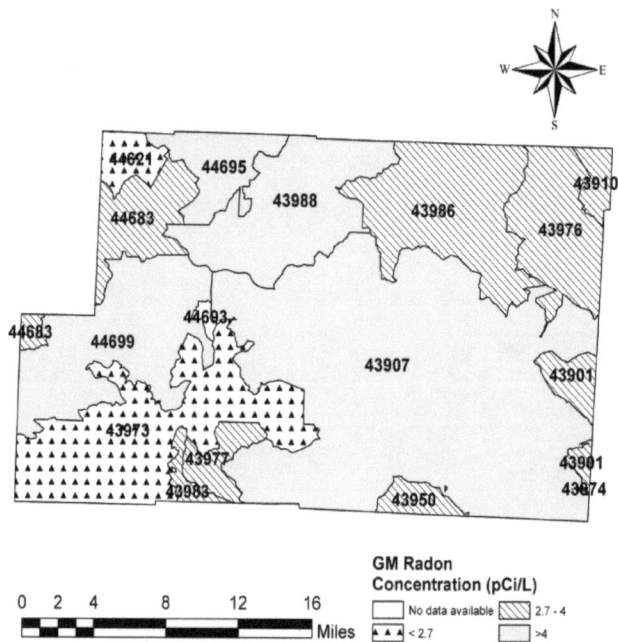

Figure 34a: GM ^{222}Rn concentrations in zip codes of Harrison county based on the WHO-USEPA classification.

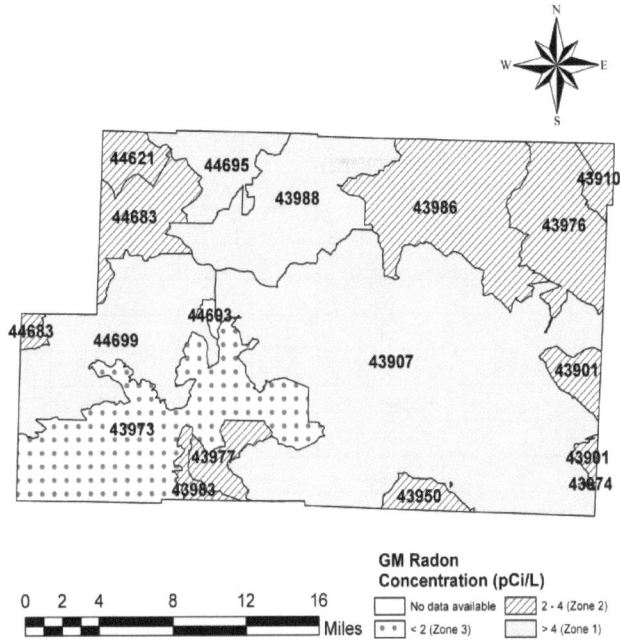

Figure 34b: GM ^{222}Rn concentrations in zip codes of Harrison county based on the USEPA classification.

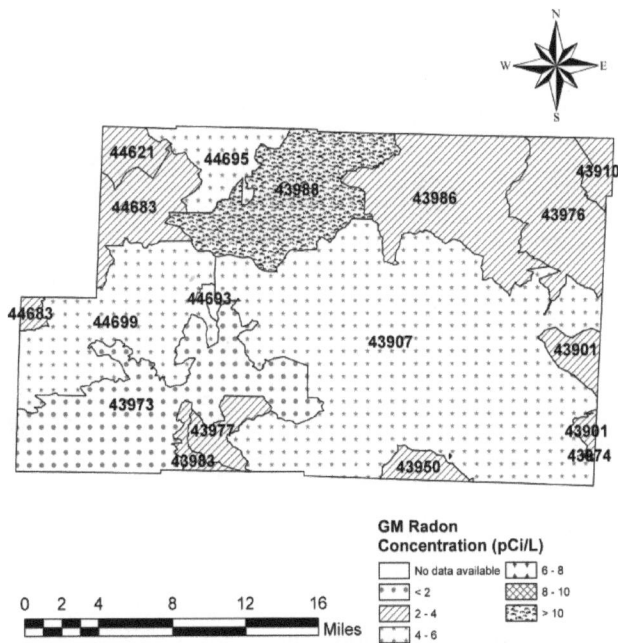

Figure 34c: GM ^{222}Rn concentrations in zip codes of Harrison county based on the 2 pCi/L breakdown classification.

Table 35: ^{222}Rn statistics for Henry county

Zip code	No.	Min.	Max.	AM	GM	SD	CV
43502*	91	0.2	19.4	5.06	3.4	4.41	0.23
43511*	15	0.1	24.1	5.45	2.12	7.84	0.4
43512*	211	0.1	21.2	3.26	2.12	3.24	0.17
43516	13	0.7	6.1	2.35	2.04	1.39	0.07
43522*	60	0.1	27.5	4.06	2.18	5.33	0.27
43523	1	2	2	2	2	0	0
43524	2	0.1	2.1	1.1	0.46	1.41	0.07
43527	12	0.6	29.4	6.14	3.11	8.26	0.42
43532	28	0.1	10.8	2.24	0.91	2.79	0.14
43534	8	1.9	13.5	5.75	4.78	3.87	0.2
43535	3	3.6	5.8	5	4.89	1.22	0.06
43545	79	0.1	19.5	3.81	2.58	3.24	0.17
43548	10	1.5	22.9	8.33	6.26	6.81	0.35
43557*	14	0.4	18.7	5.4	2.92	5.64	0.29
43567*	59	0.1	16.2	3.79	2.2	3.56	0.18
45856*	50	0.5	52.5	6.57	4.03	8.93	0.46
43550**	3	5.6	12.9	10.47	9.77	4.21	0.22

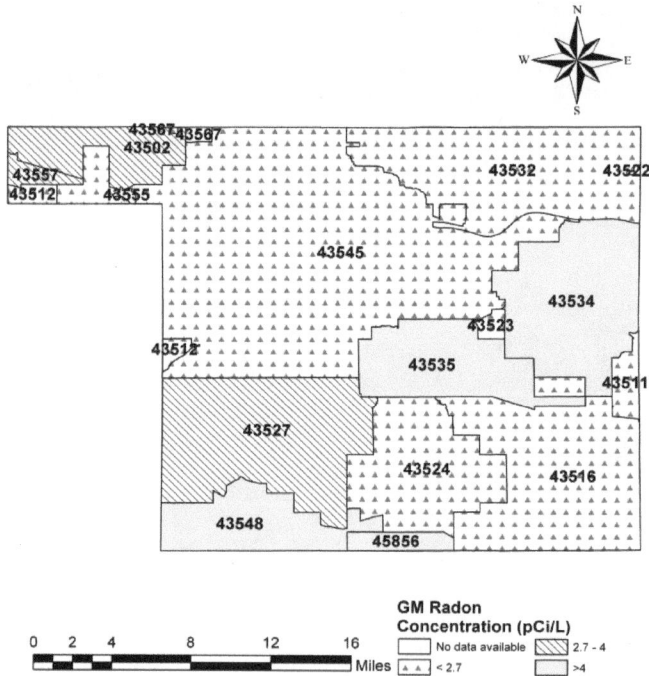

Figure 35a: GM ^{222}Rn concentrations in zip codes of Henry county based on the WHO-USEPA classification.

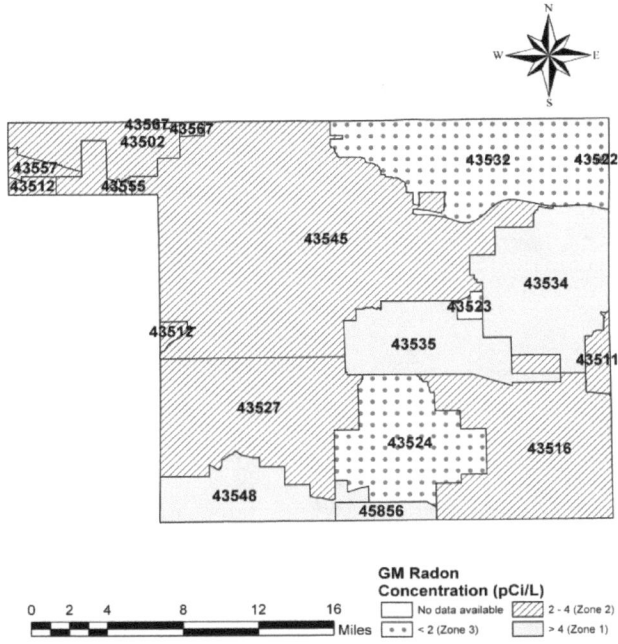

Figure 35b: GM ^{222}Rn concentrations in zip codes of Henry county based on the USEPA classification.

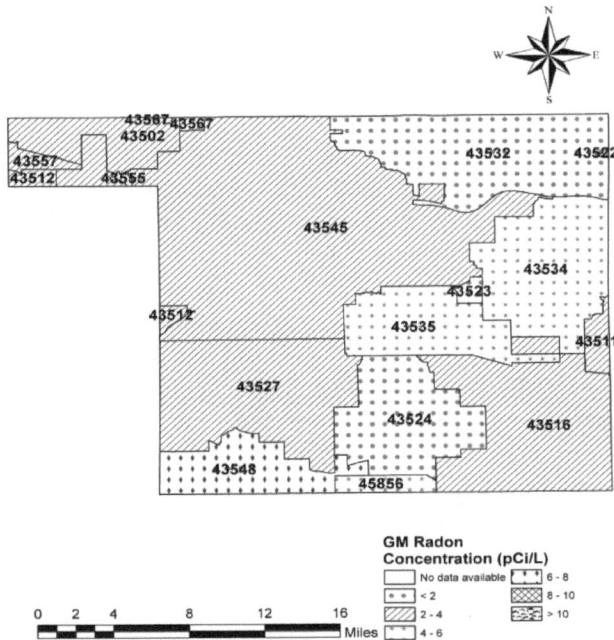

Figure 35c: GM ^{222}Rn concentrations in zip codes of Henry county based on the 2 pCi/L breakdown classification.

Table 36: ^{222}Rn statistics for Highland county

Zip code	No.	Min.	Max.	AM	GM	SD	CV
45118*	15	0.1	8	2.93	1.86	2.39	0.12
45123	20	0.2	43	8.27	4.41	10.5	0.54
45132	5	0.9	9.2	3.48	2.38	3.4	0.17
45133	126	0.1	31.1	4.14	2.49	4.51	0.23
45135	10	0.5	15.4	3.54	1.94	4.58	0.23
45142	25	0.5	9.4	4.15	3.05	2.78	0.14
45154*	19	0.5	18.3	3.42	1.88	4.35	0.22
45155	2	5.5	6.5	6	5.98	0.71	0.04
45159*	37	0.1	30.9	6.73	2.94	7.69	0.39
45171*	27	0.4	9.8	3.01	2.05	2.64	0.14
45172	1	0.1	0.1	0.1	0.1	0	0
45612*	48	0.2	28.4	5.95	2.28	7.52	0.39
45660*	65	0.1	42.5	5.09	1.71	8.41	0.43
45679*	16	0.7	6.9	3.61	3.14	1.81	0.09
45697*	22	0.1	15.6	4.63	2.41	4.79	0.25

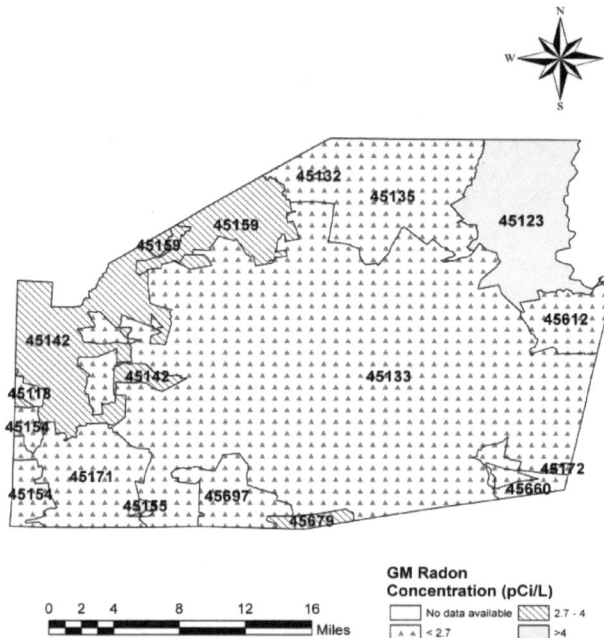

Figure 36a: GM ^{222}Rn concentrations in zip codes of Highland county based on the WHO-USEPA classification.

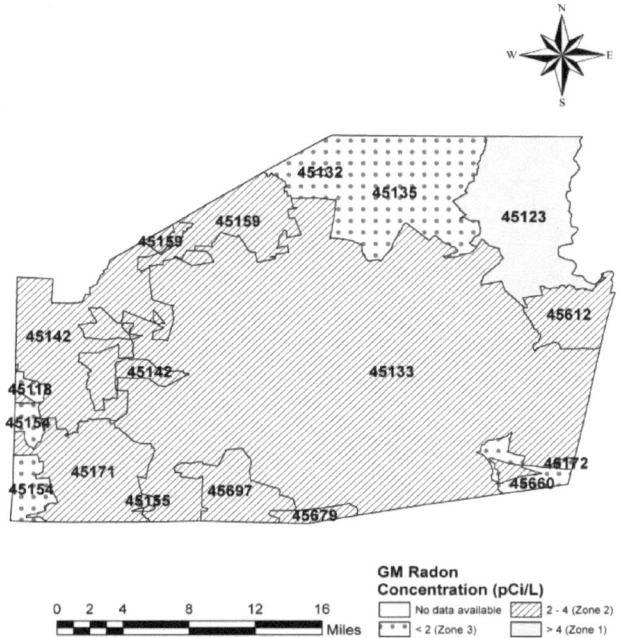

Figure 36b: GM ^{222}Rn concentrations in zip codes of Highland county based on the USEPA classification.

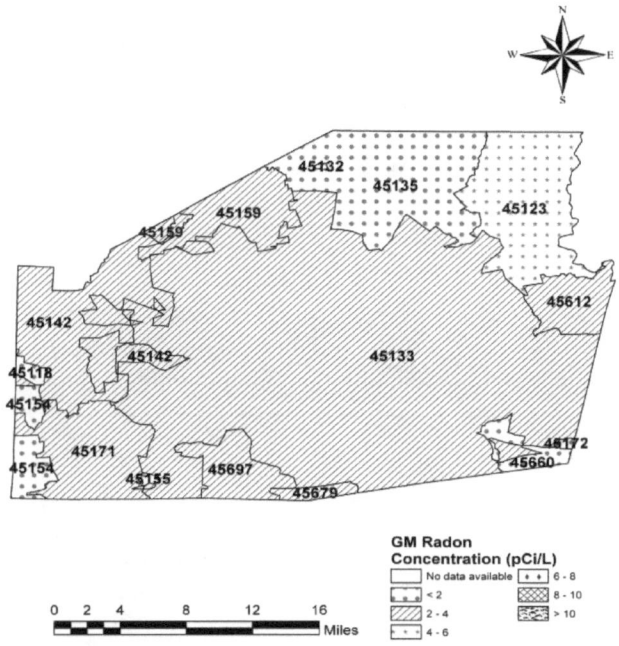

Figure 36c: GM ^{222}Rn concentrations in zip codes of Highland county based on the 2 pCi/L breakdown classification.

Table 37: ^{222}Rn statistics for Hocking county

Zip code	No.	Min.	Max.	AM	GM	SD	CV
43102*	18	0.2	33.4	8.13	3.79	9.88	0.51
43107*	32	0.8	260	16.24	5.71	46.2	2.37
43111	1	21.2	21.2	21.2	21.2	0	0
43127	2	1.6	15.2	8.4	4.93	9.62	0.49
43135	26	0.8	93.2	11.76	6.68	18.43	0.94
43138	169	0.1	99.5	6.96	3.63	9.68	0.5
43144	1	23.3	23.3	23.3	23.3	0	0
43149	49	0.3	238	21.55	6.55	47.35	2.42
43152	10	0.1	10.2	3.48	2.01	3.02	0.15
43155*	21	0.8	40.2	9.25	4.28	12.24	0.63
43158	1	7.5	7.5	7.5	7.5	0	0
43766*	4	1.3	2.4	1.8	1.75	0.47	0.02
45622*	2	0.9	2.3	1.6	1.44	0.99	0.05
45654*	4	1.1	5.8	2.63	2.13	2.14	0.11
45764*	22	0.6	9.2	4.13	3.15	2.82	0.14

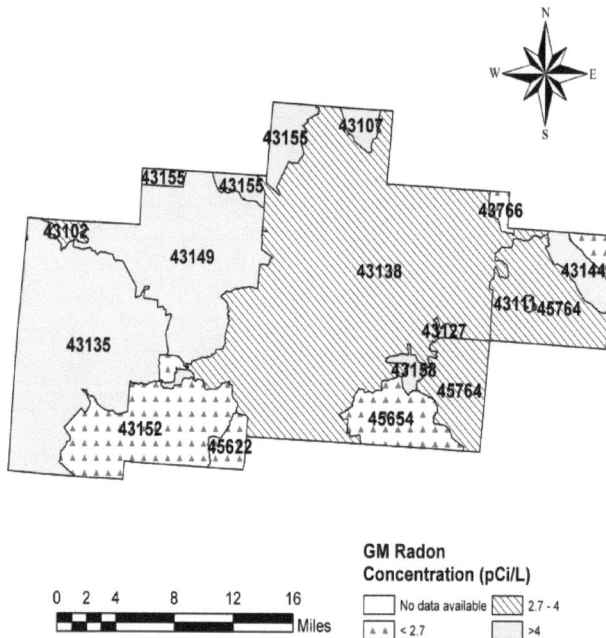

Figure 37a: GM ^{222}Rn concentrations in zip codes of Hocking county based on the WHO-USEPA classification.

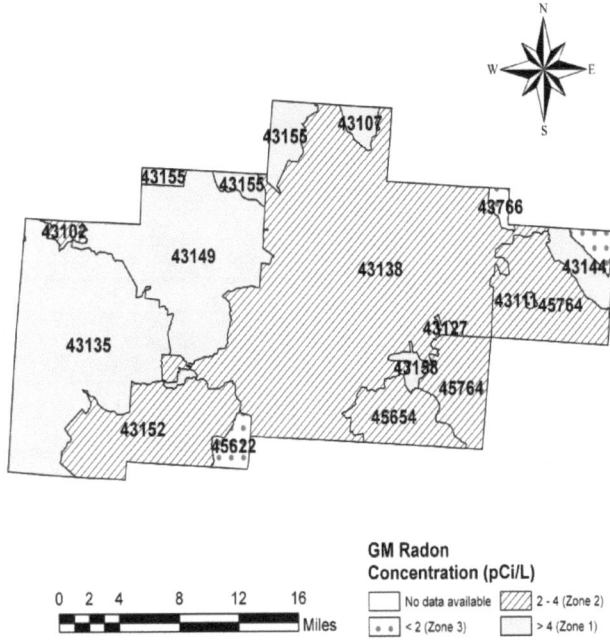

Figure 37b: GM ^{222}Rn concentrations in zip codes of Hocking county based on the USEPA classification.

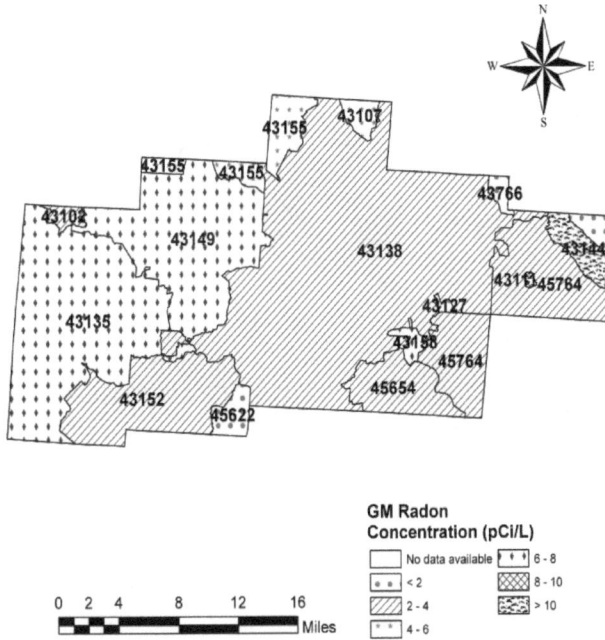

Figure 37c: GM ^{222}Rn concentrations in zip codes of Hocking county based on the 2 pCi/L breakdown classification.

Table 38: ^{222}Rn statistics for Holmes county

Zip code	No.	Min.	Max.	AM	GM	SD	CV
43006	4	1	15.9	10.35	6.83	7.12	0.36
43804	2	0.8	2.5	1.65	1.41	1.2	0.06
44610	5	1.4	7.8	4.44	3.67	2.8	0.14
44611	8	1.8	9.2	4.33	3.81	2.42	0.12
44624[*]	16	1.1	58	8.66	4.16	15.11	0.77
44627[*]	3	2.2	17.5	7.3	4.39	8.83	0.45
44628	16	0.8	23.6	6.16	3.09	7.64	0.39
44633	7	0.1	35.2	8.43	2.06	12.55	0.64
44637	19	0.6	140.3	27.02	9.77	43.48	2.23
44638	7	0.9	11.8	4.79	3.75	3.56	0.18
44654	77	0.1	43.7	7.24	3.74	9.54	0.49
44661	1	11.1	11.1	11.1	11.1	0	0
44676[*]	18	1.1	164.2	20.16	5.05	45.49	2.33
44681[*]	29	0.1	88.2	8.7	3.76	16.84	0.86
44687	1	3.8	3.8	3.8	3.8	0	0
44842[*]	52	0.1	29.9	6.49	3.53	6.88	0.35
44617[**]	1	0.6	0.6	0.6	0.6	0	0
44660[**]	2	1.1	1.7	1.4	1.37	0.42	0.02

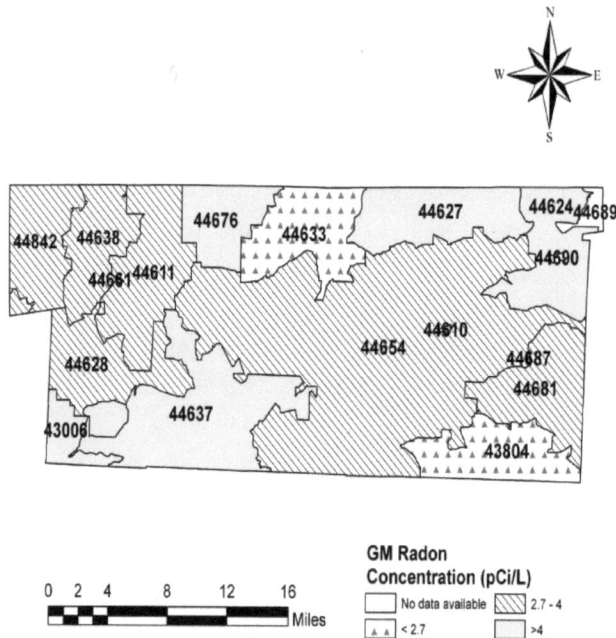

Figure 38a: GM ^{222}Rn concentrations in zip codes of Holmes county based on the WHO-USEPA classification.

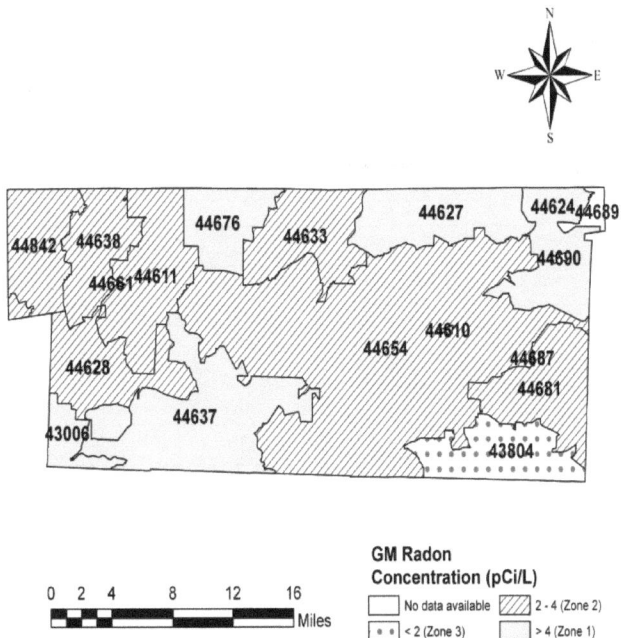

Figure 38b: GM ^{222}Rn concentrations in zip codes of Holmes county based on the USEPA classification.

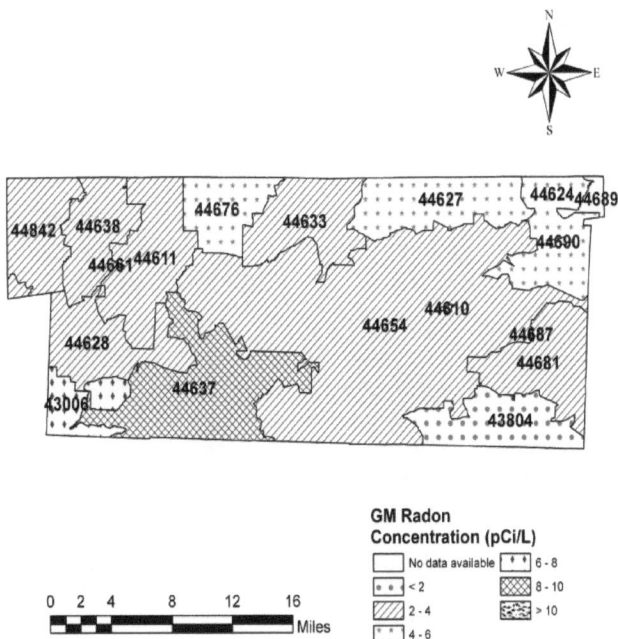

Figure 38c: GM ^{222}Rn concentrations in zip codes of Holmes county based on the 2 pCi/L breakdown classification.

Table 39: ^{222}Rn statistics for Huron county

Zip code	No.	Min.	Max.	AM	GM	SD	CV
44807*	25	0.2	44.7	7.13	3.1	10.74	0.55
44811*	262	0.1	159	15.21	6.34	24.33	1.25
44826	15	0.6	15.8	5.37	3.42	5.12	0.26
44837	17	0.7	50.8	8.52	4.2	13.38	0.69
44846*	101	0.1	81.6	9.22	4.5	13.68	0.7
44847	132	0.1	149.7	23.18	9.05	35.72	1.83
44850	1	1	1	1	1	0	0
44851	32	0.9	13.7	5.01	4.14	3.28	0.17
44855	6	3.5	22.4	8.22	6.47	7.21	0.37
44857	335	0.1	104.1	6.41	3.23	10.19	0.52
44865*	15	2.3	38.1	10.89	7.47	11.62	0.6
44878*	14	0.4	16.6	4.61	3.35	4.13	0.21
44889	44	0.1	24.8	4.74	2.88	4.77	0.24
44890	80	0.3	139.3	8.69	3.92	21.85	1.12
44888**	4	1.2	7.6	4.55	3.76	2.64	0.14

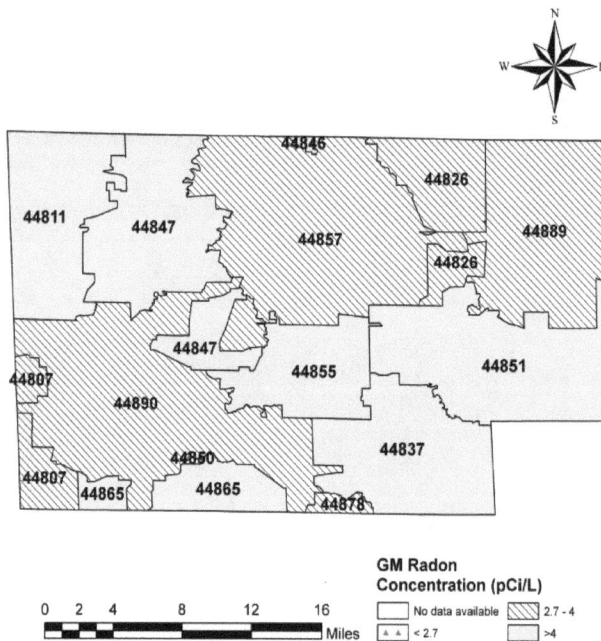

Figure 39a: GM ^{222}Rn concentrations in zip codes of Huron county based on the WHO-USEPA classification.

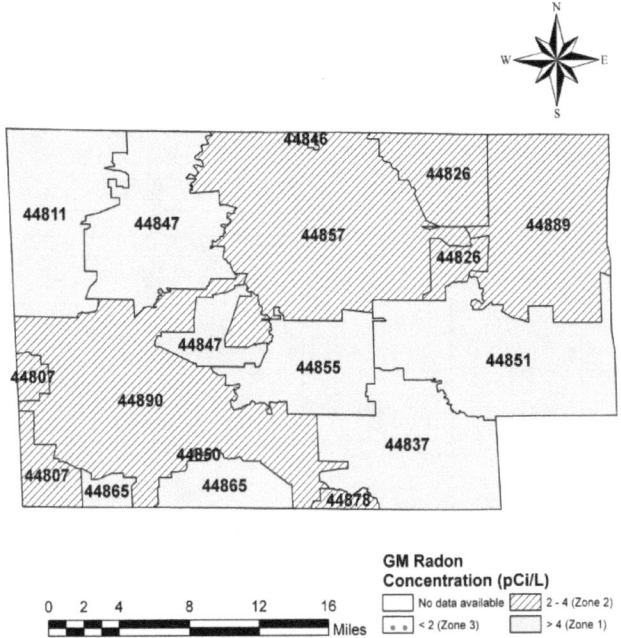

Figure 39b: GM ^{222}Rn concentrations in zip codes of Huron county based on the USEPA classification.

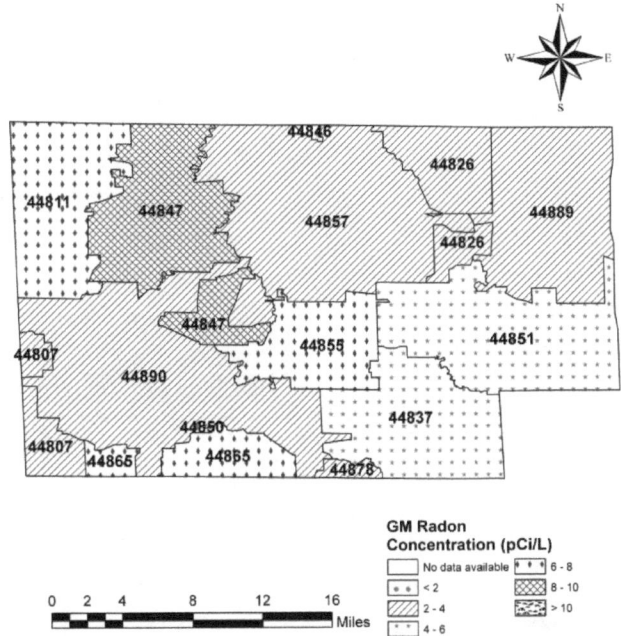

Figure 39c: GM ^{222}Rn concentrations in zip codes of Huron county based on the 2 pCi/L breakdown classification.

Table 40: ^{222}Rn statistics for Jackson county

Zip code	No.	Min.	Max.	AM	GM	SD	CV
45601*	1058	0.1	223.4	9.61	5.1	15.04	0.77
45613*	19	0.1	37.9	4.12	1.18	8.54	0.44
45621	3	0.1	10.5	3.67	0.75	5.92	0.3
45634*	8	0.7	16.4	4.98	2.77	5.45	0.28
45640	99	0.1	20.3	3.43	2.21	3.2	0.16
45656	14	0.6	12.4	3.62	2.3	3.89	0.2
45672*	4	1.9	13.5	5	3.39	5.68	0.29
45685*	2	1.7	1.7	1.7	1.7	0	0
45692	30	0.2	24.3	3.93	1.79	6.01	0.31

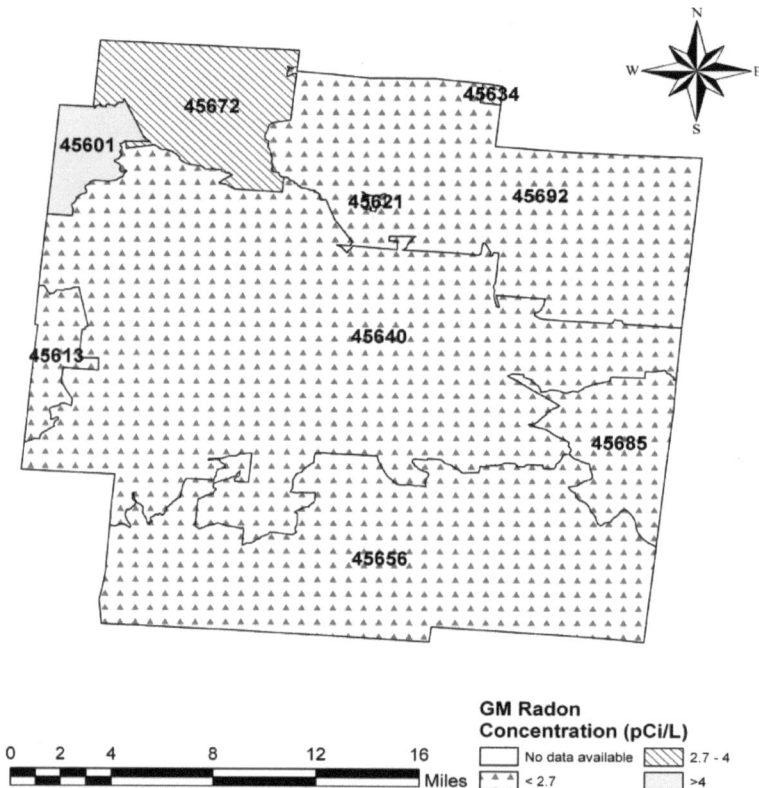

Figure 40a: GM ^{222}Rn concentrations in zip codes of Jackson county based on the WHO-USEPA classification.

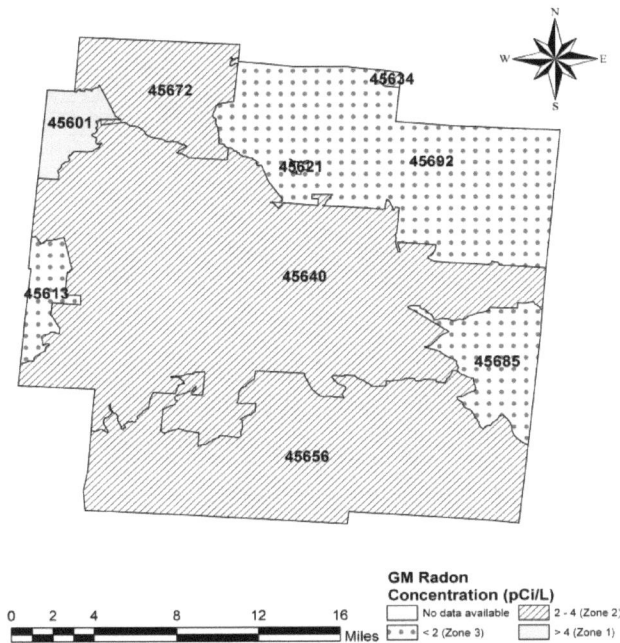

Figure 40b: GM ^{222}Rn concentrations in zip codes of Jackson county based on the USEPA classification.

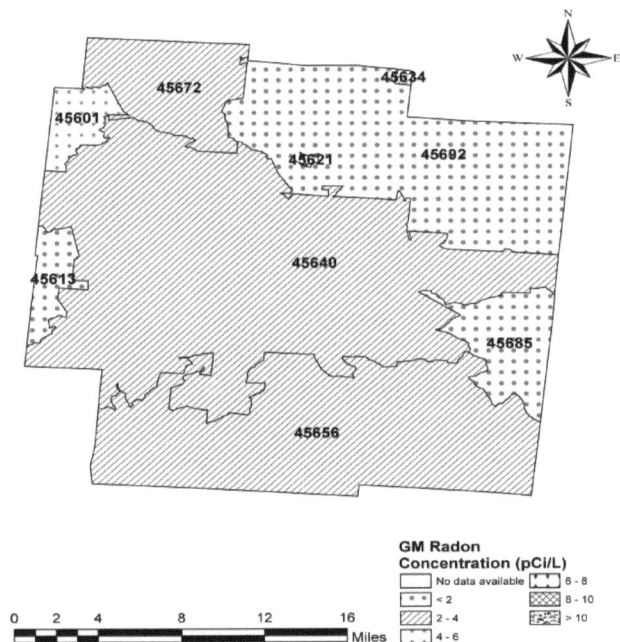

Figure 40c: GM ^{222}Rn concentrations in zip codes of Jackson county based on the 2 pCi/L breakdown classification.

Table 41: ^{222}Rn statistics for Jefferson county

Zip code	No.	Min.	Max.	AM	GM	SD	CV
43901	5	1.1	8.9	4.74	3.46	3.6	0.18
43903[*]	19	0.1	23	3.58	1.85	5.22	0.27
43907[*]	43	1.2	42.6	9.52	5.37	11.78	0.6
43908	3	1.9	3.5	2.47	2.37	0.9	0.05
43910	23	0.6	44.7	9.69	3.73	14.14	0.72
43913	5	1.5	5.8	3.92	3.31	2.19	0.11
43917	26	2.3	65.3	11.88	7.91	13.93	0.71
43925	1	10.5	10.5	10.5	10.5	0	0
43926	2	2	6.1	4.05	3.49	2.9	0.15
43930	4	3.5	39	15.05	9.08	16.74	0.86
43932	4	1.9	36.7	14.93	9.39	15.08	0.77
43938	17	0.5	21.4	5.19	3.18	5.58	0.29
43943	36	0.2	180	29.74	9.69	40.64	2.08
43944	11	0.5	9.3	4.05	2.53	3.35	0.17
43945[*]	29	0.6	177.8	20.52	5.57	41.09	2.1
43948	1	0.3	0.3	0.3	0.3	0	0
43952	186	0.1	927.6	10.17	3.5	67.92	3.48
43953	61	0.2	61.2	8.35	4.16	11.45	0.59
43961	1	1.9	1.9	1.9	1.9	0	0
43963	7	0.9	11.4	3.87	2.84	3.63	0.19
43964	51	0.5	132.2	13.79	4.64	31.68	1.62
43971	3	0.8	6.7	3.37	2.41	3.02	0.15

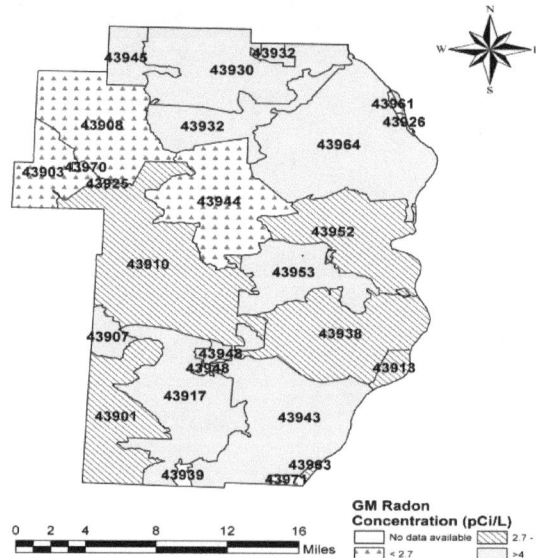

Figure 41a: GM ^{222}Rn concentrations in zip codes of Jefferson county based on the WHO-USEPA classification.

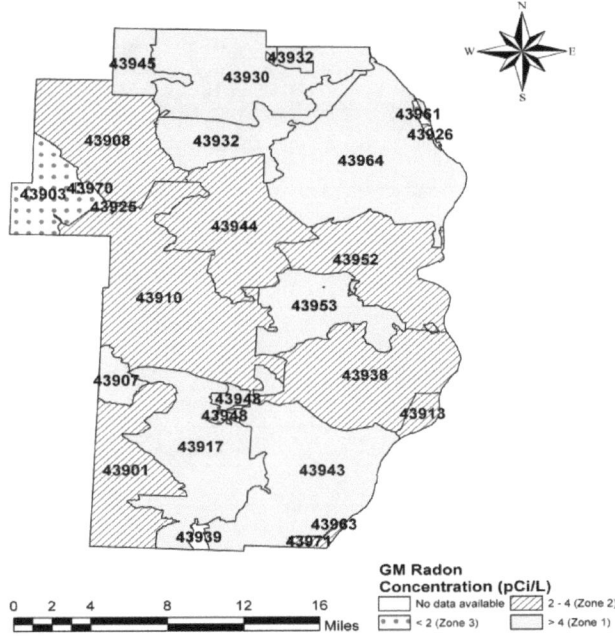

Figure 41b: GM ^{222}Rn concentrations in zip codes of Jefferson county based on the USEPA classification.

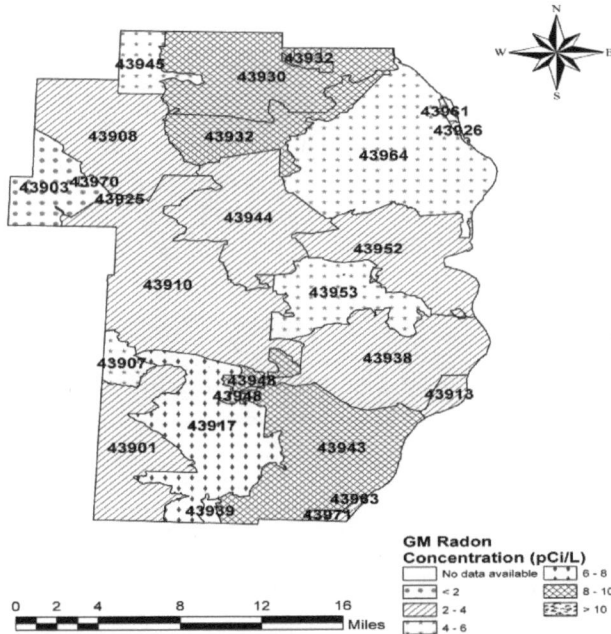

Figure 41c: GM ^{222}Rn concentrations in zip codes of Jefferson county based on the 2 pCi/L breakdown classification.

Table 42: ^{222}Rn statistics for Knox county

Zip code	No.	Min.	Max.	AM	GM	SD	CV
43005	6	0.9	14	7.73	5.46	5.38	0.28
43006*	4	1	15.9	10.35	6.83	7.12	0.36
43011	154	0.1	61.7	8.61	5.53	9.87	0.51
43014	120	0.1	239.3	19.33	6.06	39.09	2
43019	410	0.1	189.5	9.45	3.38	19.71	1.01
43022	254	0.1	553.7	29.08	7.58	67.88	3.48
43028	1568	0.1	844.2	39.53	11	69.26	3.55
43037	4	1	1.2	1.1	1.1	0.12	0.01
43050	1677	0.1	742.7	14.53	5.48	32.72	1.68
43080*	204	0.1	146.3	14.43	6.4	24.59	1.26
43822*	84	0.2	131.1	11.75	5.7	18.43	0.94
43843*	11	1.6	60.5	10.39	4.74	17.77	0.91
44628*	16	0.8	23.6	6.16	3.09	7.64	0.39
44813*	209	0.2	318.8	20.72	8.99	38.36	1.96
44822*	56	1.2	264	36.45	9.52	63.76	3.26

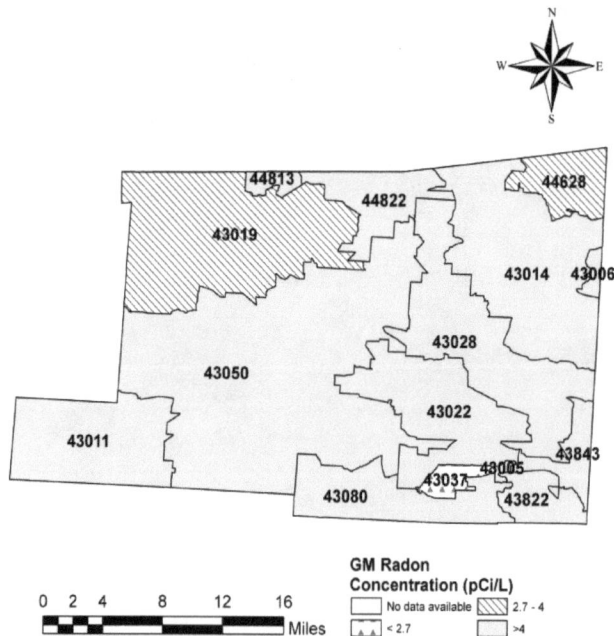

Figure 42a: GM ^{222}Rn concentrations in zip codes of Knox county based on the WHO-USEPA classification.

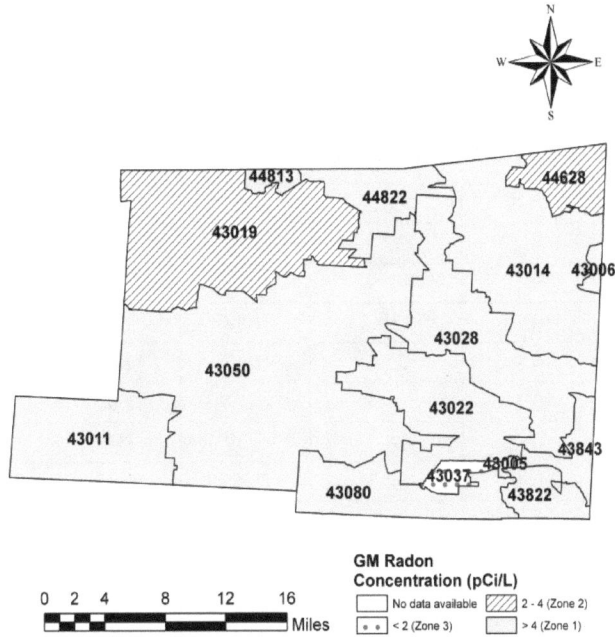

Figure 42b: GM ^{222}Rn concentrations in zip codes of Knox county based on the USEPA classification.

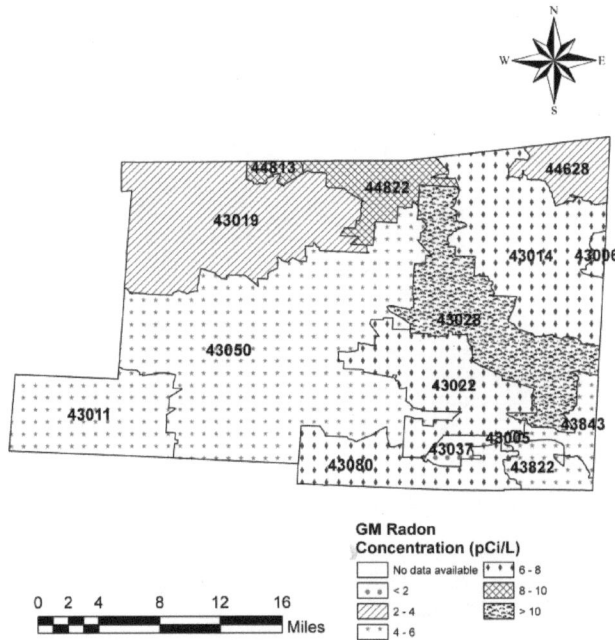

Figure 42c: GM ^{222}Rn concentrations in zip codes of Knox county based on the 2 pCi/L breakdown classification.

Table 43: ^{222}Rn statistics for Lake county

Zip code	No.	Min.	Max.	AM	GM	SD	CV
44024[*]	319	0.1	19.6	2.83	2.07	2.51	0.13
44041[*]	87	0.1	14	2.18	1.16	2.66	0.14
44045	6	0.1	5.1	1.65	0.56	2.14	0.11
44057	152	0.1	37.8	3.97	1.86	6.77	0.35
44060	899	0.1	49.9	3.78	2.06	5.86	0.3
44077	870	0.1	139.3	4.38	2.38	7.51	0.38
44081	124	0.1	22.6	4.43	2.72	4.52	0.23
44086[*]	25	0.1	27.5	3.72	2.07	5.59	0.29
44092	137	0.1	24.7	2.32	1.42	2.94	0.15
44094	528	0.1	33.3	2.76	1.55	3.61	0.18
44095	203	0.1	9.3	1.56	1.07	1.52	0.08
44061[**]	3	0.5	20	7.1	2	11.17	0.57
44096[**]	1	2.5	2.5	2.5	2.5	0	0
44097[**]	2	1	3.4	2.2	1.84	1.7	0.09

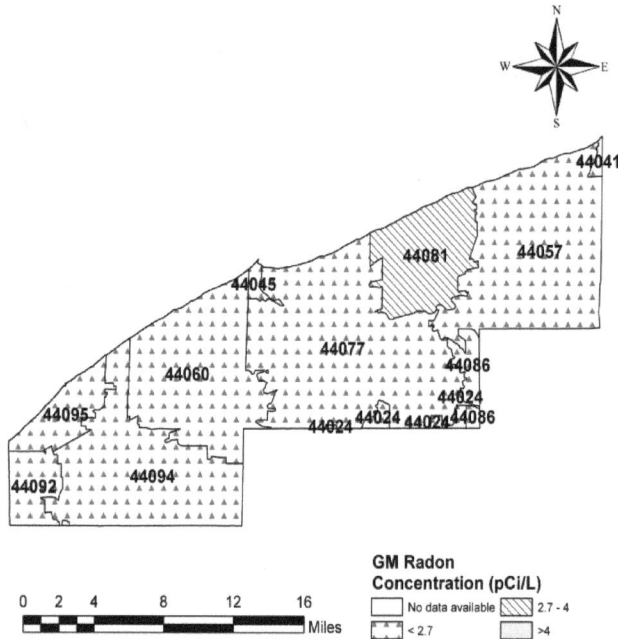

Figure 43a: GM ^{222}Rn concentrations in zip codes of Lake county based on the WHO-USEPA classification.

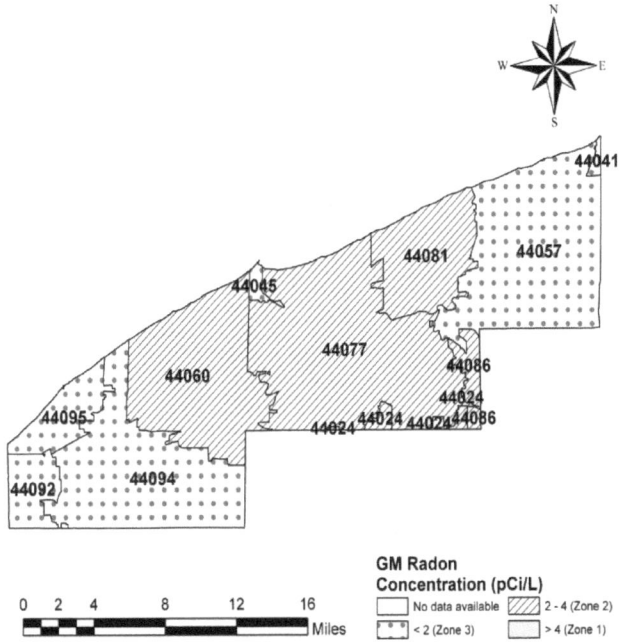

Figure 43b: GM ^{222}Rn concentrations in zip codes of Lake county based on the USEPA classification.

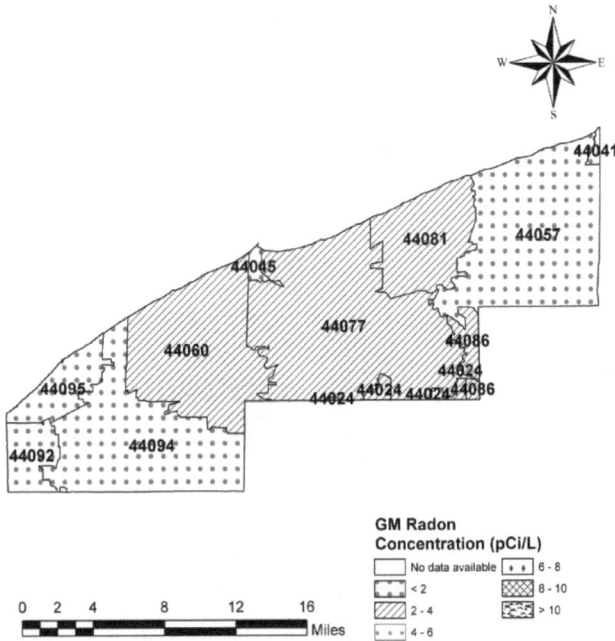

Figure 43c: GM ^{222}Rn concentrations in zip codes of Lake county based on the 2 pCi/L breakdown classification.

Table 44: ^{222}Rn statistics for Lawrence county

Zip code	No.	Min.	Max.	AM	GM	SD	CV
45619	19	0.1	7.6	2.16	1.39	1.98	0.1
45623*	5	0.6	4.9	1.8	1.32	1.78	0.09
45638	52	0.1	33	3.64	2	5.33	0.27
45645	6	0.7	1.9	0.95	0.88	0.47	0.02
45656*	14	0.6	12.4	3.62	2.3	3.89	0.2
45669	60	0.1	333.4	9.39	2.16	42.88	2.2
45678	8	0.1	1	0.75	0.59	0.37	0.02
45680	26	0.6	8.1	2.28	1.7	2.01	0.1
45682*	5	1.8	2.8	2.42	2.39	0.41	0.02

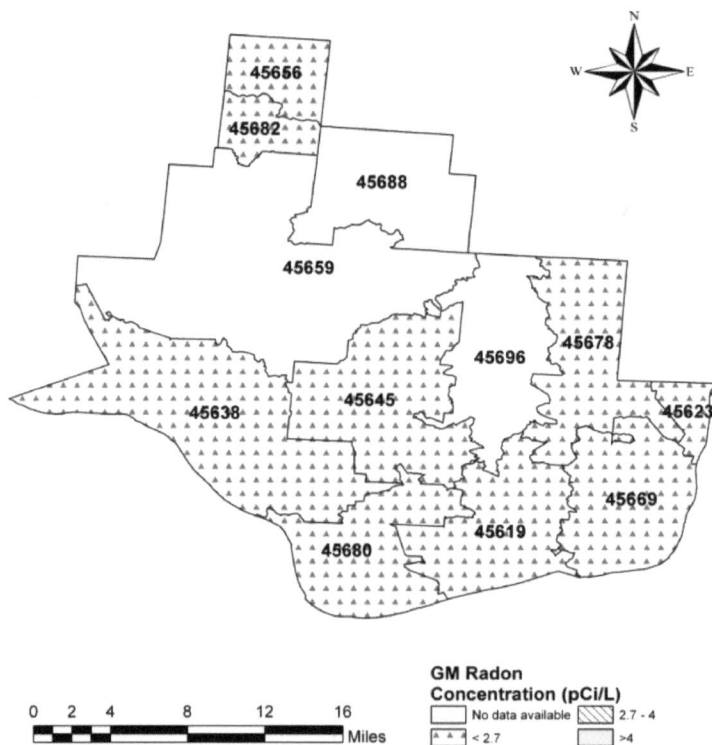

Figure 44a: GM ^{222}Rn concentrations in zip codes of Lawrence county based on the WHO-USEPA classification.

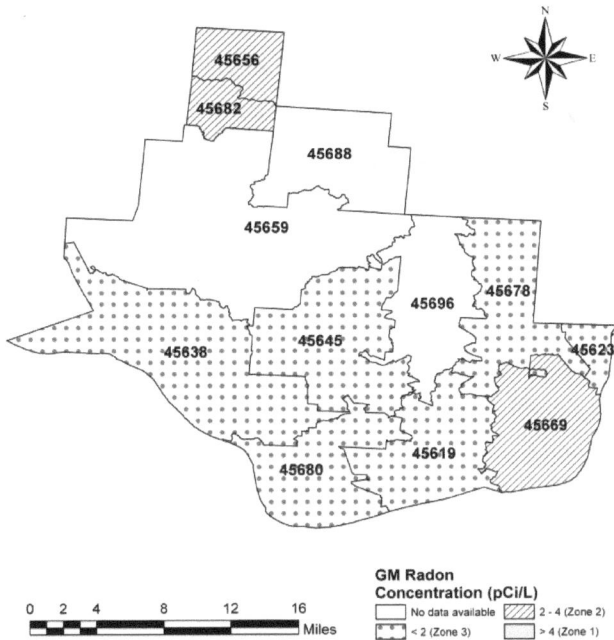

Figure 44b: GM ^{222}Rn concentrations in zip codes of Lawrence county based on the USEPA classification.

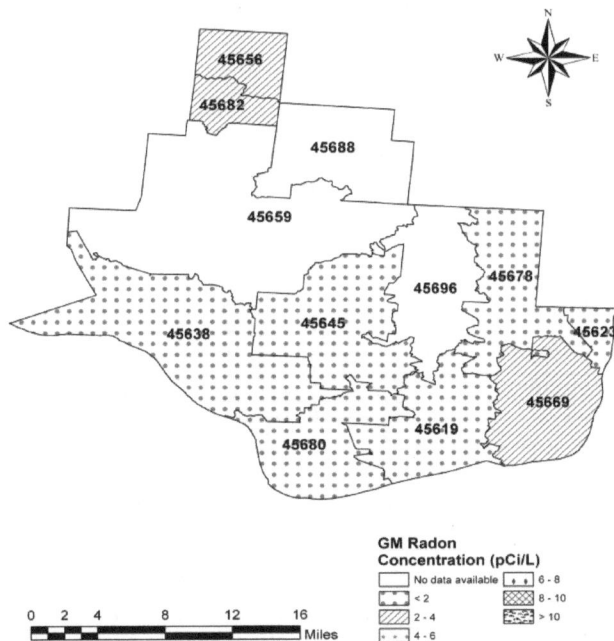

Figure 44c: GM ^{222}Rn concentrations in zip codes of Lawrence county based on the 2 pCi/L breakdown classification.

Table 45: ^{222}Rn statistics for Licking county

Zip code	No.	Min.	Max.	AM	GM	SD	CV
43001	168	0.1	317.7	19.53	6.47	52.61	2.69
43008	34	0.1	14.3	2.76	1.55	3.04	0.16
43011*	154	0.1	61.7	8.61	5.53	9.87	0.51
43013	41	0.4	25.4	7.33	4.73	6.65	0.34
43023	3339	0.1	1400	20.24	8.16	44.7	2.29
43025	283	0.1	684.5	13.08	5.65	50.58	2.59
43030	9	3.6	40.8	14.72	10.44	12.78	0.65
43031	398	0.1	49.9	6.72	4.21	7.01	0.36
43033	11	1.2	145	30.15	13.89	40.76	2.09
43046*	53	0.9	29	5.24	3.6	5.27	0.27
43055	4170	0.1	559	20.63	10.09	32.49	1.66
43056	1060	0.1	191	18.28	8.92	23.76	1.22
43062	910	0.1	470	9.39	5	21.67	1.11
43068*	951	0.1	180.5	8.6	5.03	12.12	0.62
43071	95	0.1	160.8	14.57	6.83	22.37	1.15
43074*	547	0.1	75.7	6.52	3.49	8.28	0.42
43076*	168	0.2	274.5	13.41	4.4	36.3	1.86
43080	204	0.1	146.3	14.43	6.4	24.59	1.26
43147*	1316	0.1	110.1	9.34	5.41	9.88	0.51
43721	7	2	5.9	3.6	3.32	1.57	0.08
43739*	19	2	19.4	7.67	5.78	5.56	0.28
43740	4	12.2	12.3	12.25	12.25	0.06	0
43746*	7	1.3	27	6.43	3.58	9.23	0.47
43760*	10	0.5	3.1	1.89	1.59	0.97	0.05
43822*	84	0.2	131.1	11.75	5.7	18.43	0.94
43830*	129	0.2	129.7	9.2	4	17.16	0.88
43018**	5	0.1	19.9	6.82	1.64	8.61	0.44
43027**	2	4.9	5.1	5	5	0.14	0.01
43058**	46	1.5	125.1	17.45	9	27.63	1.41
43073**	3	2.5	17.5	7.5	4.78	8.66	0.44
43093**	5	0.9	1.2	1.08	1.07	0.13	0.01

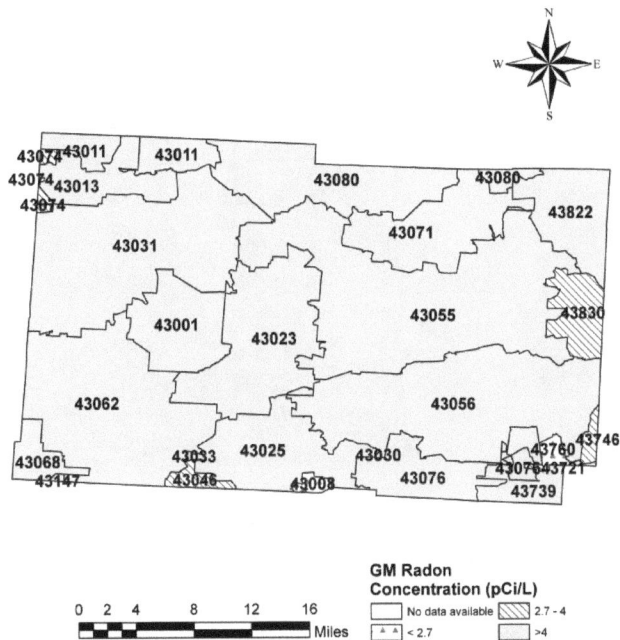

Figure 45a: GM ^{222}Rn concentrations in zip codes of Licking county based on the WHO-USEPA classification.

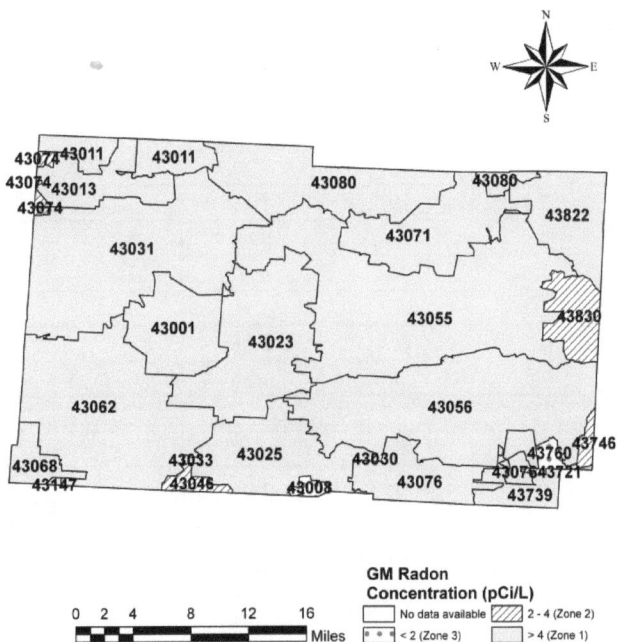

Figure 45b: GM ^{222}Rn concentrations in zip codes of Licking county based on the USEPA classification.

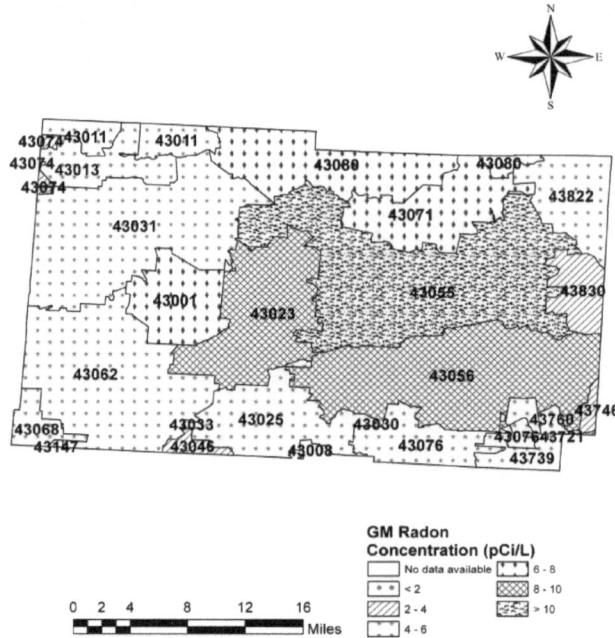

Figure 45c: GM ^{222}Rn concentrations in zip codes of Licking county based on the 2 pCi/L breakdown classification.

Table 46: ^{222}Rn statistics for Logan county

Zip code	No.	Min.	Max.	AM	GM	SD	CV
43060*	51	0.2	62.7	8.24	4.51	11.03	0.56
43310	32	0.5	29.8	7.51	5.19	6.84	0.35
43311	448	0.1	319.2	10.9	6.21	18.72	0.96
43318	59	0.8	150.3	9.64	5.33	20.02	1.03
43319	20	0.6	20.4	3.97	2.81	4.37	0.22
43324	32	0.2	34.1	8.36	4.52	7.91	0.41
43331	19	0.5	8.7	2.06	1.47	2.08	0.11
43333	13	0.3	14.6	7.36	5.48	4.05	0.21
43336	3	1.4	3.8	2.77	2.55	1.23	0.06
43343	18	0.1	38.8	14.47	6.01	14.88	0.76
43345*	6	1.2	6.4	3.08	2.37	2.39	0.12
43347	7	0.7	4.9	2.6	2.07	1.71	0.09
43348	9	0.7	11	4.33	3.01	3.61	0.18
43357	207	0.1	75	16	8.17	17.07	0.87
43358	24	0.5	12.6	4.75	3.47	3.49	0.18
43360	66	0.4	90.2	11.66	6.23	15.94	0.82
45334*	19	0.7	17	3.7	2.54	4.05	0.21
45895*	122	0.1	34.2	6.16	3.81	5.97	0.31

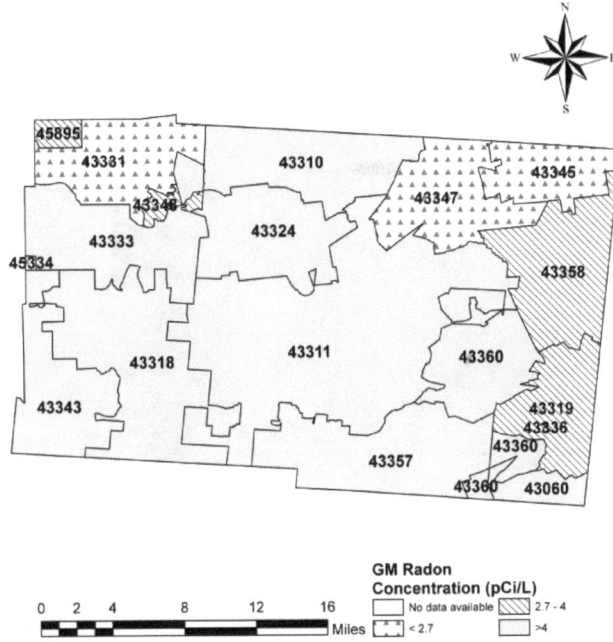

Figure 46a: GM ^{222}Rn concentrations in zip codes of Logan county based on the WHO-USEPA classification.

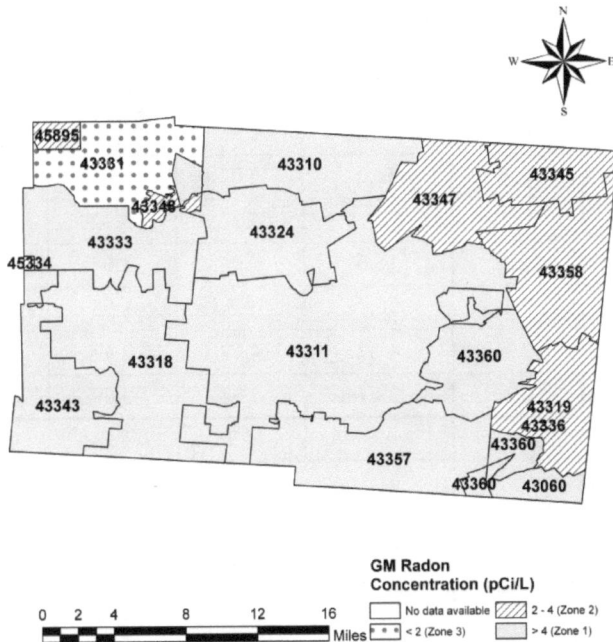

Figure 46b: GM ^{222}Rn concentrations in zip codes of Logan county based on the USEPA classification.

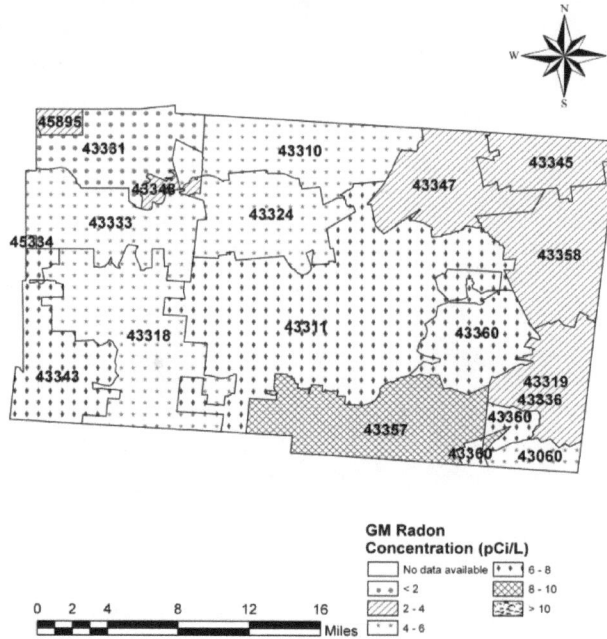

Figure 46c: GM ^{222}Rn concentrations in zip codes of Logan county based on the 2 pCi/L breakdown classification.

Table 47: ^{222}Rn statistics for Lorain county

Zip code	No.	Min.	Max.	AM	GM	SD	CV
44001	594	0.1	42.8	2.96	1.2	4.87	0.25
44011	738	0.1	52.7	5.02	3.04	6.22	0.32
44012	1910	0.1	618.2	5.33	3.12	15.05	0.77
44028	75	0.2	16.6	4.94	3.68	3.79	0.19
44035	384	0.1	70.1	3.3	1.93	4.84	0.25
44039	465	0.1	40.2	4.12	2.47	4.72	0.24
44044	127	0.2	28.7	4.02	2.69	3.82	0.2
44049	1	1.3	1.3	1.3	1.3	0	0
44050	33	0.1	51.2	7.02	4.03	9.23	0.47
44052	130	0.1	20.2	3.01	1.47	3.55	0.18
44053	156	0.1	32.8	5.16	3.09	5.22	0.27
44054	117	0.1	27.2	5.19	3.13	4.99	0.26
44055	42	0.1	23.4	5.05	2.88	5.02	0.26
44074	120	0.1	54.1	3.87	2.47	5.59	0.29

Zip code	No.	Min.	Max.	AM	GM	SD	CV
44089*	270	0.1	89.6	7.21	4.03	10.7	0.55
44090	75	0.1	28.3	5.56	3.41	5.72	0.29
44253*	33	1	23.9	4.69	3.33	4.8	0.25
44256*	1446	0.1	60.3	3.93	2.73	4.35	0.22
44275*	11	0.6	14.1	5.28	3.31	4.93	0.25
44280*	52	0.5	36.1	7.66	5.01	8.4	0.43
44851*	32	0.9	13.7	5.01	4.14	3.28	0.17
44880*	13	0.6	17.7	5.42	3.18	5.48	0.28
44889*	44	0.1	24.8	4.74	2.88	4.77	0.24
44036**	8	0.2	4.7	2.49	1.79	1.66	0.09

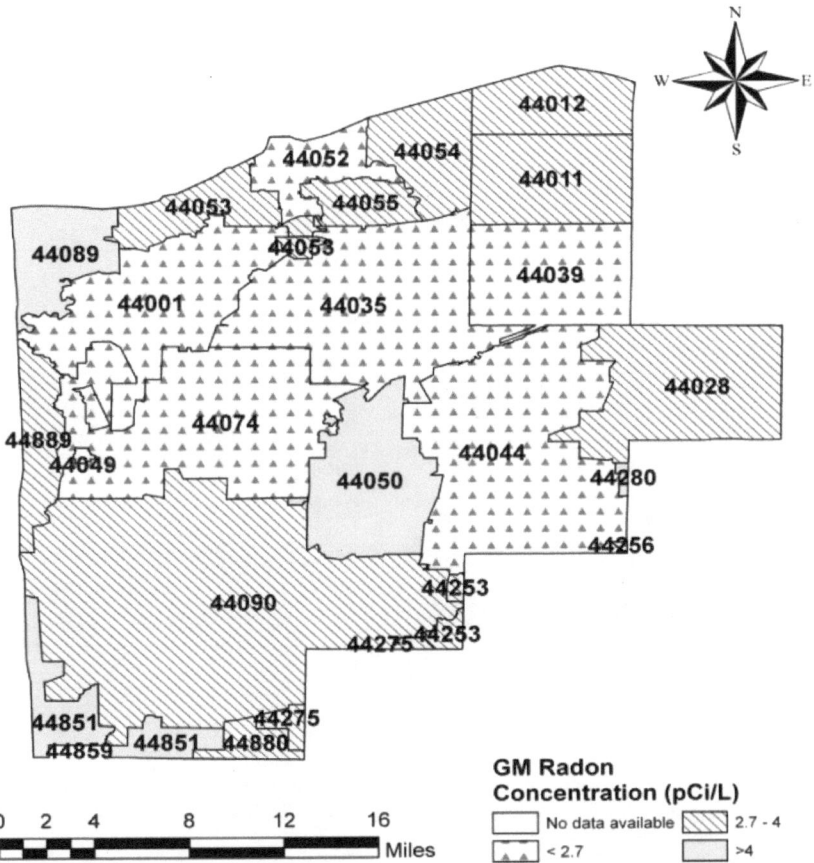

Figure 47a: GM ^{222}Rn concentrations in zip codes of Lorain county based on the WHO-USEPA classification.

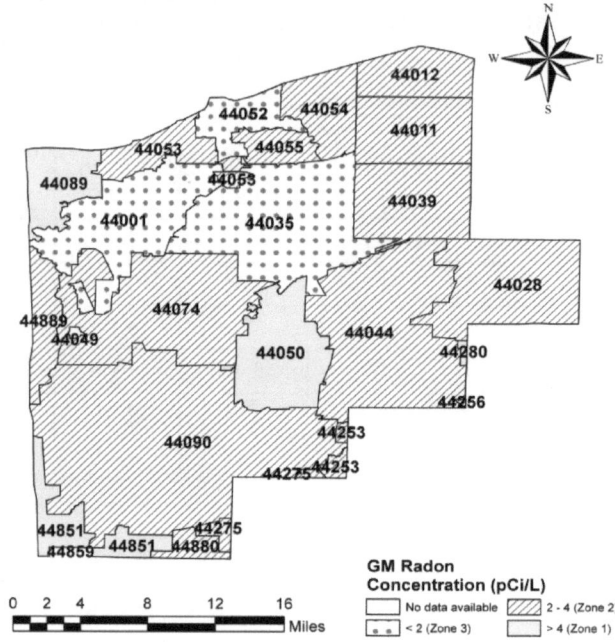

Figure 47b: GM [222]Rn concentrations in zip codes of Lorain county based on the USEPA classification.

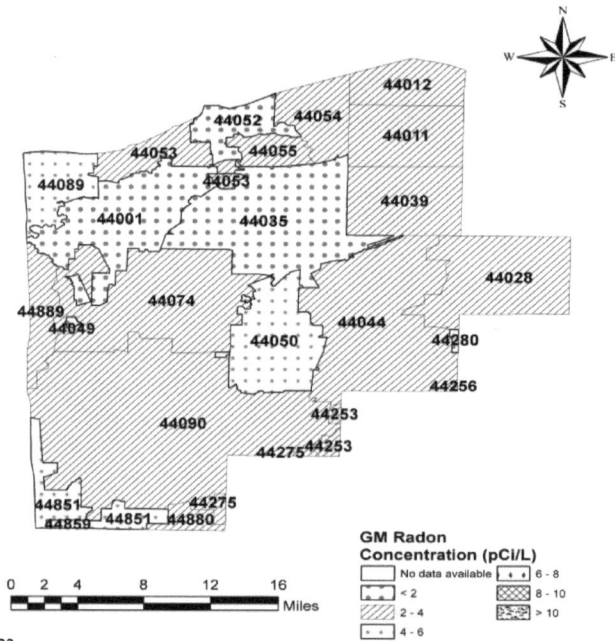

Figure 47c: GM [222]Rn concentrations in zip codes of Lorain county based on the 2 pCi/L breakdown classification.

Table 48: ^{222}Rn statistics for Lucas county

Zip code	No.	Min.	Max.	AM	GM	SD	CV
43412	30	0.1	260	13.3	3.4	46.76	2.39
43434	1	2.8	2.8	2.8	2.8	0	0
43445*	10	0.1	7.8	3.06	1.93	2.31	0.12
43504	14	0.5	16.6	5.57	3.72	5.06	0.26
43522*	60	0.1	27.5	4.06	2.18	5.33	0.27
43528	306	0.1	17.5	2.63	1.85	2.57	0.13
43532*	28	0.1	10.8	2.24	0.91	2.79	0.14
43537	846	0.1	125.5	5.19	2.93	6.98	0.36
43542	138	0.1	19.6	4.08	2.4	3.53	0.18
43547	2	0.6	0.8	0.7	0.69	0.14	0.01
43558*	85	0.1	20.4	2.75	1.61	3.26	0.17
43560	816	0.1	70.3	3.3	1.99	5.13	0.26
43566	268	0.1	71.2	5.36	2.94	6.55	0.34
43571	107	0.1	18.8	3.28	1.97	3.45	0.18
43604	42	0.1	13.8	2.76	1.29	3.36	0.17
43605	58	0.1	12.1	3.29	2.09	2.8	0.14
43606	310	0.1	25.5	2.38	1.58	2.61	0.13
43607	46	0.1	17.7	3.39	1.96	4.09	0.21
43608	93	0.1	18.6	4.48	3.28	3.33	0.17
43609	49	0.3	26.2	2.79	1.77	4.13	0.21
43610	11	0.2	6	2.29	1.57	1.84	0.09
43611	113	0.1	24.4	4.67	2.63	4.7	0.24
43612	154	0.1	15.5	2.54	1.75	2.37	0.12
43613	209	0.1	19.2	1.86	1.23	2.2	0.11
43614	461	0.1	103.9	5.04	2.65	8.2	0.42
43615	362	0.1	16.4	2.61	1.64	2.6	0.13
43616	139	0.1	46.3	4.81	2.87	5.52	0.28
43617	156	0.1	26.1	2.1	1.64	2.28	0.12
43620	10	0.4	9.4	3.38	2.32	2.99	0.15
43623	193	0.1	50.8	2.08	1.37	3.8	0.19
43601**	2	0.6	11.2	5.9	2.59	7.5	0.38
43635**	1	6.6	6.6	6.6	6.6	0	0
43652**	1	3.1	3.1	3.1	3.1	0	0
43697**	1	2.5	2.5	2.5	2.5	0	0
43699**	7	0.1	1	0.29	0.2	0.32	0.02

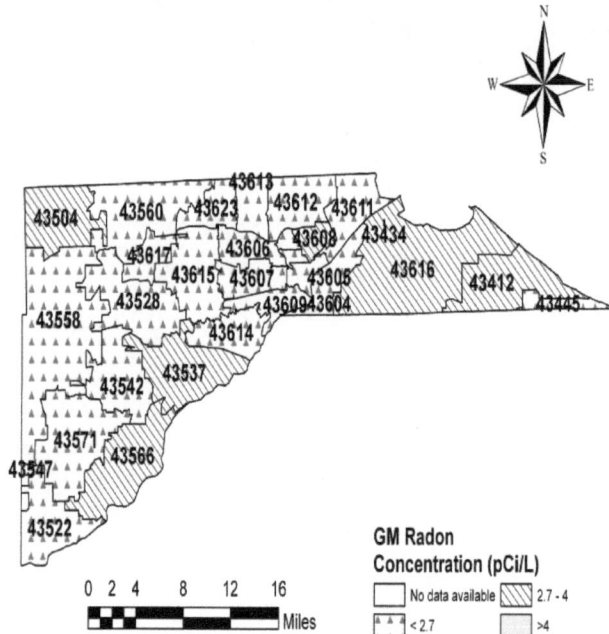

Figure 48a: GM ^{222}Rn concentrations in zip codes of Lucas county based on the WHO-USEPA classification.

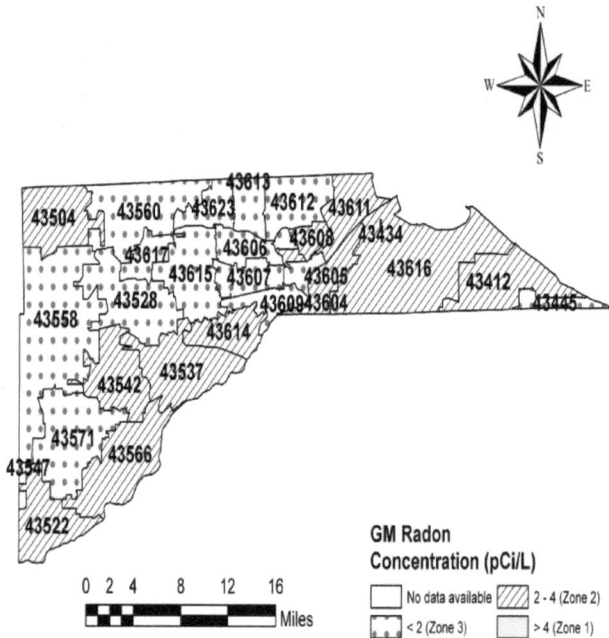

Figure 48b: GM ^{222}Rn concentrations in zip codes of Lucas county based on the USEPA classification.

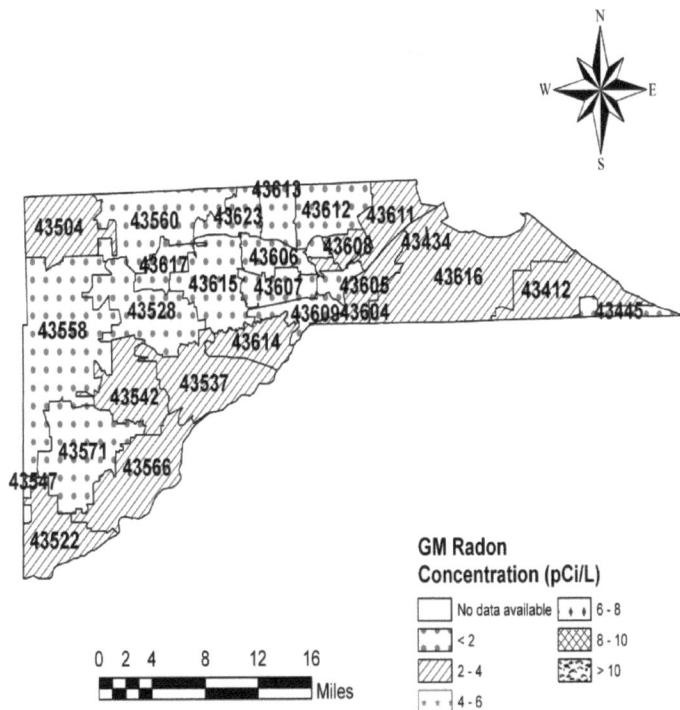

Figure 48c: GM ^{222}Rn concentrations in zip codes of Lucas county based on the 2 pCi/L breakdown classification.

Table 49: ^{222}Rn statistics for Madison county

Zip code	No.	Min.	Max.	AM	GM	SD	CV
43026*	1847	0.1	151	9.33	5.69	9.99	0.51
43029	18	0.2	13.3	5.46	3.63	3.47	0.18
43044*	110	0.1	41.5	7.51	4.32	7.59	0.39
43064	329	0.5	70.4	8.28	5.12	10.09	0.52
43119*	393	0.1	63.9	8.33	5.44	7.39	0.38
43140	397	0.1	82.1	7.87	4.88	9.39	0.48
43143	58	0.1	28.5	5.21	3.37	4.88	0.25
43146*	143	0.5	43.1	9.05	6.17	8.39	0.43
43151	4	1	9.3	6.3	4.75	3.72	0.19
43153	8	0.9	10.9	5.1	4.1	3.25	0.17
43162	142	0.5	54.4	8.54	6.2	7.9	0.4
45369*	88	0.1	21.3	5.09	3.26	4.74	0.24

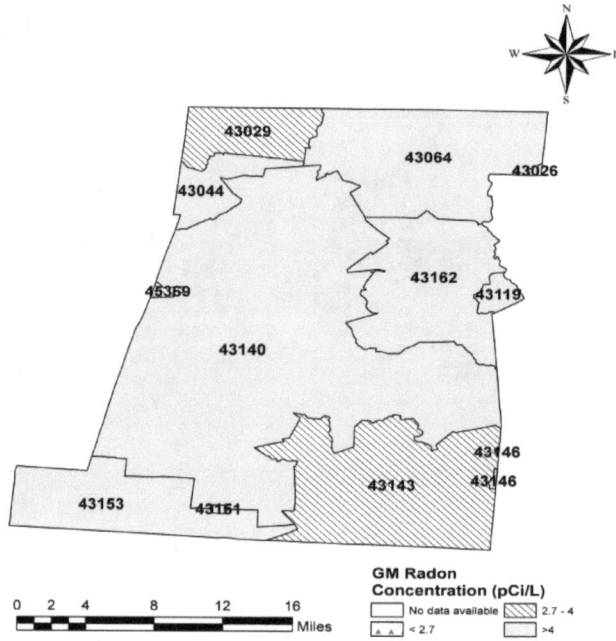

Figure 49a: GM ^{222}Rn concentrations in zip codes of Madison county based on the WHO-USEPA classification.

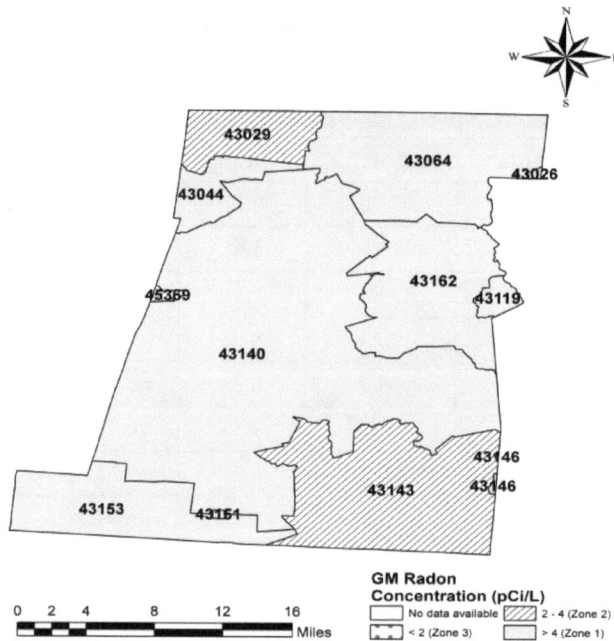

Figure 49b: GM ^{222}Rn concentrations in zip codes of Madison county based on the USEPA classification.

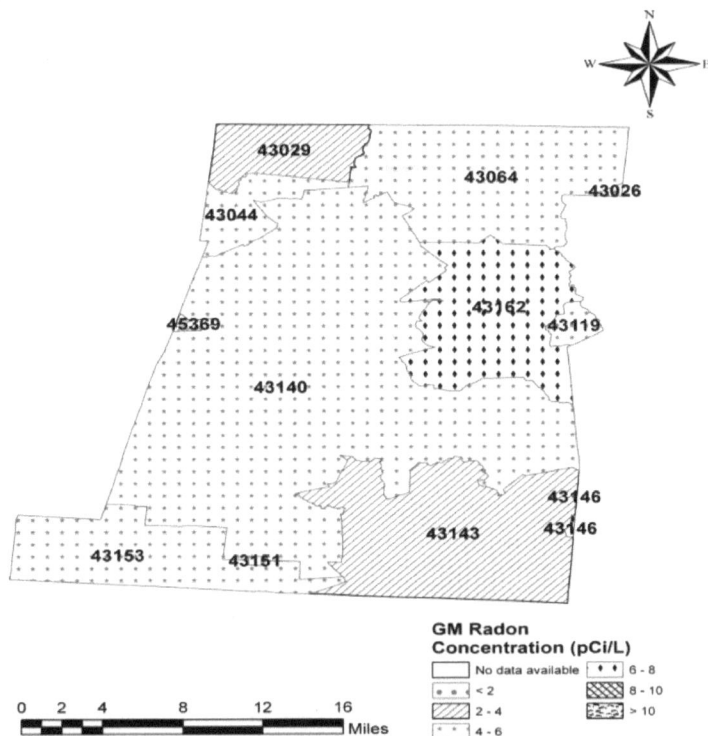

Figure 49c: GM ^{222}Rn concentrations in zip codes of Madison county based on the 2 pCi/L breakdown classification.

Table 50: ^{222}Rn statistics for Mahoning county

Zip code	No.	Min.	Max.	AM	GM	SD	CV
44401	33	0.1	7.8	2.19	1.4	1.72	0.09
44405	43	0.1	16.9	2.93	1.68	3.12	0.16
44406	415	0.1	16.3	2.99	2.13	2.51	0.13
44408*	157	0.1	234.2	6.31	3.07	19.41	0.99
44412*	27	0.7	11.3	3.5	2.4	3.1	0.16
44425*	152	0.1	40.1	3.11	2.05	3.96	0.2
44429	14	0.3	4.9	2.39	1.93	1.41	0.07
44436	21	0.4	15.8	3.06	2.2	3.16	0.16
44437*	14	0.1	6.2	2.21	1.72	1.41	0.07
44440*	32	0.1	57	6.63	2.44	12.56	0.64
44442	35	1	12.2	3.4	2.65	2.75	0.14

Zip code	No.	Min.	Max.	AM	GM	SD	CV
44443	31	0.1	207	27.88	6.65	50.99	2.61
44444[*]	64	0.1	13.9	2.47	1.74	2.22	0.11
44449	10	0.3	4.8	2.07	1.52	1.4	0.07
44451	46	0.2	12.9	2.07	1.47	2.02	0.1
44452	46	0.3	29.1	4.43	2.96	5.27	0.27
44454	10	1.1	26.8	6.03	3.82	7.6	0.39
44460[*]	226	0.1	76.7	6.56	2.95	12.28	0.63
44471	46	0.3	12.9	2.17	1.64	2.09	0.11
44481[*]	87	0.1	11.1	2.63	1.53	2.48	0.13
44490[*]	2	3.5	4	3.75	3.74	0.35	0.02
44502	30	0.1	9.9	1.81	1.19	1.82	0.09
44503	17	0.4	50	8.95	4.75	11.92	0.61
44504	16	0.5	6.8	1.56	1.26	1.46	0.07
44505	113	0.1	9.6	2.62	2.02	1.75	0.09
44506	6	0.7	8.4	4.23	2.98	2.98	0.15
44507	16	0.3	3.6	1.03	0.77	0.91	0.05
44509	49	0.1	5.7	1.88	1.27	1.48	0.08
44510	8	0.5	3.8	2.26	1.83	1.27	0.07
44511	166	0.1	32.6	2.81	1.74	3.32	0.17
44512	406	0.1	20	2.65	2	1.99	0.1
44514	369	0.1	12.1	2.84	2.03	2.2	0.11
44515	251	0.1	11.7	2.52	1.85	1.98	0.1
44601[*]	214	0.1	16.3	2.62	1.89	2.21	0.11
44609[*]	17	0.3	5.3	1.43	1.06	1.31	0.07
44672	10	0.1	2	0.95	0.72	0.62	0.03
44416[**]	2	1.6	1.9	1.75	1.74	0.21	0.01
44422[**]	2	0.4	1.7	1.05	0.82	0.92	0.05
44501[**]	11	0.6	4.9	2.17	1.81	1.43	0.07
44513[**]	4	1.4	3.9	2.83	2.65	1.04	0.05
44555[**]	113	0.1	16.6	2.58	1	3.89	0.2
44619[**]	5	0.1	10.1	3.04	0.76	4.32	0.22

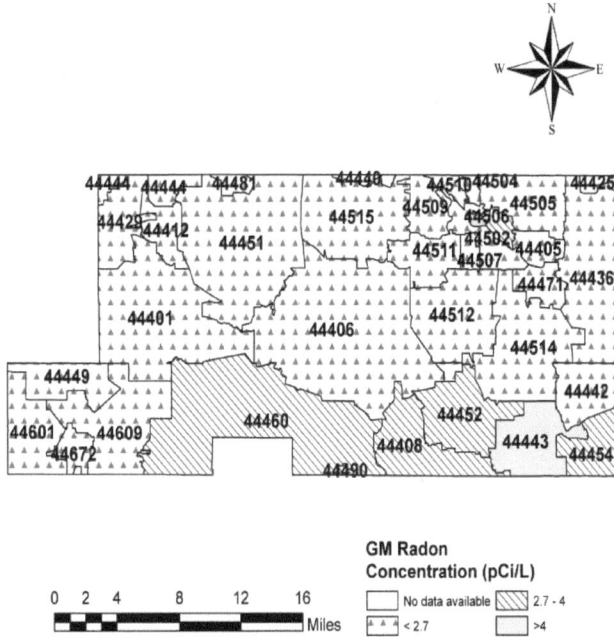

Figure 50a: GM ^{222}Rn concentrations in zip codes of Mahoning county based on the WHO-USEPA classification.

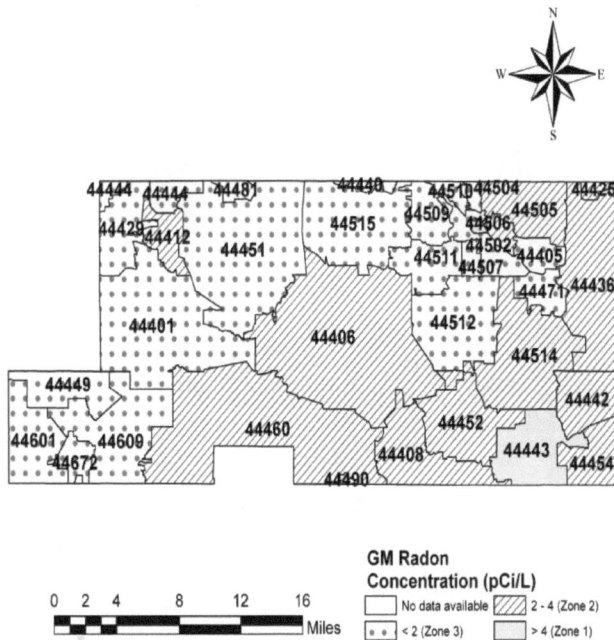

Figure 50b: GM ^{222}Rn concentrations in zip codes of Mahoning county based on the USEPA classification.

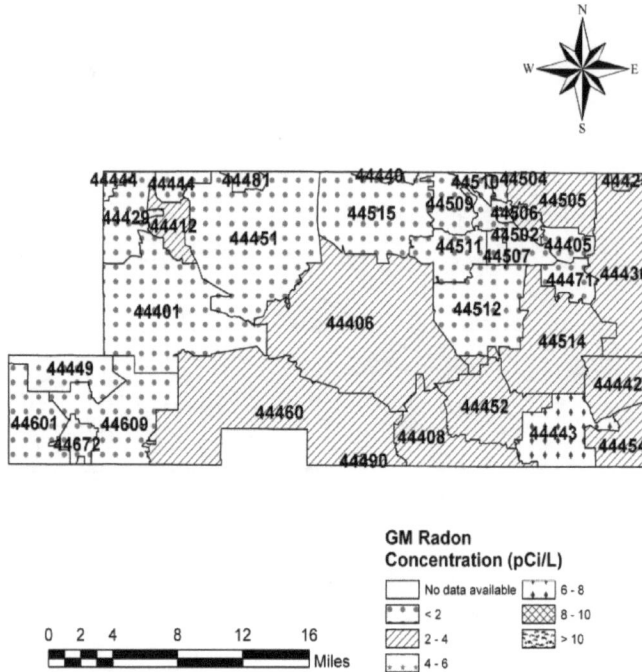

Figure 50c: GM ^{222}Rn concentrations in zip codes of Mahoning county based on the 2 pCi/L breakdown classification.

Table 51: ^{222}Rn statistics for Marion county

Zip code	No.	Min.	Max.	AM	GM	SD	CV
43302	911	0.1	67.1	6.25	3.47	7.2	0.37
43314	83	0.1	26.3	4.65	3.38	4.12	0.21
43315[*]	116	0.1	77.2	7.48	3.85	12.18	0.62
43322	2	2.5	7.8	5.15	4.42	3.75	0.19
43332	18	2	30.3	12.28	8.78	8.91	0.46
43337	7	0.8	8.6	4.06	3.05	2.96	0.15
43341	12	1.4	19.3	7.21	5.5	5.96	0.31
43342	58	0.6	21.3	9.15	6.79	5.92	0.3
43344[*]	153	0.1	61.8	5.54	3.14	7.7	0.39
43356	33	0.7	28.8	6.16	3.94	6.15	0.32
44833[*]	493	0.1	140.2	7.75	4.41	12.72	0.65
44849[*]	52	0.2	20.9	4.15	2.88	4.36	0.22
43301[**]	2	5.3	5.3	5.3	5.3	0	0
43306[**]	3	0.5	6.8	2.77	1.5	3.5	0.18
43335[**]	1	6.1	6.1	6.1	6.1	0	0

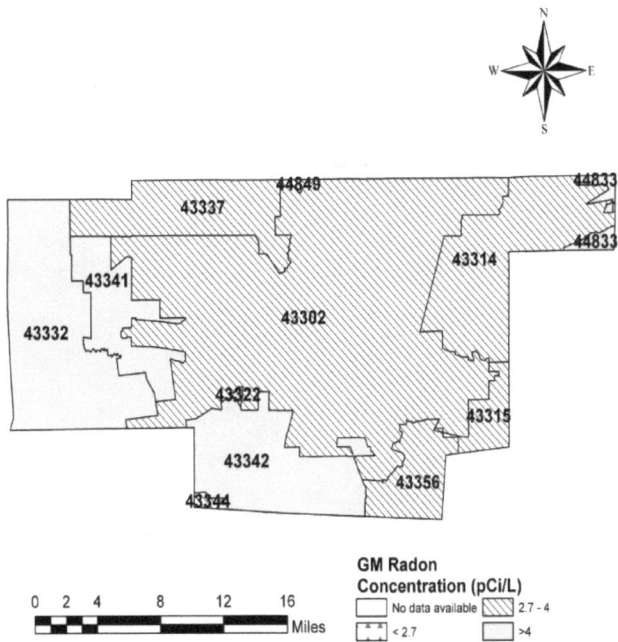

Figure 51a: GM ^{222}Rn concentrations in zip codes of Marion county based on the WHO-USEPA classification.

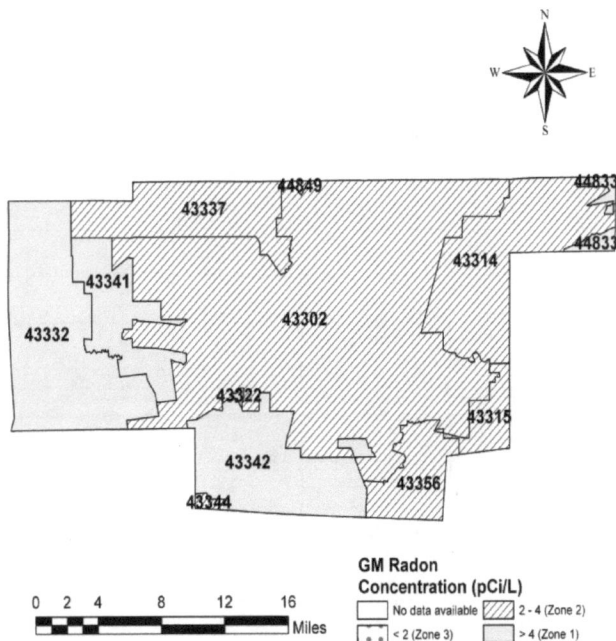

Figure 51b: GM ^{222}Rn concentrations in zip codes of Marion county based on the USEPA classification.

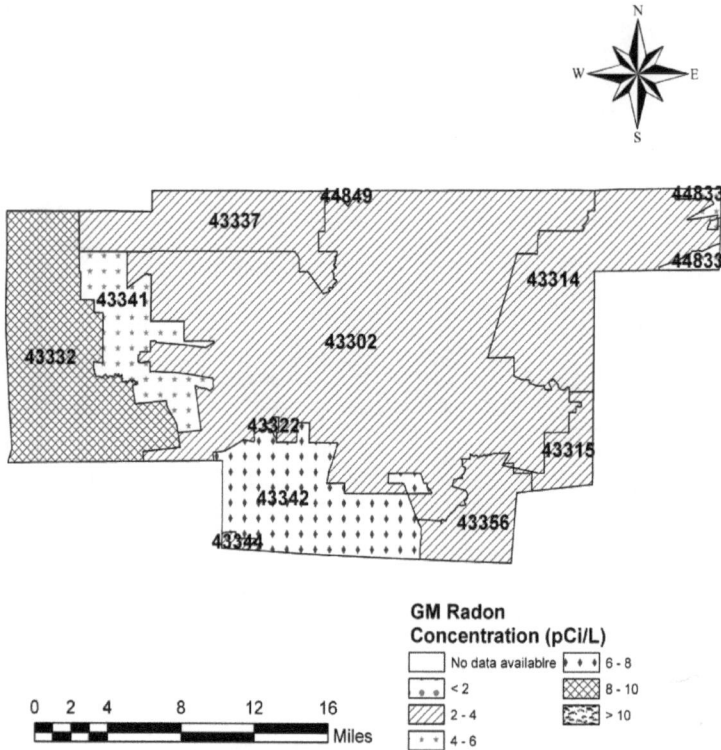

Figure 51c: GM ^{222}Rn concentrations in zip codes of Marion county based on the 2 pCi/L breakdown classification.

Table 52: ^{222}Rn statistics for Medina county

Zip code	No.	Min.	Max.	AM	GM	SD	CV
44136[*]	560	0.1	656.2	5.12	2.16	38.94	1.99
44203[*]	482	0.1	30	4.21	2.74	4.16	0.21
44212	545	0.1	169.3	3.38	2.22	7.54	0.39
44214[*]	24	0.3	5.6	2.9	2.29	1.67	0.09
44215	21	0.9	21.2	4	2.69	4.55	0.23
44217[*]	28	0.1	90.4	9.45	2.88	18.33	0.94
44230[*]	99	0.3	77.5	7.22	4.34	9.86	0.5
44233	172	0.1	22.1	3.11	2.26	2.67	0.14
44235	21	0.8	44.2	6.23	3.89	9.32	0.48
44251	38	0.6	11.6	4.22	3.59	2.5	0.13
44253	33	1	23.9	4.69	3.33	4.8	0.25

Zip code	No.	Min.	Max.	AM	GM	SD	CV
44254	50	0.5	31.5	5.18	3.31	5.9	0.3
44256	1446	0.1	60.3	3.93	2.73	4.35	0.22
44270[*]	43	0.1	14.7	3.9	2.6	3.37	0.17
44273	93	0.4	48.4	6.49	4.08	8.96	0.46
44274	8	0.3	6.4	2.65	1.78	2.19	0.11
44275	11	0.6	14.1	5.28	3.31	4.93	0.25
44280	52	0.5	36.1	7.66	5.01	8.4	0.43
44281	688	0.1	52.5	4.71	2.86	5.69	0.29
44287[*]	39	0.1	17	4.78	2.85	4.49	0.23
44321[*]	554	0.1	591.6	7.19	3.26	26.23	1.34
44333[*]	621	0.1	50.8	6.67	3.75	8.28	0.42
44880[*]	13	0.6	17.7	5.42	3.18	5.48	0.28
44258[**]	9	1.6	13.6	4.33	3.23	4.16	0.21
44282[**]	3	0.8	9.1	3.97	2.44	4.49	0.23

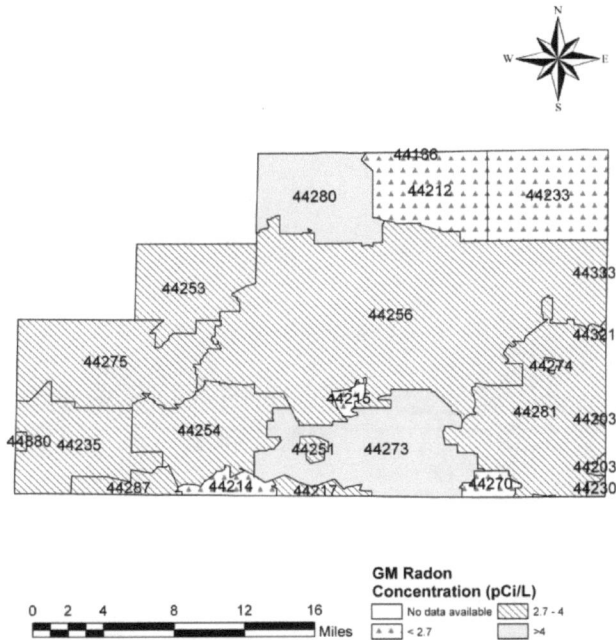

Figure 52a: GM ^{222}Rn concentrations in zip codes of Medina county based on the WHO-USEPA classification.

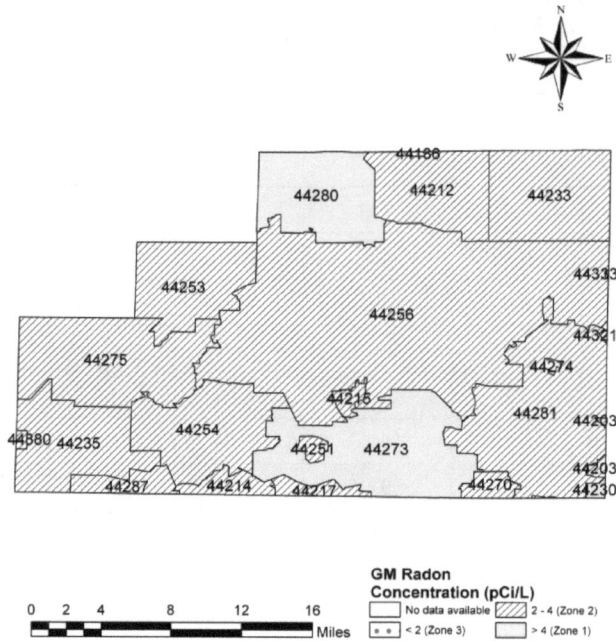

Figure 52b: GM ^{222}Rn concentrations in zip codes of Medina county based on the USEPA classification.

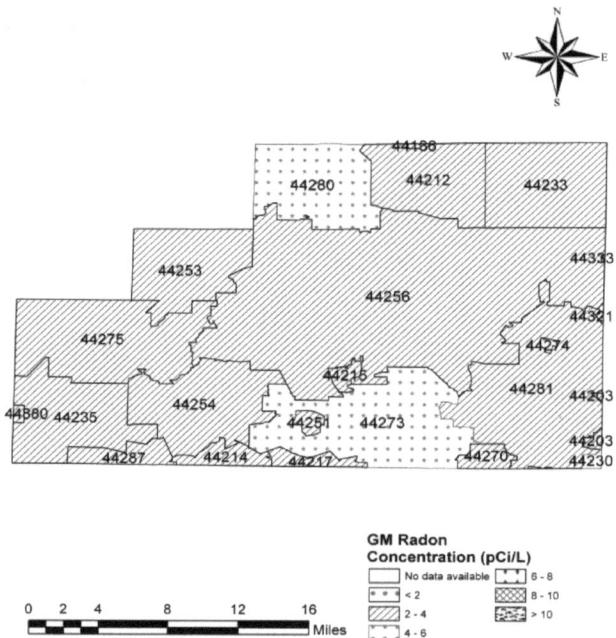

Figure 52c: GM ^{222}Rn concentrations in zip codes of Medina county based on the 2 pCi/L breakdown classification.

Table 53: ^{222}Rn statistics for Meigs county

Zip code	No.	Min.	Max.	AM	GM	SD	CV
45620*	5	1.6	4.1	2.78	2.65	0.92	0.05
45686*	9	0.5	3.7	1.59	1.41	0.88	0.05
45710*	53	0.3	12.4	3.18	2.06	3.11	0.16
45723*	11	0.3	4.2	2.24	1.66	1.4	0.07
45741	2	3.2	4.1	3.65	3.62	0.64	0.03
45743	2	1	2.8	1.9	1.67	1.27	0.07
45760	12	0.5	4	1.82	1.48	1.19	0.06
45769	29	0.1	4.2	1.92	1.32	1.33	0.07
45770	2	0.9	1	0.95	0.95	0.07	0
45771	13	0.7	6.3	2.6	2.19	1.58	0.08
45772	3	0.1	2.9	1.93	0.93	1.59	0.08
45775	3	1.1	1.7	1.43	1.41	0.31	0.02
45776	6	0.2	6.4	2.02	0.96	2.48	0.13
45779	4	0.3	3.9	1.45	0.9	1.67	0.09
45720**	1	3.8	3.8	3.8	3.8	0	0

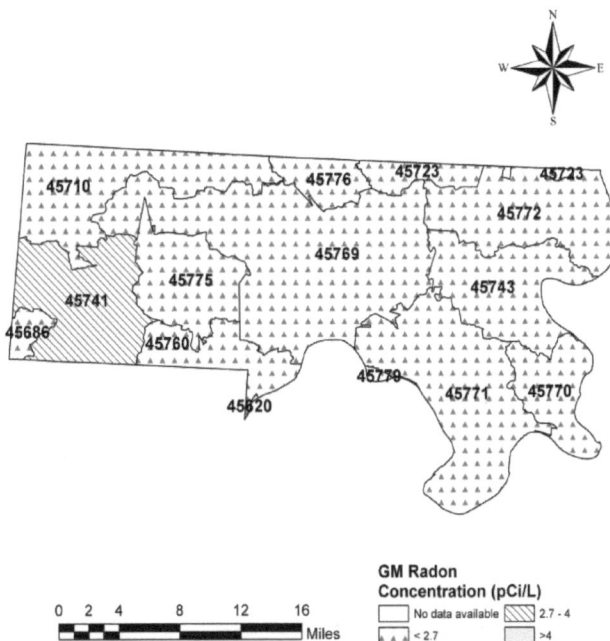

Figure 53a: GM ^{222}Rn concentrations in zip codes of Meigs county based on the WHO-USEPA classification.

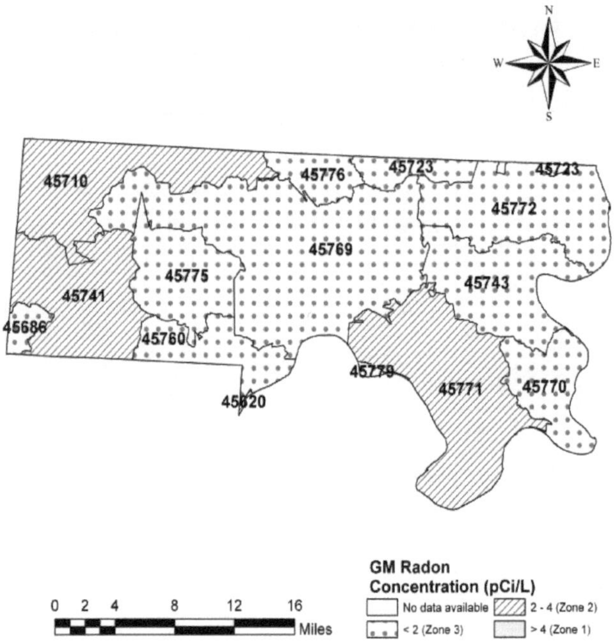

Figure 53b: GM ^{222}Rn concentrations in zip codes of Meigs county based on the USEPA classification.

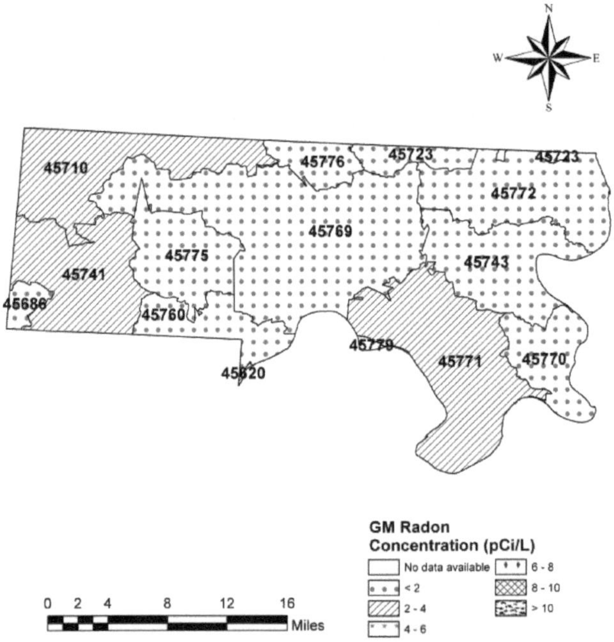

Figure 53c: GM ^{222}Rn concentrations in zip codes of Meigs county based on the 2 pCi/L breakdown classification.

Table 54: ^{222}Rn statistics for Mercer county

Zip code	No.	Min.	Max.	AM	GM	SD	CV
45310	8	0.7	13	5.96	4.16	4.33	0.22
45822	203	0.1	48.6	7.15	4.25	7.92	0.41
45826	1	5.8	5.8	5.8	5.8	0	0
45828	71	0.7	17.6	5.24	3.98	3.95	0.2
45846	42	0.1	38.5	5.88	3.37	7.2	0.37
45860	65	0.5	56.1	9.43	5.43	12.03	0.62
45862	7	0.4	16.1	5.23	2.33	6.48	0.33
45869*	71	0.4	24.9	6.51	4.18	6.03	0.31
45882	17	0.1	13.3	3.48	1.83	4.14	0.21
45883	59	0.1	53	7.34	4.27	8.33	0.43
45885*	75	0.1	31.1	6.12	3.35	6.65	0.34
45887*	14	0.5	14.3	4.41	2.69	4.43	0.23
45898*	8	1.8	7.2	4.61	4.08	2.26	0.12

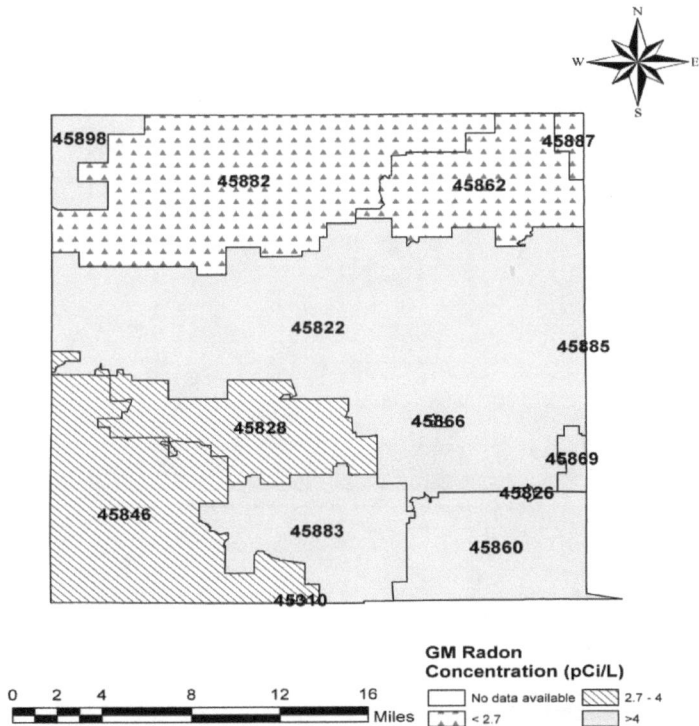

Figure 54a: GM ^{222}Rn concentrations in zip codes of Mercer county based on the WHO-USEPA classification.

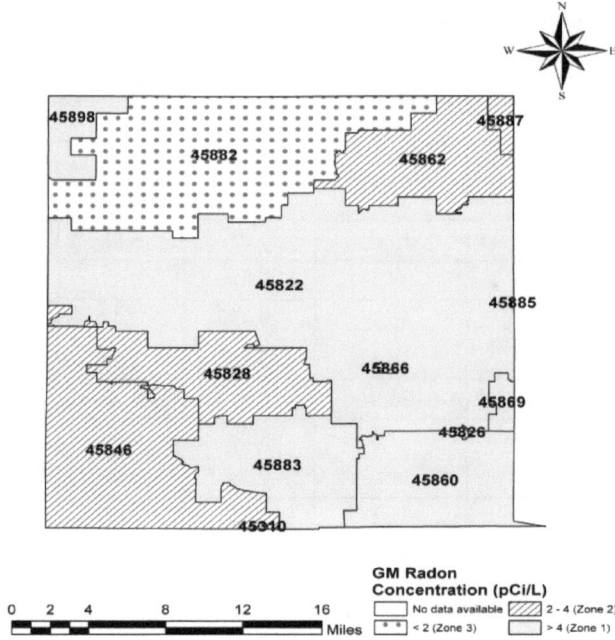

Figure 54b: GM ^{222}Rn concentrations in zip codes of Mercer county based on the USEPA classification.

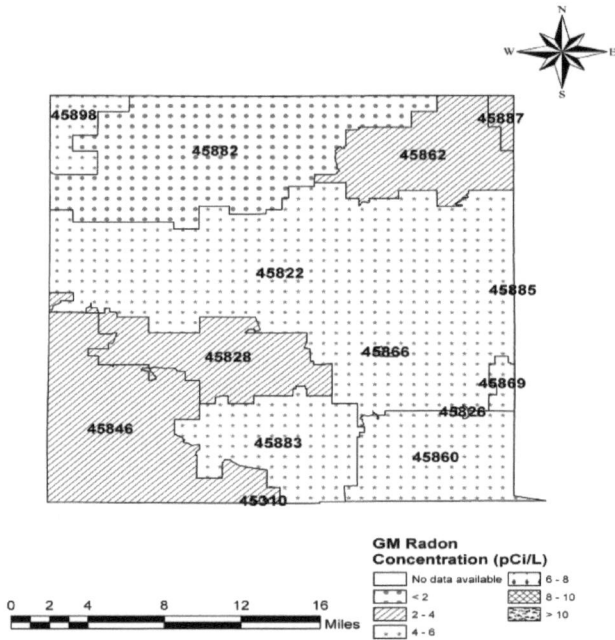

Figure 54c: GM ^{222}Rn concentrations in zip codes of Mercer county based on the 2 pCi/L breakdown classification.

Table 55: ^{222}Rn statistics for Miami county

Zip code	No.	Min.	Max.	AM	GM	SD	CV
45304[*]	136	0.1	49	7.09	3.95	7.55	0.39
45308[*]	66	0.1	71	8.43	4.69	12.25	0.63
45312	43	0.1	59.2	9.64	6.01	10.29	0.53
45317[*]	16	0.2	19.6	8.84	6.2	5.7	0.29
45318	88	0.1	252	11.42	5.54	27.38	1.4
45322[*]	655	0.1	69.6	6.06	3.42	7.32	0.37
45326	27	0.1	20.1	6.49	3.7	5.56	0.28
45337	49	0.2	18.1	6.04	4.54	4.44	0.23
45339	55	0.5	40.4	11.1	6.91	10.84	0.55
45344[*]	440	0.1	150.1	11.3	5.75	16.58	0.85
45356	454	0.1	160	8.59	5.09	12.29	0.63
45359	62	0.5	105	12.71	5.96	21.84	1.12
45361	2	6.3	6.3	6.3	6.3	0	0
45371	787	0.1	204	7.69	4.54	11.83	0.61
45373	1402	0.1	223	8.44	5.09	10.72	0.55
45383	167	0.1	62.9	8.16	4.75	10.13	0.52
45424[*]	1187	0.1	83.2	5.16	2.89	6.83	0.35
45374[**]	1	9.2	9.2	9.2	9.2	0	0

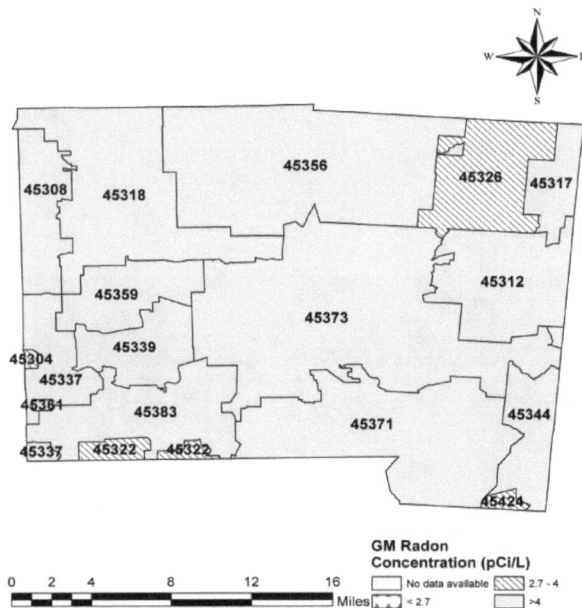

Figure 55a: GM ^{222}Rn concentrations in zip codes of Miami county based on the WHO-USEPA classification.

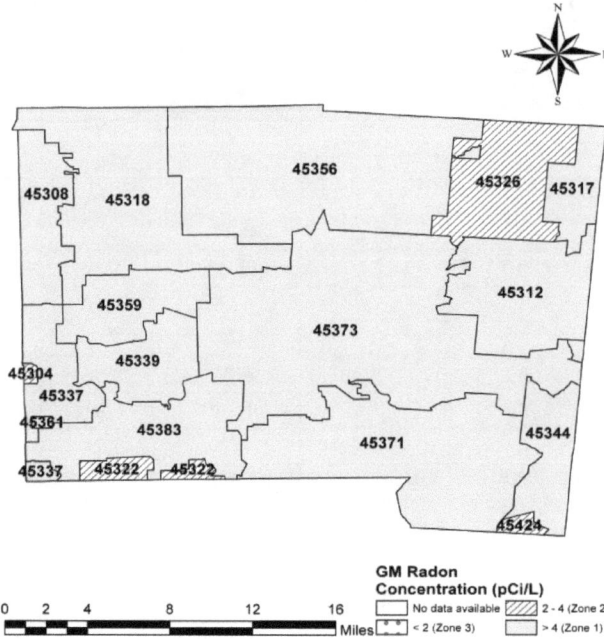

Figure 55b: GM ^{222}Rn concentrations in zip codes of Miami county based on the USEPA classification.

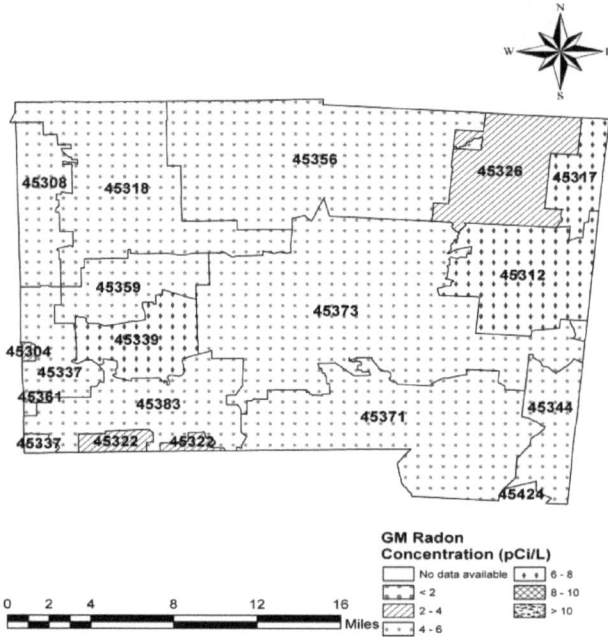

Figure 55c: GM ^{222}Rn concentrations in zip codes of Miami county based on the 2 pCi/L breakdown classification.

Table 56: ^{222}Rn statistics for Monroe county

Zip code	No.	Min.	Max.	AM	GM	SD	CV
43716	39	0.6	51.2	8.27	4.27	10.97	0.56
43747	5	0.4	6.1	3.74	2.11	3.01	0.15
43754	4	0.1	3.9	1.73	0.9	1.67	0.09
43773*	5	0.6	9.2	3.28	2.21	3.42	0.18
43788*	2	1	1.3	1.15	1.14	0.21	0.01
43793	24	0.1	8.6	3.33	2.38	2.36	0.12
43915	9	0.1	39.3	8.21	3.28	12.19	0.62
43931	1	0.8	0.8	0.8	0.8	0	0
43942*	18	0.5	15.3	5.57	3.06	5.32	0.27
43946	2	0.9	1.7	1.3	1.24	0.57	0.03
45734	2	1.5	1.7	1.6	1.6	0.14	0.01
45745*	5	1.1	3.9	2	1.74	1.23	0.06
45767	10	0.3	9.35	2.19	1.38	2.67	0.14
45789	1	2.9	2.9	2.9	2.9	0	0
43752**	1	1.1	1.1	1.1	1.1	0	0

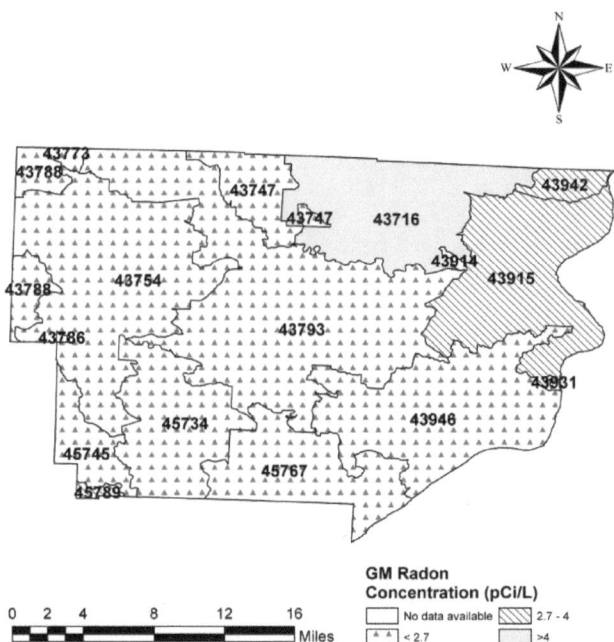

Figure 56a: GM ^{222}Rn concentrations in zip codes of Monroe county based on the WHO-USEPA classification.

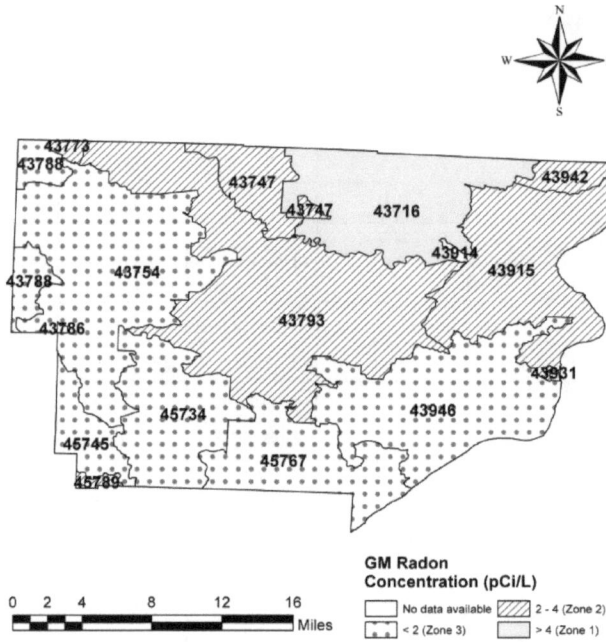

Figure 56b: GM [222]Rn concentrations in zip codes of Monroe county based on the USEPA classification.

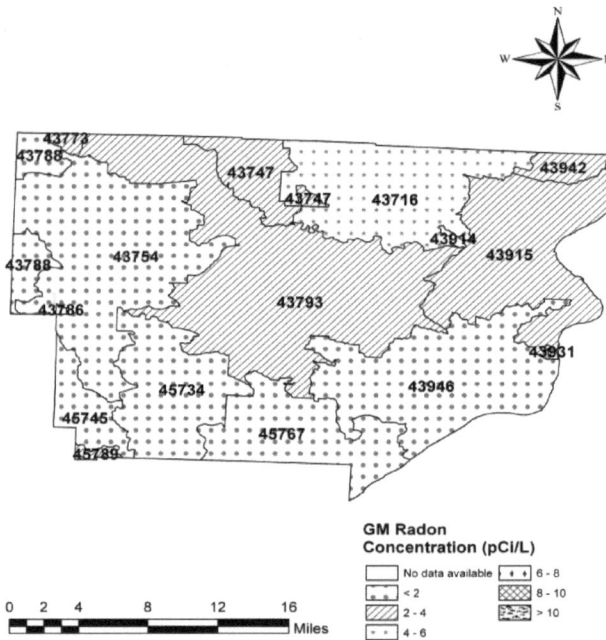

Figure 56c: GM [222]Rn concentrations in zip codes of Monroe county based on the 2 pCi/L breakdown classification.

Table 57: ^{222}Rn statistics for Montgomery county

Zip code	No.	Min.	Max.	AM	GM	SD	CV	Zip code	No.	Min.	Max.	AM	GM	SD	CV
45005[*]	580	0.1	70.7	6.82	3.56	8.76	0.45	45418	120	0.1	21.5	3.38	2.09	3.93	0.2
45066[*]	1044	0.1	58.1	4.5	2.89	4.82	0.25	45419	1394	0.1	220.1	5.49	3.28	10.18	0.52
45309	428	0.1	267.8	7.08	3.65	14.87	0.76	45420	1100	0.1	76.5	4.87	3.06	5.91	0.3
45315	183	0.1	67.3	5.36	3.38	6.5	0.33	45424	1187	0.1	83.2	5.16	2.89	6.83	0.35
45322	655	0.1	69.6	6.06	3.42	7.32	0.37	45426	427	0.1	161	5.72	2.97	12.79	0.65
45325	62	0.5	27.7	6.78	4.54	6.61	0.34	45428	19	0.1	4.2	1.77	1.16	1.3	0.07
45327	233	0.1	86.1	9.23	5.43	11.52	0.59	45429	2059	0.1	440.7	5.53	3.3	11.77	0.6
45342	926	0.1	52.7	4.21	2.52	4.91	0.25	45431[*]	975	0.1	117	5.65	3.5	6.94	0.36
45344[*]	440	0.1	150.1	11.3	5.75	16.58	0.85	45432[*]	864	0.1	117	5.85	3.7	6.68	0.34
45345	145	0.1	40.1	5.92	3.86	5.99	0.31	45433[*]	31	0.1	268.8	44.31	3.62	83.01	4.25
45354	22	0.1	18.3	3.96	1.13	5.58	0.29	45439	183	0.1	56.1	8.34	5.07	8.23	0.42
45371[*]	787	0.1	204	7.69	4.54	11.83	0.61	45440	1520	0.1	359	5.35	3.2	12.18	0.62
45377	538	0.1	40.8	4.58	2.9	4.38	0.22	45449	412	0.1	51.6	5.12	2.76	6.49	0.33
45381[*]	123	0.1	90.3	8.12	4.28	11.83	0.61	45458	1266	0.1	35.5	3.82	2.27	3.94	0.2
45402	127	0.1	84.8	6.57	2.87	10.23	0.52	45459	2159	0.1	153	5.56	3.22	7.93	0.41
45403	248	0.1	65	5.05	3.31	6.06	0.31	45401[**]	97	0.1	36.5	11.63	6.07	9.59	0.49
45404	134	0.2	48.2	8.74	6.05	7.92	0.41	45413[**]	4	2.7	11.4	6.38	5.62	3.64	0.19
45405	462	0.1	49.6	5.78	3.56	5.78	0.3	45422[**]	8	0.1	10.9	4.05	1.82	3.55	0.18
45406	441	0.1	32.9	3.69	2.44	3.7	0.19	45423[**]	4	0.3	16.9	5.58	1.92	7.79	0.4
45409	462	0.1	39.9	5.97	3.57	5.73	0.29	45437[**]	6	0.9	11.8	3.77	2.52	4.07	0.21
45410	374	0.1	35	3.79	2.55	3.79	0.19	45441[**]	5	0.3	13.5	5.36	1.65	6.77	0.35
45414	720	0.1	410	7.41	4.06	16.99	0.87	45448[**]	14	0.5	6.1	1.8	1.26	1.9	0.1
45415	727	0.1	416.3	7.65	4.35	17.3	0.89	45469[**]	6	0.9	7.3	3.1	2.58	2.18	0.11
45416	152	0.1	61.2	5.5	2.69	8.82	0.45	45475[**]	2	4.3	15.5	9.9	8.16	7.92	0.41
45417	95	0.1	60.1	5.46	2.28	8.82	0.45	45490[**]	3	1	2.1	1.37	1.28	0.64	0.03

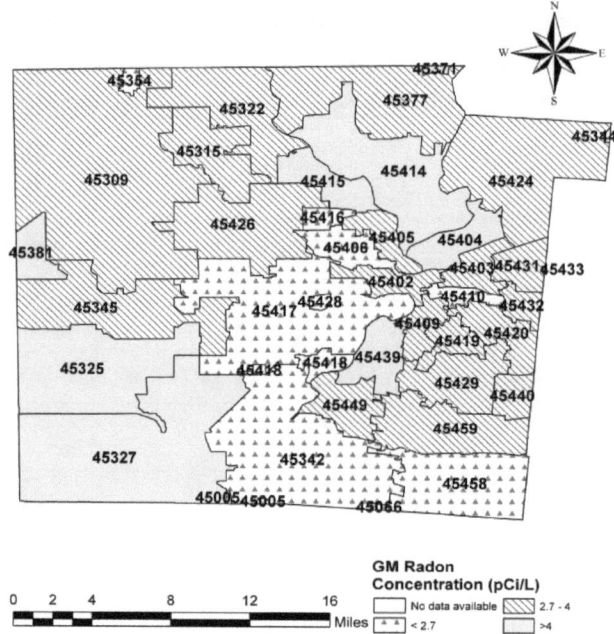

Figure 57a: GM ^{222}Rn concentrations in zip codes of Montgomery county based on the WHO-USEPA classification.

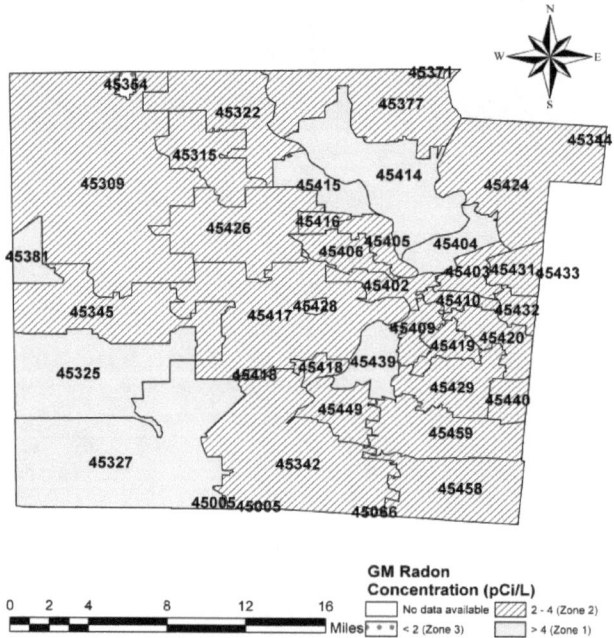

Figure 57b: GM ^{222}Rn concentrations in zip codes of Montgomery county based on the USEPA classification.

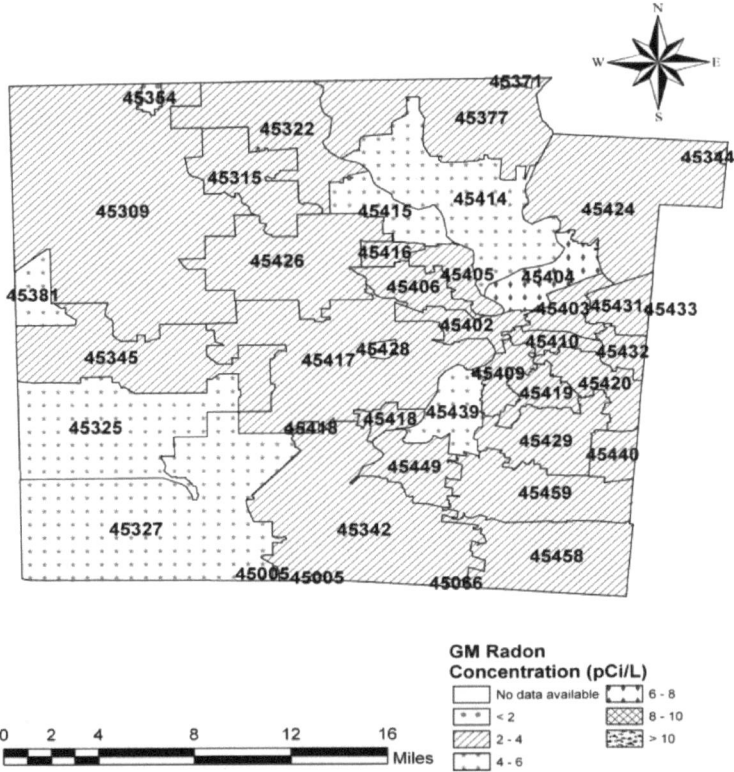

Figure 57c: GM ^{222}Rn concentrations in zip codes of Montgomery county based on the 2 pCi/L breakdown classification.

Table 58: ^{222}Rn statistics for Morgan county

Zip code	No.	Min.	Max.	AM	GM	SD	CV
43724*	32	0.1	10.3	2.79	1.68	2.51	0.13
43728	3	1.7	5	2.9	2.57	1.82	0.09
43731*	15	0.6	6.5	3.37	2.82	1.86	0.1
43756	24	0.1	12.4	3.33	2.44	2.57	0.13
43758	12	0.2	11.6	4.67	3.07	3.74	0.19
43787	11	0.7	33.9	7.81	4.02	10.14	0.52
45711*	12	0.4	4.5	1.96	1.36	1.67	0.09
45715*	12	2.2	27	6.94	5.45	6.62	0.34
45732*	17	0.1	7.9	3.21	2	2.69	0.14
45786*	17	0.4	19.6	3.84	2.48	4.71	0.24

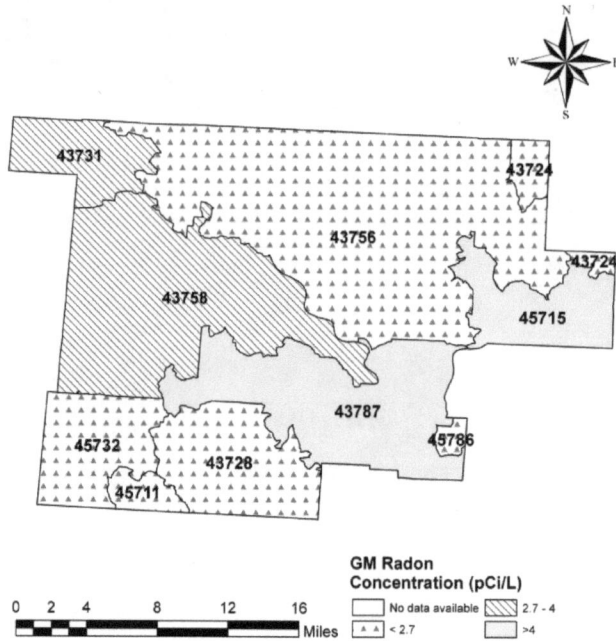

Figure 58a: GM ^{222}Rn concentrations in zip codes of Morgan county based on the WHO-USEPA classification.

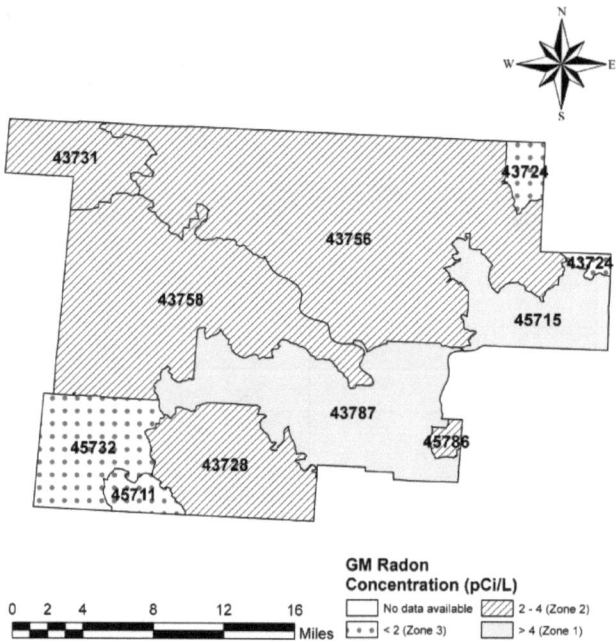

Figure 58b: GM ^{222}Rn concentrations in zip codes of Morgan county based on the USEPA classification.

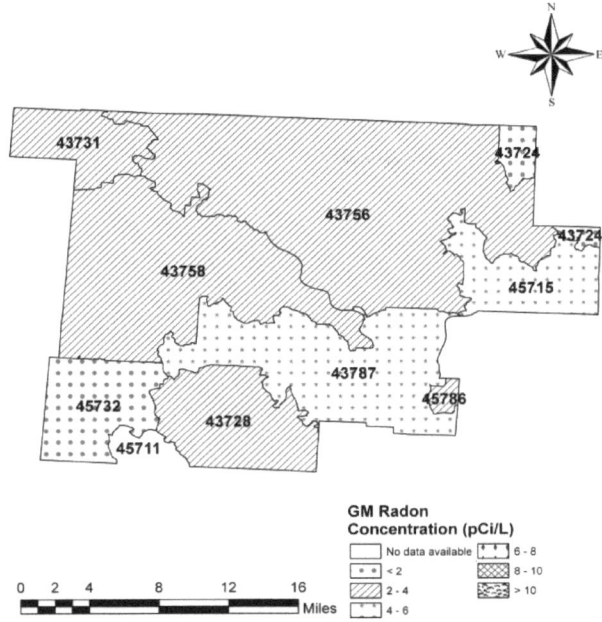

Figure 58c: GM ^{222}Rn concentrations in zip codes of Morgan county based on the 2 pCi/L breakdown classification.

Table 59: ^{222}Rn statistics for Morrow county

Zip code	No.	Min.	Max.	AM	GM	SD	CV
43003*	44	0.1	50.1	6.1	2.83	9.28	0.48
43011*	154	0.1	61.7	8.61	5.53	9.87	0.51
43019*	410	0.1	189.5	9.45	3.38	19.71	1.01
43050*	1677	0.1	742.7	14.53	5.48	32.72	1.68
43314*	83	0.1	26.3	4.65	3.38	4.12	0.21
43315	116	0.1	77.2	7.48	3.85	12.18	0.62
43317	4	1.4	14.8	9.68	7.13	5.87	0.3
43320	24	0.6	32.4	6.84	3.68	8.43	0.43
43321	8	0.2	14.6	4.96	2.25	5.1	0.26
43334	94	0.1	62.9	9.01	4.68	12.88	0.66
43338	168	0.1	53.1	6.52	3.51	8.51	0.44
43356*	33	0.7	28.8	6.16	3.94	6.15	0.32
44813*	209	0.2	318.8	20.72	8.99	38.36	1.96
44833*	493	0.1	140.2	7.75	4.41	12.72	0.65
44903*	494	0.1	150.4	8.02	4.43	12.43	0.64
44904*	798	0.1	292.3	9.75	4.76	17.45	0.89
43325**	2	4.9	6.4	5.65	5.6	1.06	0.05
43349**	6	0.7	12.9	4.93	3.11	4.93	0.25
43350**	2	5.1	10.7	7.9	7.39	3.96	0.2

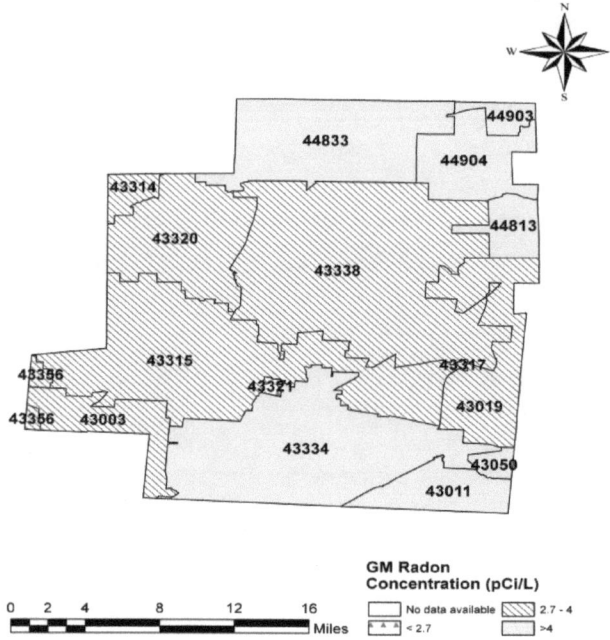

Figure 59a: GM ^{222}Rn concentrations in zip codes of Morrow county based on the WHO-USEPA classification.

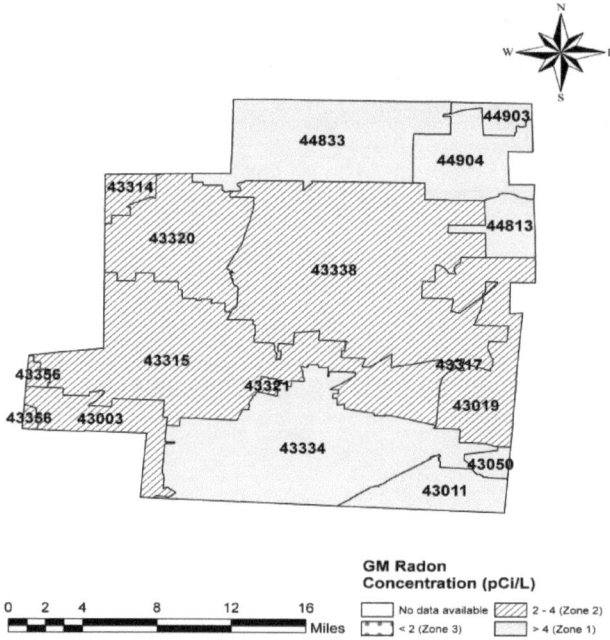

Figure 59b: GM ^{222}Rn concentrations in zip codes of Morrow county based on the USEPA classification.

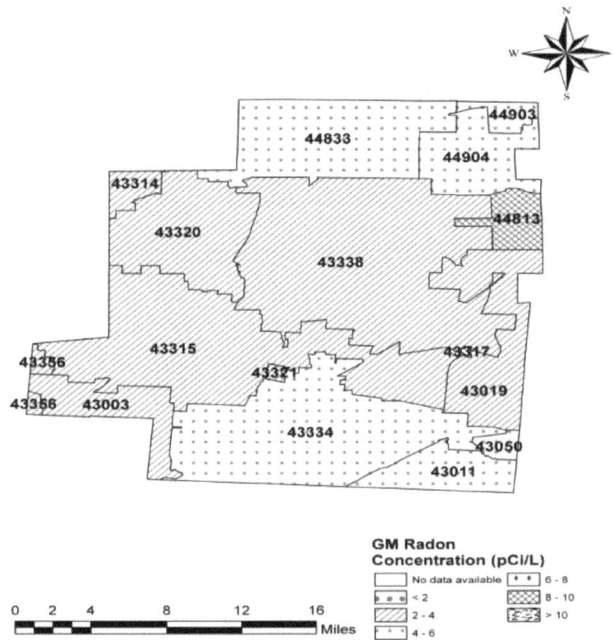

Figure 59c: GM ^{222}Rn concentrations in zip codes of Morrow county based on the 2 pCi/L breakdown classification.

Table 60: ^{222}Rn statistics for Muskingum county

Zip code	No.	Min.	Max.	AM	GM	SD	CV
43701	584	0.1	118.3	6.54	3.73	10.79	0.55
43720	4	1.1	5.3	3.7	3.15	1.95	0.1
43727	10	0.5	7.1	3.68	2.67	2.6	0.13
43732*	13	0.1	8.6	2.93	1.66	3.04	0.16
43734	8	0.8	13.9	4.96	3.38	4.74	0.24
43735	5	2.6	14.2	8.82	7.47	5	0.26
43738	2	1.5	11.7	6.6	4.19	7.21	0.37
43740*	4	12.2	12.3	12.25	12.25	0.06	0
43746	7	1.3	27	6.43	3.58	9.23	0.47
43760*	10	0.5	3.1	1.89	1.59	0.97	0.05
43762	112	0.1	129	6.7	3.21	13.62	0.7
43767	12	1.2	58.5	12.53	5.21	19.61	1
43771	8	0.2	15.4	6.56	3.35	6.33	0.32

Zip code	No.	Min.	Max.	AM	GM	SD	CV
43777	5	0.6	7.5	4.8	3.56	2.94	0.15
43802	4	0.8	6.9	3.4	2.62	2.57	0.13
43812[*]	180	0.1	95.6	11.78	5.17	18.64	0.95
43821	30	0.2	14.6	4.16	2.9	3.15	0.16
43822	84	0.2	131.1	11.75	5.7	18.43	0.94
43830	129	0.2	129.7	9.2	4	17.16	0.88
43842	1	1.9	1.9	1.9	1.9	0	0
43702[**]	13	0.2	19.1	4.56	2.96	4.78	0.24

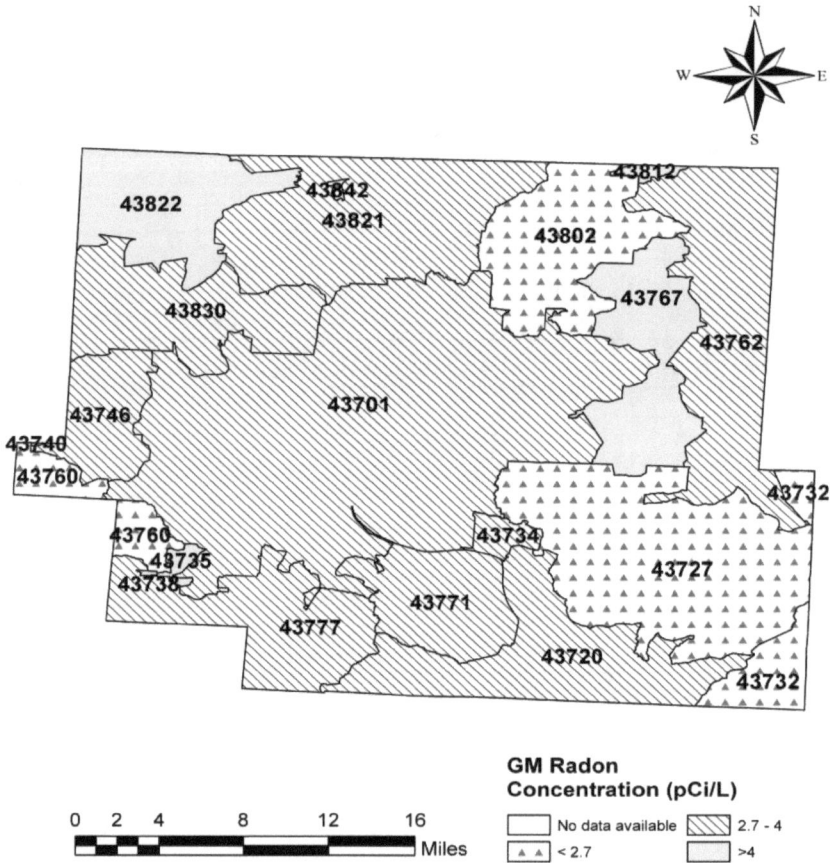

Figure 60a: GM ^{222}Rn concentrations in zip codes of Muskingum county based on the WHO-USEPA classification.

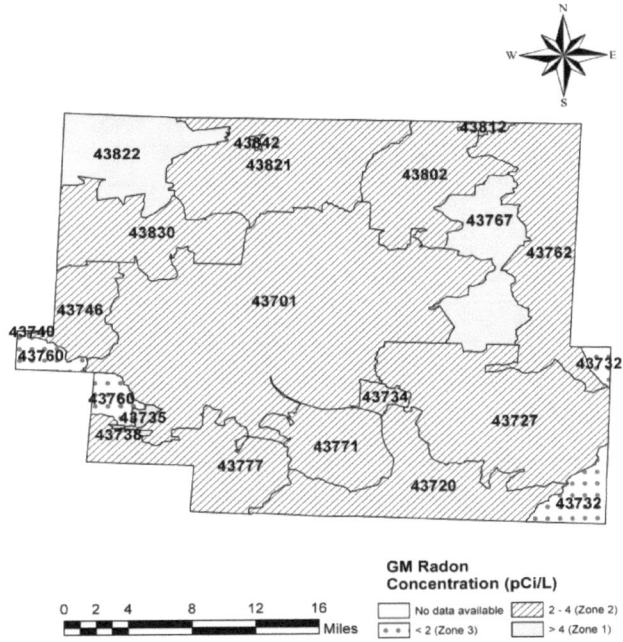

Figure 60b: GM ^{222}Rn concentrations in zip codes of Muskingum county based on the USEPA classification.

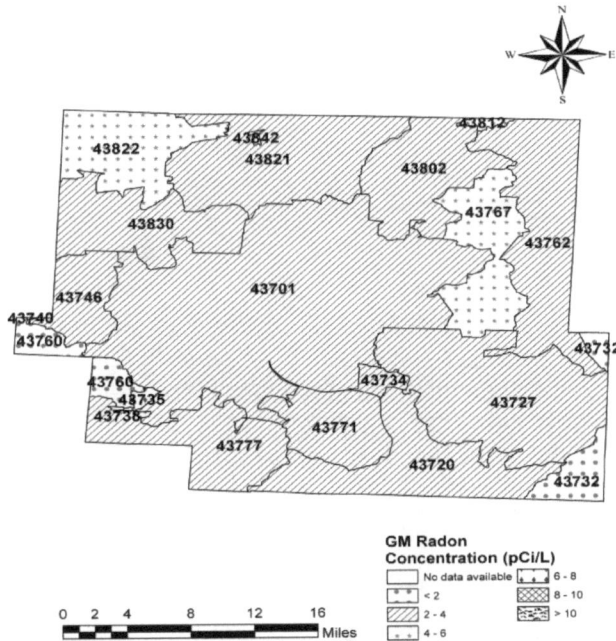

Figure 60c: GM ^{222}Rn concentrations in zip codes of Muskingum county based on the 2 pCi/L breakdown classification.

Table 61: ^{222}Rn statistics for Noble county

Zip code	No.	Min.	Max.	AM	GM	SD	CV
43724	32	0.1	10.3	2.79	1.68	2.51	0.13
43732*	13	0.1	8.6	2.93	1.66	3.04	0.16
43772*	13	0.1	8.4	2.91	1.37	3.02	0.15
43773*	5	0.6	9.2	3.28	2.21	3.42	0.18
43778*	3	1.7	5.9	3.7	3.27	2.11	0.11
43779	2	1.5	2.7	2.1	2.01	0.85	0.04
43780*	9	0.5	4.6	1.58	1.04	1.73	0.09
43788	2	1	1.3	1.15	1.14	0.21	0.01
45715*	12	2.2	27	6.94	5.45	6.62	0.34
45727	3	0.7	3.4	1.7	1.34	1.48	0.08
45744*	25	1	10.7	3.89	3	3.03	0.16
45745*	5	1.1	3.9	2	1.74	1.23	0.06
45746*	3	0.5	3.4	1.7	1.27	1.51	0.08

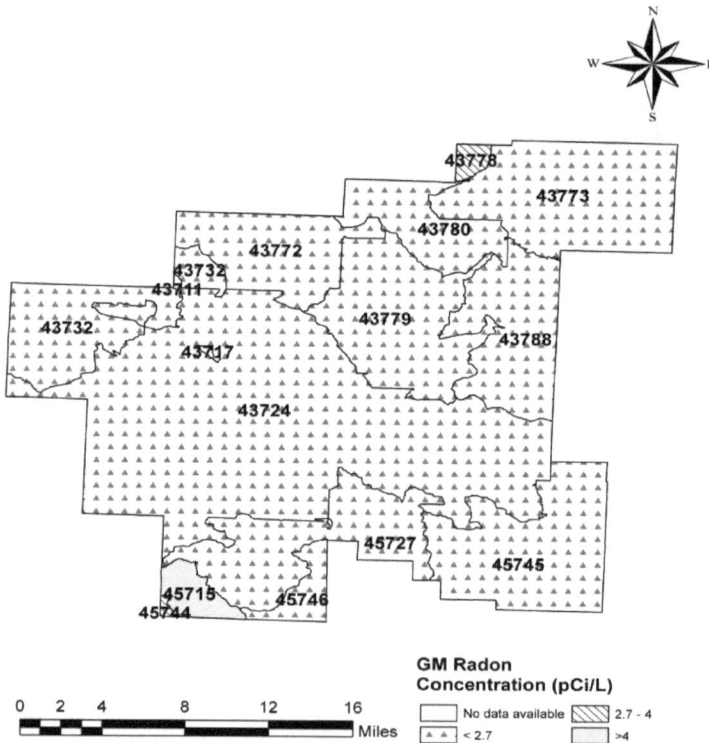

Figure 61a: GM ^{222}Rn concentrations in zip codes of Noble county based on the WHO-USEPA classification.

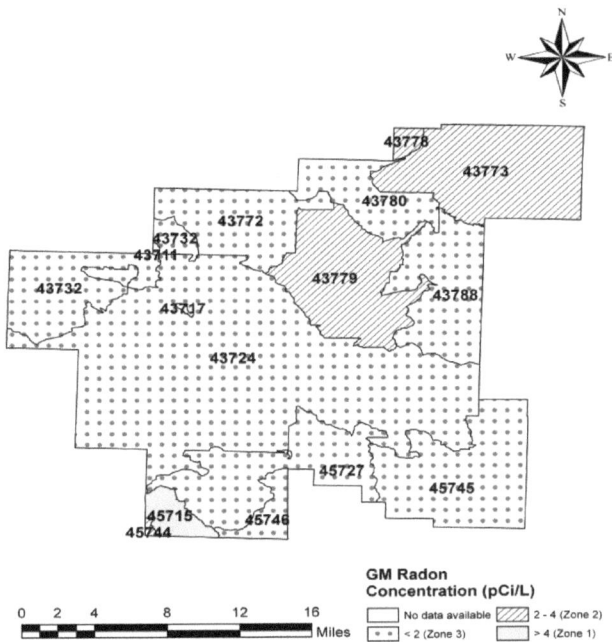

Figure 61b: GM ^{222}Rn concentrations in zip codes of Noble county based on the USEPA classification.

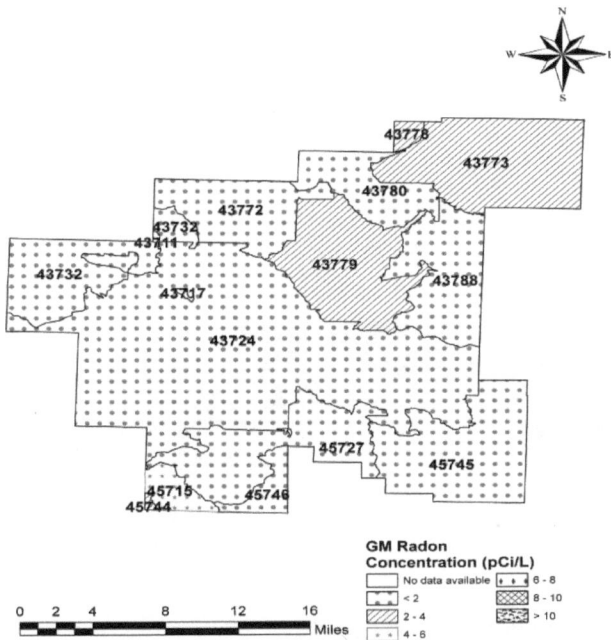

Figure 61c: GM ^{222}Rn concentrations in zip codes of Noble county based on the 2 pCi/L breakdown classification.

Table 62: ^{222}Rn statistics for Ottawa county

Zip code	No.	Min.	Max.	AM	GM	SD	CV
43412*	30	0.1	260	13.3	3.4	46.76	2.39
43416	23	1.1	17.2	5.96	4.75	4.07	0.21
43430	54	0.4	27.8	4.84	3.22	4.99	0.26
43432	13	0.8	6.5	2.08	1.65	1.67	0.09
43439	1	4.5	4.5	4.5	4.5	0	0
43440	84	0.1	30.2	5.13	2.74	5.52	0.28
43445	10	0.1	7.8	3.06	1.93	2.31	0.12
43446	2	12.4	12.8	12.6	12.6	0.28	0.01
43447*	42	0.2	17.2	4.98	2.96	4.8	0.25
43449	59	0.1	14.1	3.08	2	3	0.15
43452	265	0.1	320	5.44	2.43	20.18	1.03
43468	5	2.2	14.8	6.14	4.95	4.99	0.26
43469*	28	0.1	46.1	6.34	3.65	8.48	0.43

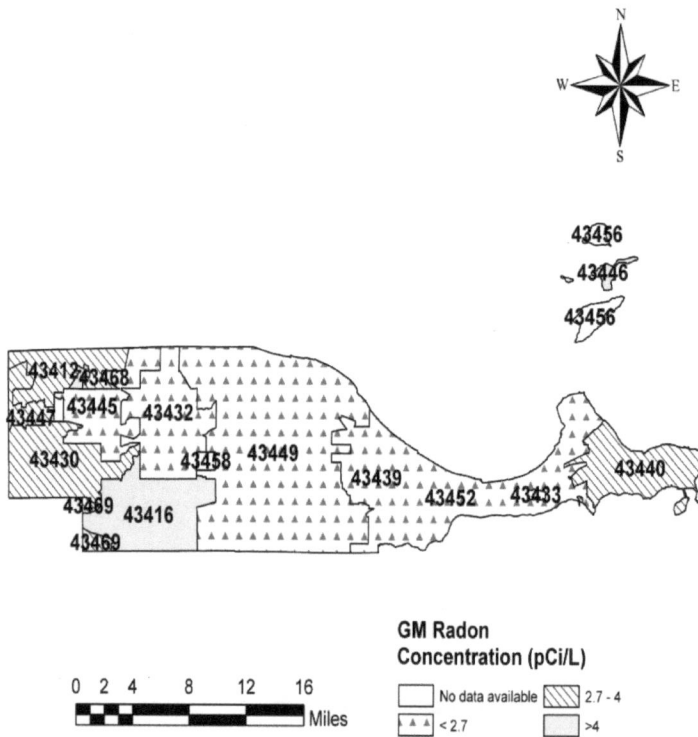

Figure 62a: GM ^{222}Rn concentrations in zip codes of Ottawa county based on the WHO-USEPA classification.

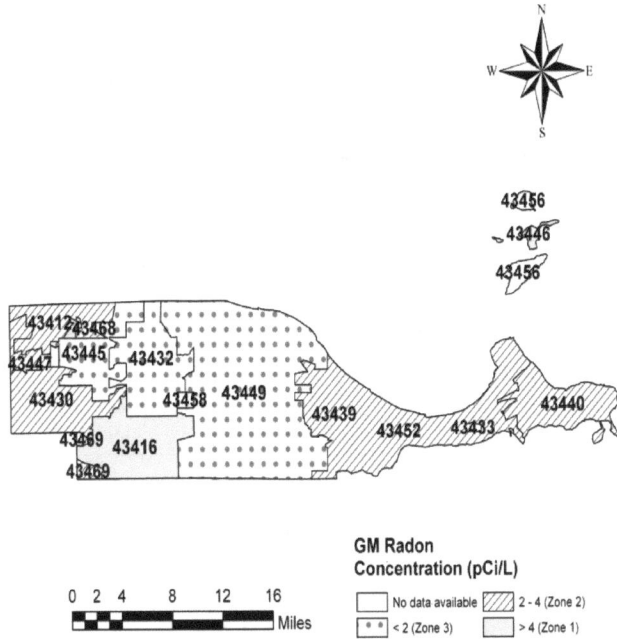

Figure 62b: GM ^{222}Rn concentrations in zip codes of Ottawa county based on the USEPA classification.

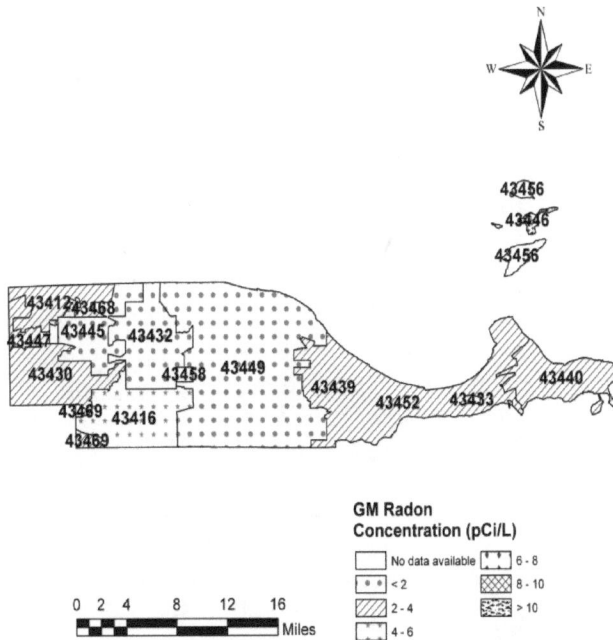

Figure 62c: GM ^{222}Rn concentrations in zip codes of Ottawa county based on the 2 pCi/L breakdown classification.

Table 63: ^{222}Rn statistics for Paulding county

Zip code	No.	Min.	Max.	AM	GM	SD	CV
43512*	211	0.1	21.2	3.26	2.12	3.24	0.17
43526*	12	0.1	5.6	3.32	2.52	1.62	0.08
45813	15	1	7.6	3.17	2.64	1.99	0.1
45821	1	3	3	3	3	0	0
45827*	10	0.4	11.7	4.77	2.71	4.25	0.22
45832*	10	0.5	11.3	4.84	2.89	4.27	0.22
45849	4	0.3	0.8	0.5	0.47	0.22	0.01
45851	3	0.1	3	1.97	0.94	1.62	0.08
45855	1	4	4	4	4	0	0
45861	1	0.5	0.5	0.5	0.5	0	0
45873	5	0.6	3.1	1.74	1.38	1.25	0.06
45879	31	0.1	13.2	5.03	2.91	4.53	0.23
45880	11	0.3	9.5	2.34	1.52	2.63	0.13
45886*	1	0.9	0.9	0.9	0.9	0	0

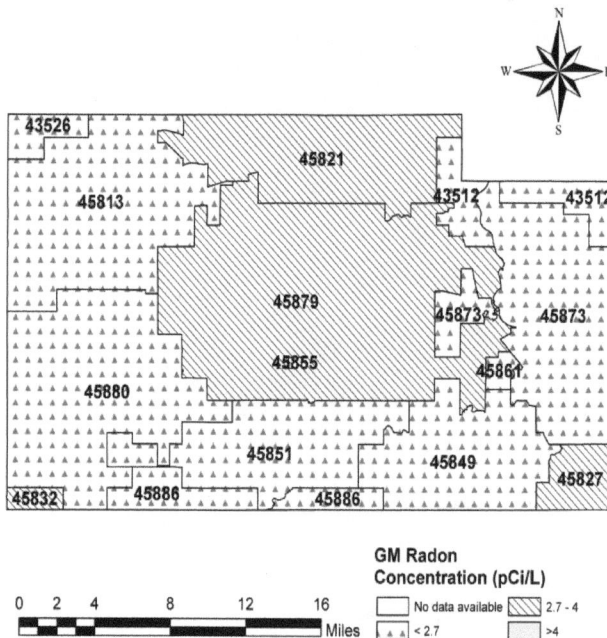

Figure 63a: GM ^{222}Rn concentrations in zip codes of Paulding county based on the WHO-USEPA classification.

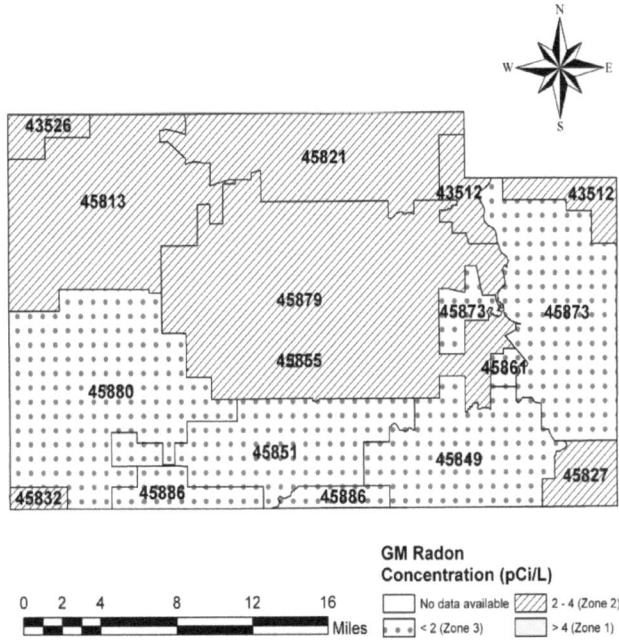

Figure 63b: GM ^{222}Rn concentrations in zip codes of Paulding county based on the USEPA classification.

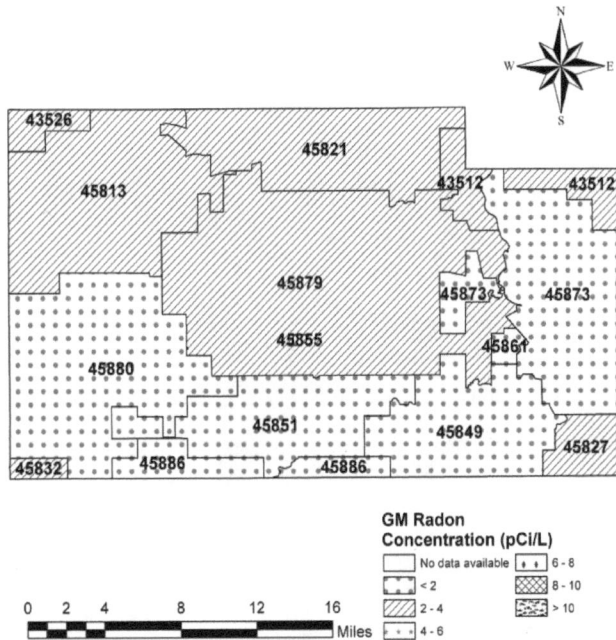

Figure 63c: GM ^{222}Rn concentrations in zip codes of Paulding county based on the 2 pCi/L breakdown classification.

Table 64: ^{222}Rn statistics for Perry county

Zip code	No.	Min.	Max.	AM	GM	SD	CV
43076	168	0.2	274.5	13.41	4.4	36.3	1.86
43107*	32	0.8	260	16.24	5.71	46.2	2.37
43138*	169	0.1	99.5	6.96	3.63	9.68	0.5
43148*	24	0.1	45.7	8.63	5.28	9.74	0.5
43150*	6	0.3	5.9	2.45	1.49	2.39	0.12
43730	8	0.6	15.3	4.09	2.3	4.88	0.25
43731	15	0.6	6.5	3.37	2.82	1.86	0.1
43739	19	2	19.4	7.67	5.78	5.56	0.28
43748	20	0.1	11.1	2.66	1.68	2.51	0.13
43760	10	0.5	3.1	1.89	1.59	0.97	0.05
43761	2	1.3	3.8	2.55	2.22	1.77	0.09
43764	43	0.8	15.8	4.95	3.73	3.8	0.19
43766	4	1.3	2.4	1.8	1.75	0.47	0.02
43777*	5	0.6	7.5	4.8	3.56	2.94	0.15
43782	3	0.6	9.1	6.27	3.68	4.91	0.25
43783	25	1	42	6.28	4.12	8.35	0.43
45732*	17	0.1	7.9	3.21	2	2.69	0.14

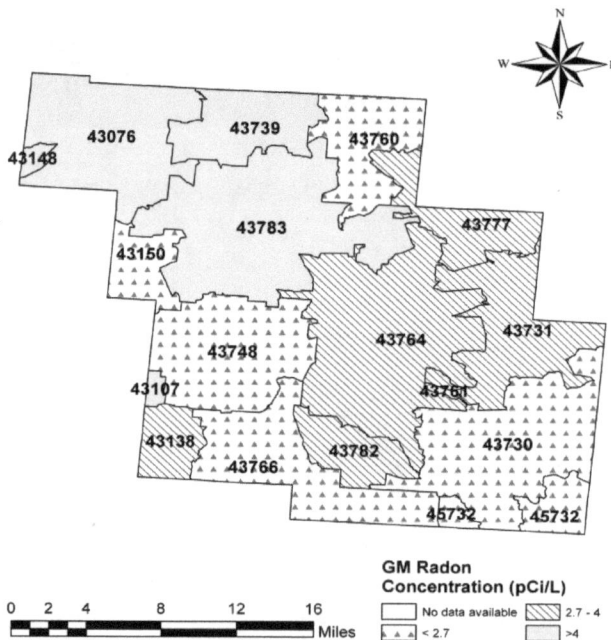

Figure 64a: GM ^{222}Rn concentrations in zip codes of Perry county based on the WHO-USEPA classification.

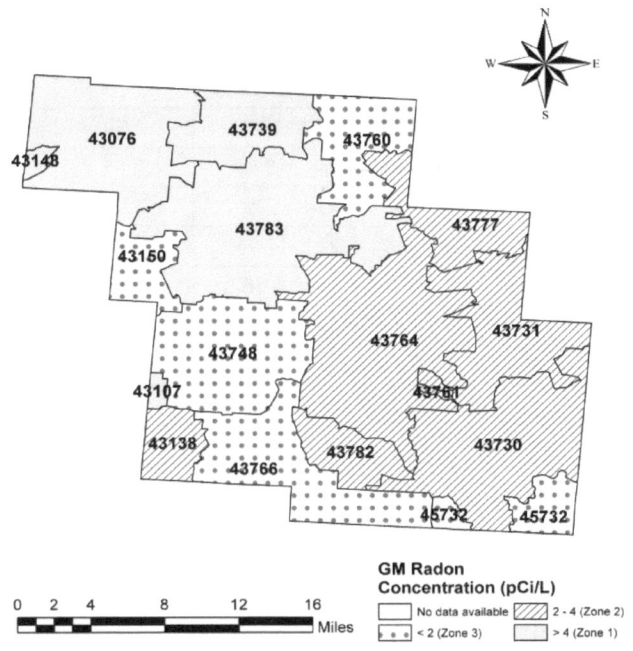

Figure 64b: GM ^{222}Rn concentrations in zip codes of Perry county based on the USEPA classification.

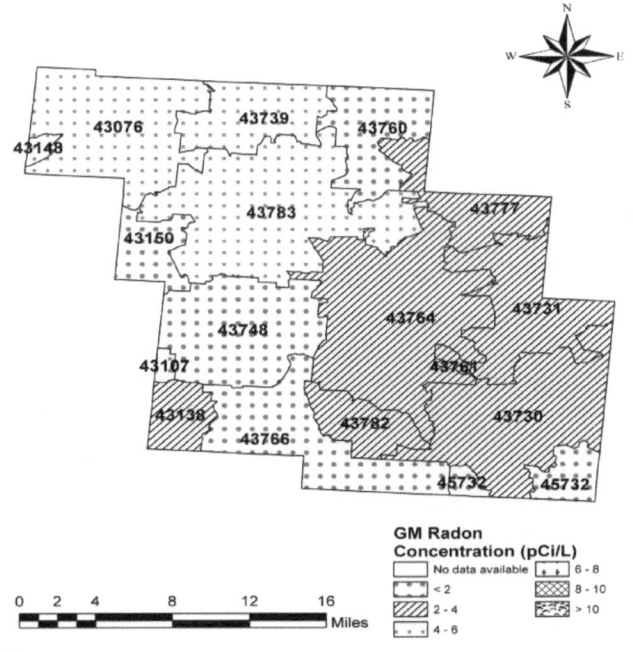

Figure 64c: GM ^{222}Rn concentrations in zip codes of Perry county based on the 2 pCi/L breakdown classification.

Table 65: ^{222}Rn statistics for Pickaway county

Zip code	No.	Min.	Max.	AM	GM	SD	CV
43102*	18	0.2	33.4	8.13	3.79	9.88	0.51
43103	159	0.1	102.5	15.65	8.55	18.32	0.94
43110*	438	0.1	112	10.56	6.45	11.49	0.59
43113	342	0.2	86.6	11.96	7.89	12.53	0.64
43115*	28	0.1	16.1	4.16	2.58	3.78	0.19
43116	10	3.3	12.5	7.39	6.75	3.16	0.16
43117	5	2.5	8.6	5.58	5.15	2.31	0.12
43125*	253	0.1	94.8	13.06	8.2	13.75	0.7
43135*	26	0.8	93.2	11.76	6.68	18.43	0.94
43137*	21	0.9	100.5	24.05	13.41	27.44	1.41
43143*	58	0.1	28.5	5.21	3.37	4.88	0.25
43145	4	2.9	41.1	16.88	11.32	16.68	0.85
43146	143	0.5	43.1	9.05	6.17	8.39	0.43
43154*	20	0.3	19.1	7.31	4.86	5.79	0.3
43164	9	2.3	45.8	9.8	5.82	13.87	0.71
45644*	25	0.6	27.7	7	4.47	7.06	0.36

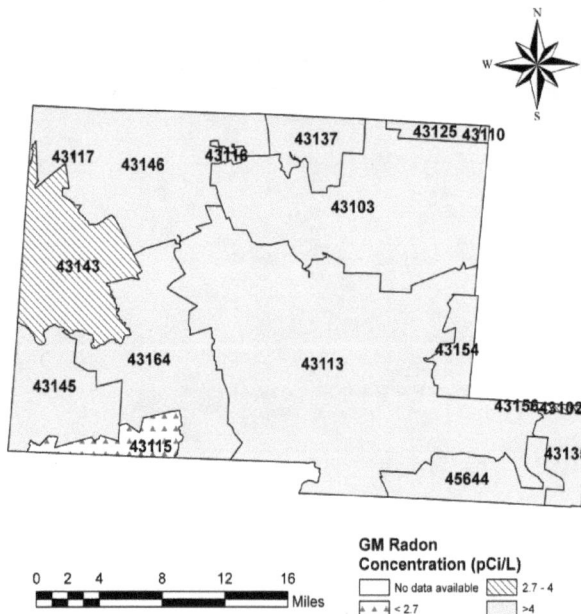

Figure 65a: GM ^{222}Rn concentrations in zip codes of Pickaway county based on the WHO-USEPA classification.

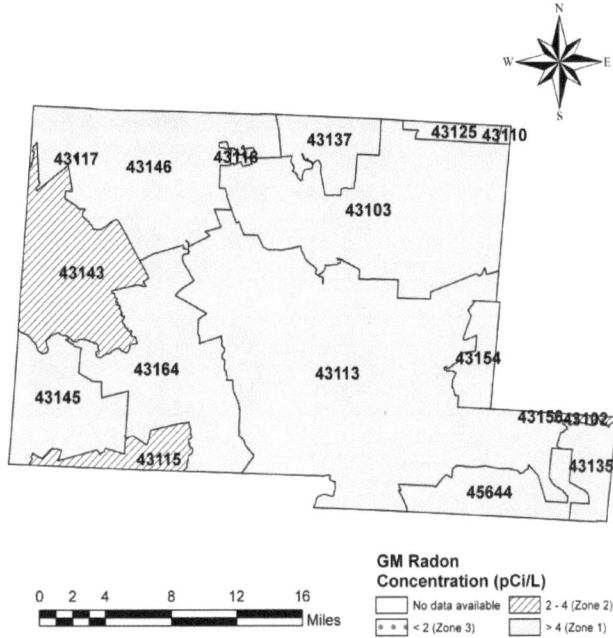

Figure 65b: GM ^{222}Rn concentrations in zip codes of Pickaway county based on the USEPA classification.

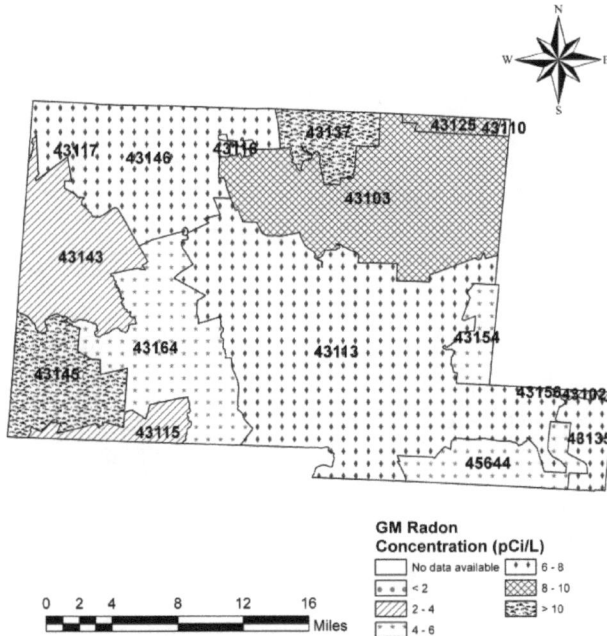

Figure 65c: GM ^{222}Rn concentrations in zip codes of Pickaway county based on the 2 pCi/L breakdown classification.

Table 66: ^{222}Rn statistics for Pike county

Zip code	No.	Min.	Max.	AM	GM	SD	CV
45133*	126	0.1	31.1	4.14	2.49	4.51	0.23
45601*	1058	0.1	223.4	9.61	5.1	15.04	0.77
45612*	48	0.2	28.4	5.95	2.28	7.52	0.39
45613	19	0.1	37.9	4.12	1.18	8.54	0.44
45624	4	0.2	0.2	0.2	0.2	0	0
45646	20	0.2	73.2	17.08	4.44	22.77	1.17
45648*	50	0.1	29.5	5.73	2.62	6.58	0.34
45660*	65	0.1	42.5	5.09	1.71	8.41	0.43
45661	140	0.1	43.3	5.36	1.82	8.14	0.42
45690	825	0.1	74.9	6.61	3.31	8.36	0.43

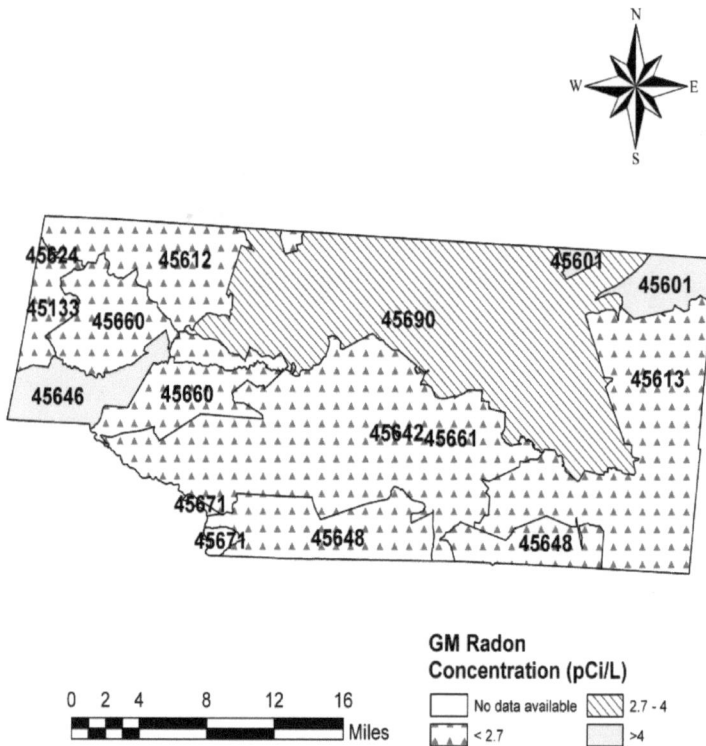

Figure 66a: GM ^{222}Rn concentrations in zip codes of Pike county based on the WHO-USEPA classification.

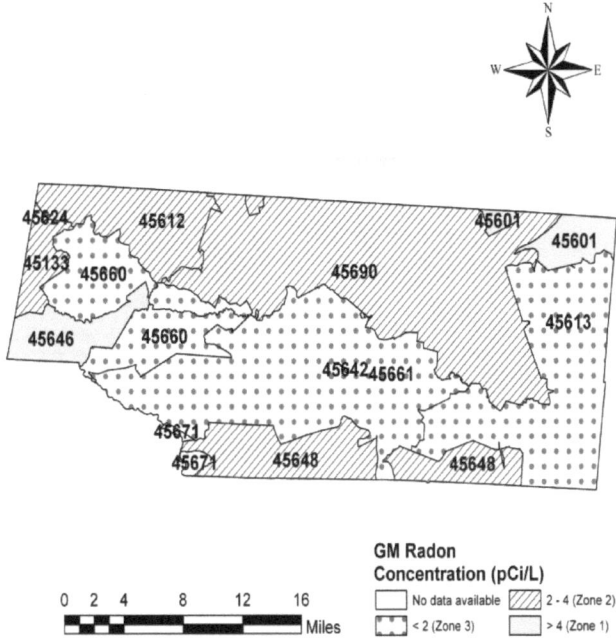

Figure 66b: GM ^{222}Rn concentrations in zip codes of Pike county based on the USEPA classification.

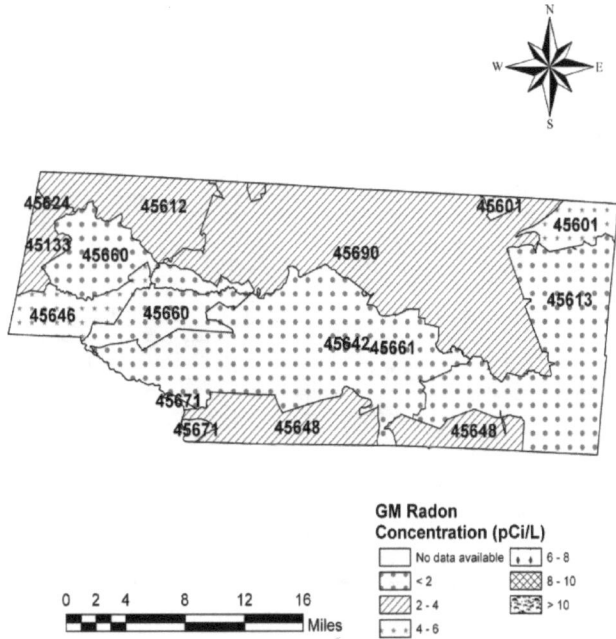

Figure 66c: GM ^{222}Rn concentrations in zip codes of Pike county based on the 2 pCi/L breakdown classification.

Table 67: ^{222}Rn statistics for Portage county

Zip code	No.	Min.	Max.	AM	GM	SD	CV
44201	38	0.2	8.9	3.11	2.25	2.41	0.12
44202	465	0.1	22.4	2.43	1.65	2.35	0.12
44224[*]	1231	0.1	58.5	4.05	2.47	4.97	0.25
44231	63	0.4	14.1	3.46	2.64	2.65	0.14
44234	36	0.5	29.8	3.48	2.2	5.37	0.27
44236[*]	1270	0.1	319	3.76	2.21	10.53	0.54
44240	503	0.1	70.1	4.33	2.56	5.53	0.28
44241	162	0.1	31.2	4.75	2.93	5.13	0.26
44243	4	0.1	5.8	3.5	1.71	2.65	0.14
44255	65	0.2	41.2	4.69	2.72	6.07	0.31
44260	197	0.1	44.9	5.71	3.38	6.09	0.31
44265	7	0.1	21.8	11.01	4.87	8.98	0.46
44266	386	0.1	25.6	2.62	1.48	3.34	0.17
44272	52	0.1	37.4	5.83	3.38	7.98	0.41
44278[*]	293	0.1	55.3	3.44	2.3	4.08	0.21
44285	2	4.7	5.5	5.1	5.08	0.57	0.03
44288	9	0.2	5.3	2.21	1.49	1.81	0.09
44312[*]	374	0.1	48.2	5.09	3.1	5.91	0.3
44411	8	0.6	9.7	3.51	2.27	3.4	0.17
44412	27	0.7	11.3	3.5	2.4	3.1	0.16
44429[*]	14	0.3	4.9	2.39	1.93	1.41	0.07
44444[*]	64	0.1	13.9	2.47	1.74	2.22	0.11
44449[*]	10	0.3	4.8	2.07	1.52	1.4	0.07
44491[*]	14	0.1	5.1	1.59	1.16	1.27	0.07
44632[*]	128	0.1	23.6	4.67	2.92	4.71	0.24
44211[**]	1	2.1	2.1	2.1	2.1	0	0
44242[**]	8	0.1	4.05	1.96	1.12	1.58	0.08

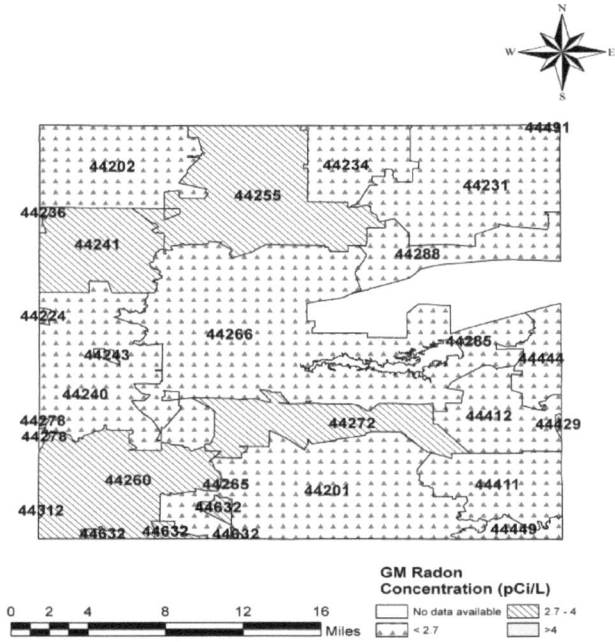

Figure 67a: GM ^{222}Rn concentrations in zip codes of Portage county based on the WHO-USEPA classification.

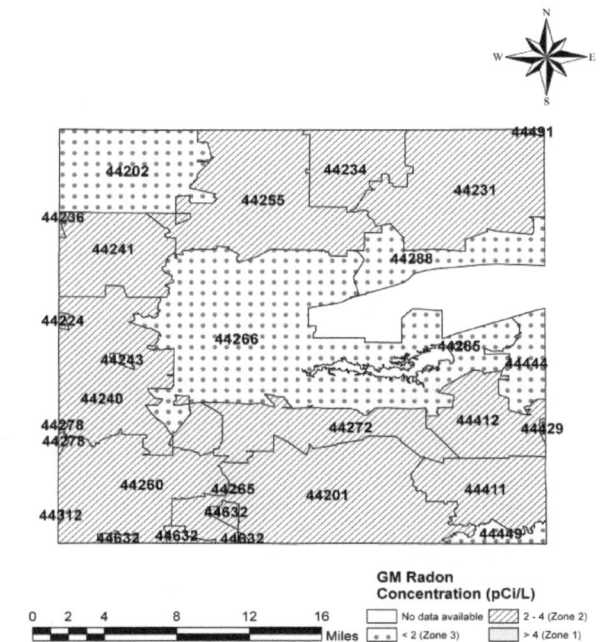

Figure 67b: GM ^{222}Rn concentrations in zip codes of Portage county based on the USEPA classification.

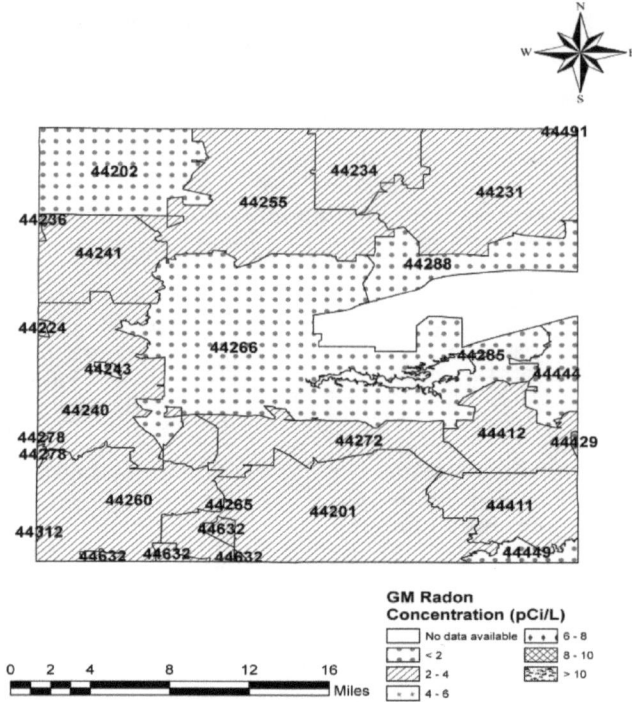

Figure 67c: GM ^{222}Rn concentrations in zip codes of Portage county based on the 2 pCi/L breakdown classification.

Table 68: ^{222}Rn statistics for Preble county

Zip code	No.	Min.	Max.	AM	GM	SD	CV
45003	10	0.9	14	6.01	4.4	4.43	0.23
45042***	640	0.1	143.8	4.51	1.96	7.76	0.4
45064***	18	0.5	30.9	6.14	2.77	9.39	0.48
45070	2	1.7	2.2	1.95	1.93	0.35	0.02
45304*	136	0.1	49	7.09	3.95	7.55	0.39
45309*	428	0.1	267.8	7.08	3.65	14.87	0.76
45311	55	0.1	41.5	5.66	3.14	7.16	0.37
45320	282	0.1	54.5	6.57	3.97	7.08	0.36
45321	12	1.2	38.1	9.1	5.23	11.11	0.57
45327*	233	0.1	86.1	9.23	5.43	11.52	0.59
45330	4	1	5.1	2.6	2.18	1.79	0.09
45338	117	0.1	57	9.57	5.48	10.35	0.53
45347	41	0.6	44.4	8.94	5.67	8.97	0.46
45378	7	1	13.8	6.41	4.38	4.92	0.25
45381	123	0.1	90.3	8.12	4.28	11.83	0.61
45382	18	0.1	45.1	10.11	4.19	11.91	0.61

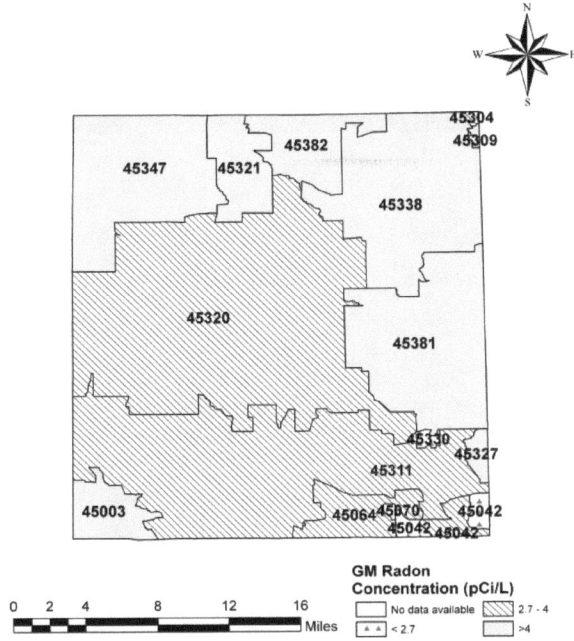

Figure 68a: GM ^{222}Rn concentrations in zip codes of Preble county based on the WHO-USEPA classification.

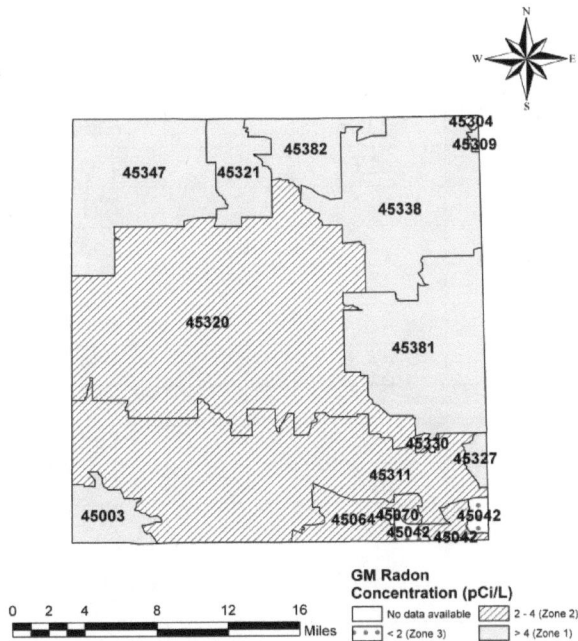

Figure 68b: GM ^{222}Rn concentrations in zip codes of Preble county based on the USEPA classification.

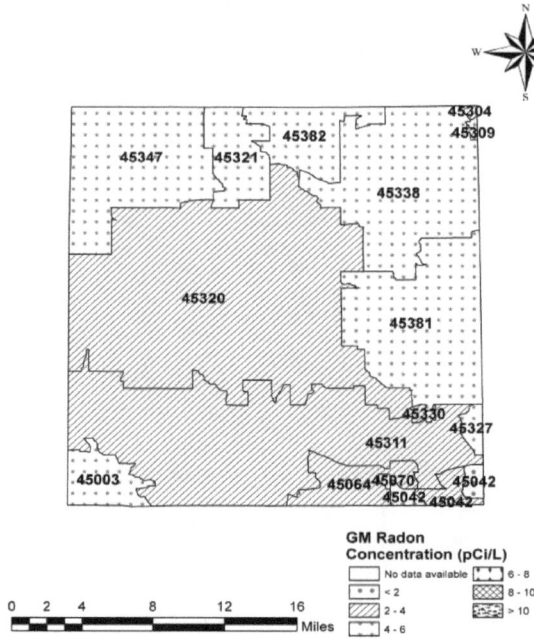

Figure 68c: GM ^{222}Rn concentrations in zip codes of Preble county based on the 2 pCi/L breakdown classification.

Table 69: ^{222}Rn statistics for Putnam county

Zip code	No.	Min.	Max.	AM	GM	SD	CV
43548[*]	10	1.5	22.9	8.33	6.26	6.81	0.35
45817[*]	38	0.5	62.6	7.51	4.02	12.79	0.65
45827	10	0.4	11.7	4.77	2.71	4.25	0.22
45830	25	0.1	35.9	9.16	5.43	9.14	0.47
45831	12	0.6	12.1	3.78	2.7	3.21	0.16
45833[*]	81	0.2	47.5	7.93	4.78	7.51	0.38
45844	36	1	28.7	7.38	4.99	6.97	0.36
45849[*]	4	0.3	0.8	0.5	0.47	0.22	0.01
45853	9	1	9.6	5.28	4.48	2.58	0.13
45856	50	0.5	52.5	6.57	4.03	8.93	0.46
45875	94	0.1	17.5	4.3	2.92	3.81	0.19
45876	3	2.1	15	6.57	4.34	7.31	0.37
45877	20	0.2	28.1	9.7	6.49	7.83	0.4
45848[**]	7	0.6	6.3	3.67	2.61	2.39	0.12
45893[**]	1	7.8	7.8	7.8	7.8	0	0

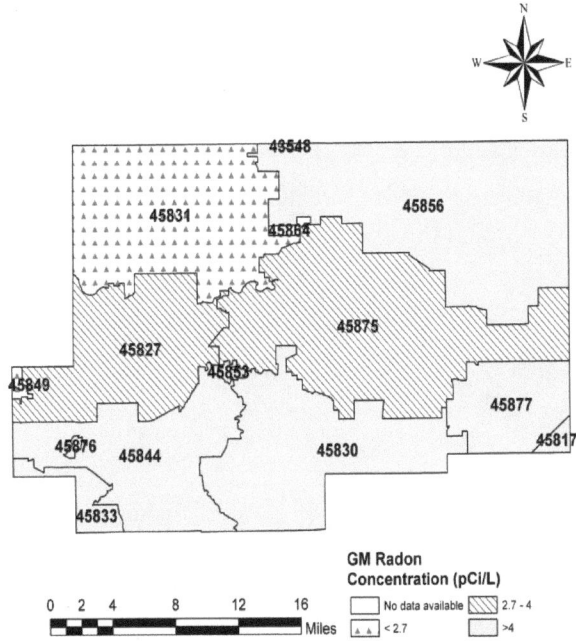

Figure 69a: GM ^{222}Rn concentrations in zip codes of Putnam county based on the WHO-USEPA classification.

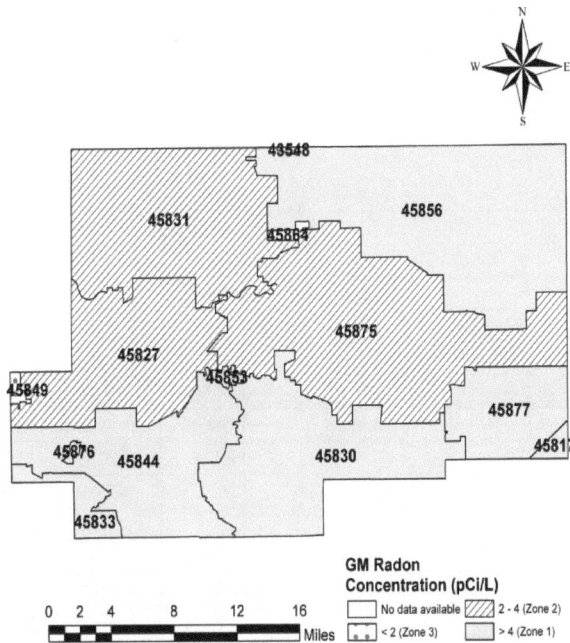

Figure 69b: GM ^{222}Rn concentrations in zip codes of Putnam county based on the USEPA classification.

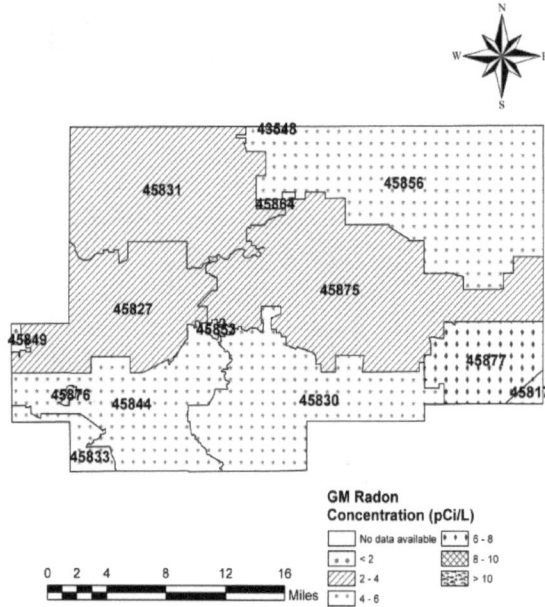

Figure 69c: GM ^{222}Rn concentrations in zip codes of Putnam county based on the 2 pCi/L breakdown classification.

Table 70: ^{222}Rn statistics for Richland county

Zip code	No.	Min.	Max.	AM	GM	SD	CV
43019[*]	410	0.1	189.5	9.45	3.38	19.71	1.01
44805[*]	451	0.1	155.8	8.17	3.62	17.7	0.91
44813	209	0.2	318.8	20.72	8.99	38.36	1.96
44822	56	1.2	264	36.45	9.52	63.76	3.26
44827[*]	181	0.1	56.1	3.18	1.25	5.67	0.29
44833[*]	493	0.1	140.2	7.75	4.41	12.72	0.65
44837[*]	17	0.7	50.8	8.52	4.2	13.38	0.69
44843	35	0.1	179	24.3	8.47	42.04	2.15
44864[*]	42	0.9	115	15.46	7.68	23.56	1.21
44865	15	2.3	38.1	10.89	7.47	11.62	0.6
44875	169	0.1	32.6	5.68	4.02	5.21	0.27
44878	14	0.4	16.6	4.61	3.35	4.13	0.21
44901	57	0.1	43.4	4.6	2.39	7.81	0.4
44902	37	0.5	10.2	2.58	1.81	2.52	0.13
44903	494	0.1	150.4	8.02	4.43	12.43	0.64
44904	798	0.1	292.3	9.75	4.76	17.45	0.89
44905	123	0.1	24.6	5.65	3.7	4.97	0.25
44906	447	0.1	44	5.51	3.4	6.03	0.31
44907	341	0.1	64.9	5.55	3.44	7.54	0.39
44862[**]	2	1.9	4.4	3.15	2.89	1.77	0.09

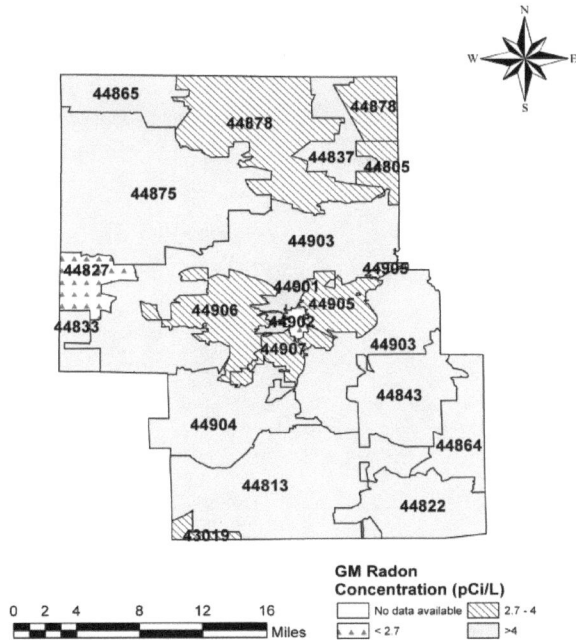

Figure 70a: GM ^{222}Rn concentrations in zip codes of Richland county based on the WHO-USEPA classification.

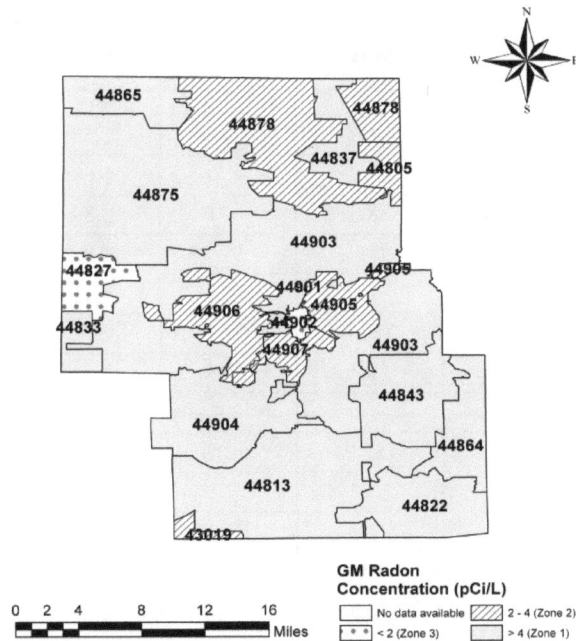

Figure 70b: GM ^{222}Rn concentrations in zip codes of Richland county based on the USEPA classification.

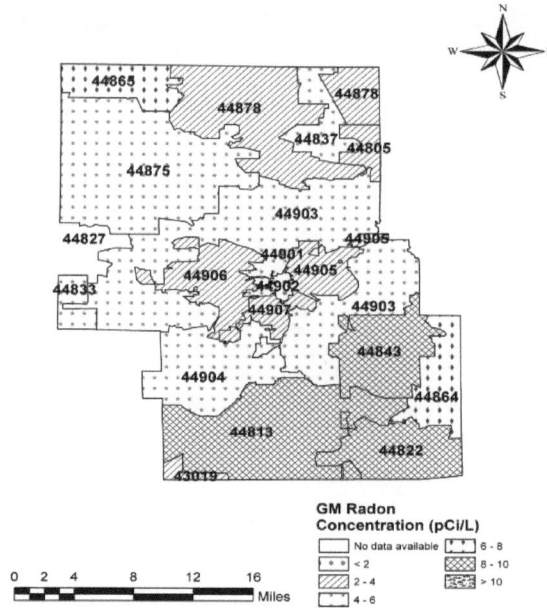

Figure 70c: GM ^{222}Rn concentrations in zip codes of Richland county based on the 2 pCi/L breakdown classification.

Table 71: ^{222}Rn statistics for Ross county

Zip code	No.	Min.	Max.	AM	GM	SD	CV
43101	1	2.1	2.1	2.1	2.1	0	0
43115	28	0.1	16.1	4.16	2.58	3.78	0.19
43135*	26	0.8	93.2	11.76	6.68	18.43	0.94
43160*	115	0.2	38.3	4.71	2.89	5.09	0.26
43164*	9	2.3	45.8	9.8	5.82	13.87	0.71
45123*	20	0.2	43	8.27	4.41	10.5	0.54
45601	1058	0.1	223.4	9.61	5.1	15.04	0.77
45612	48	0.2	28.4	5.95	2.28	7.52	0.39
45617	1	2.5	2.5	2.5	2.5	0	0
45628	61	0.1	188.7	20.13	5.91	44.25	2.27
45644	25	0.6	27.7	7	4.47	7.06	0.36
45647	25	0.7	21.4	6.92	6.09	3.59	0.18
45673	2	1.4	1.9	1.65	1.63	0.35	0.02
45681	3	4.5	52.9	24.57	15.71	25.24	1.29
45690*	825	0.1	74.9	6.61	3.31	8.36	0.43
45633**	1	15.9	15.9	15.9	15.9	0	0

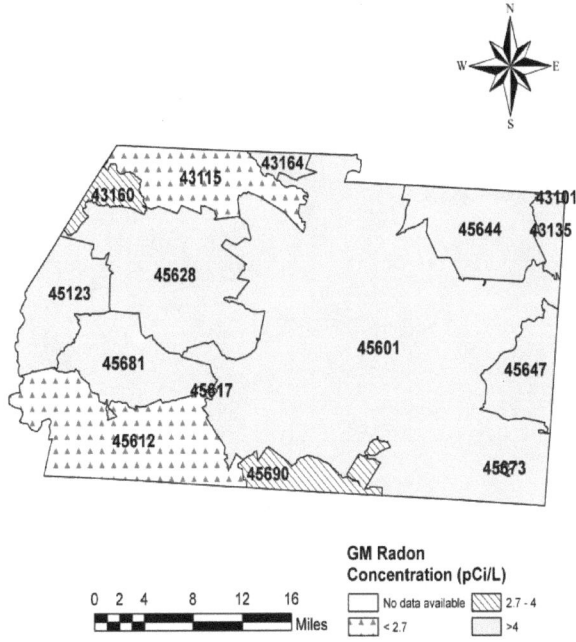

Figure 71a: GM ^{222}Rn concentrations in zip codes of Ross county based on the WHO-USEPA classification.

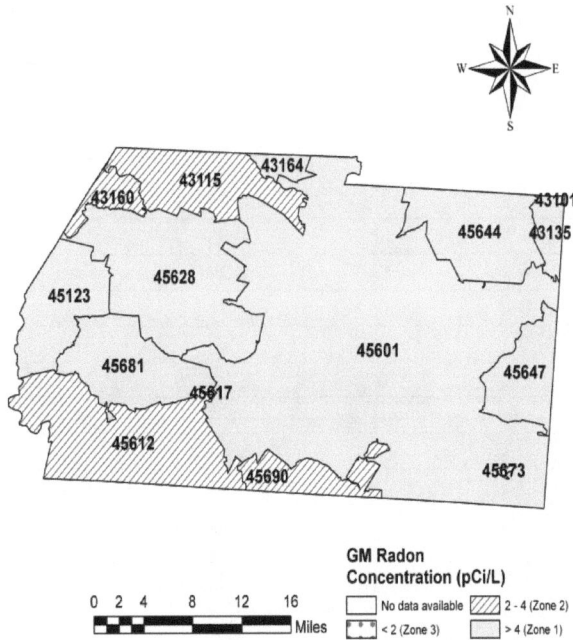

Figure 71b: GM ^{222}Rn concentrations in zip codes of Ross county based on the USEPA classification.

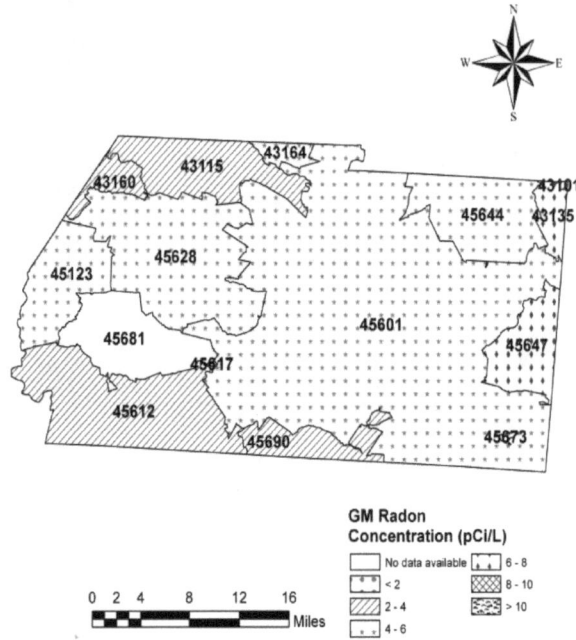

Figure 71c: GM ^{222}Rn concentrations in zip codes of Ross county based on the 2 pCi/L breakdown classification.

Table 72: ^{222}Rn statistics for Sandusky county

Zip code	No.	Min.	Max.	AM	GM	SD	CV
43406*	11	0.1	5.9	3.1	2.12	2.11	0.11
43407	6	0.1	5.1	2.55	1.57	1.87	0.1
43410	126	0.1	116.1	8.53	4.02	16.11	0.82
43416*	23	1.1	17.2	5.96	4.75	4.07	0.21
43420	354	0.1	47.7	6.29	4	7.06	0.36
43430*	54	0.4	27.8	4.84	3.22	4.99	0.26
43431	28	0.4	17	5.13	3.61	4.14	0.21
43435	18	0.1	11	3.71	2.74	2.47	0.13
43442	9	0.6	18.2	6.79	4.14	5.98	0.31
43449*	59	0.1	14.1	3.08	2	3	0.15
43457*	10	0.1	9.13	3.91	2.26	3.52	0.18
43464	7	0.2	9	2.14	0.98	3.17	0.16
43469	28	0.1	46.1	6.34	3.65	8.48	0.43
44811	262	0.1	159	15.21	6.34	24.33	1.25
44824*	129	0.1	279.2	13.43	5.23	36.71	1.88
44836*	26	0.1	20.5	5.16	2.04	6.04	0.31
44841*	10	0.1	5.2	1.69	1.01	1.59	0.08

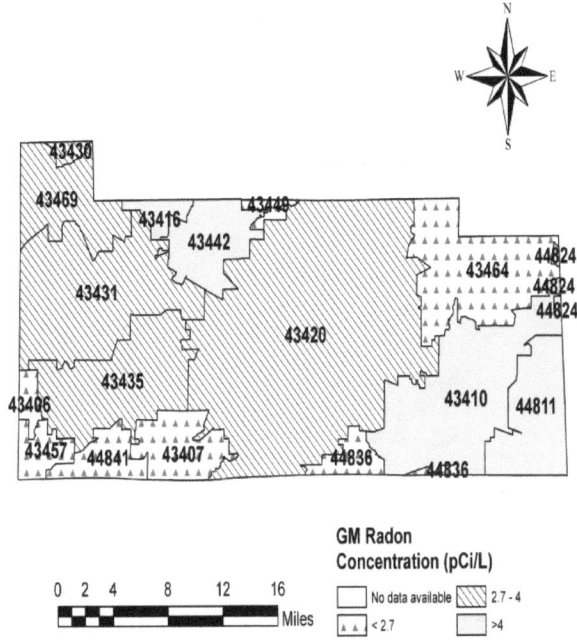

Figure 72a: GM ^{222}Rn concentrations in zip codes of Sandusky county based on the WHO-USEPA classification.

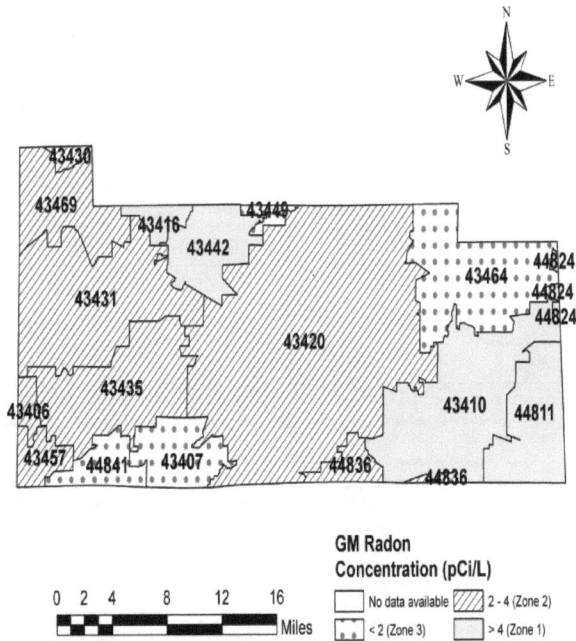

Figure 72b: GM ^{222}Rn concentrations in zip codes of Sandusky county based on the USEPA classification.

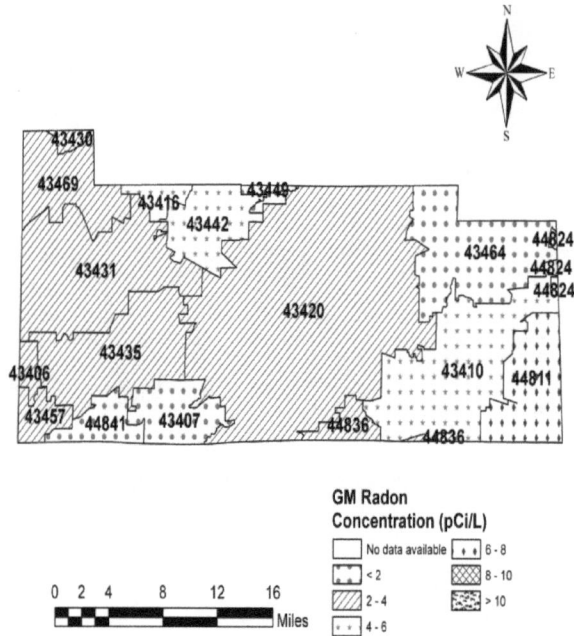

Figure 72c: GM ^{222}Rn concentrations in zip codes of Sandusky county based on the 2 pCi/L breakdown classification.

Table 73: ^{222}Rn statistics for Scioto county

Zip code	No.	Min.	Max.	AM	GM	SD	CV
45613[***]	19	0.1	37.9	4.12	1.18	8.54	0.44
45616[*]	2	1.6	4.7	3.15	2.74	2.19	0.11
45629	7	0.4	4.8	2.07	1.57	1.52	0.08
45636	1	0.6	0.6	0.6	0.6	0	0
45638[*]	52	0.1	33	3.64	2	5.33	0.27
45648	50	0.1	29.5	5.73	2.62	6.58	0.34
45652	12	0.8	6.3	3.09	2.59	1.8	0.09
45653	103	0.1	7.9	1.54	1.02	1.39	0.07
45660[*]	65	0.1	42.5	5.09	1.71	8.41	0.43
45661[***]	140	0.1	43.3	5.36	1.82	8.14	0.42
45662	98	0.1	23.4	3.84	2.65	3.93	0.2
45663	3	3.3	17.15	7.92	5.72	8	0.41
45682	5	1.8	2.8	2.42	2.39	0.41	0.02
45684	11	0.8	21.6	9.22	6.54	7.21	0.37
45694	29	0.8	9.3	3.18	2.53	2.35	0.12

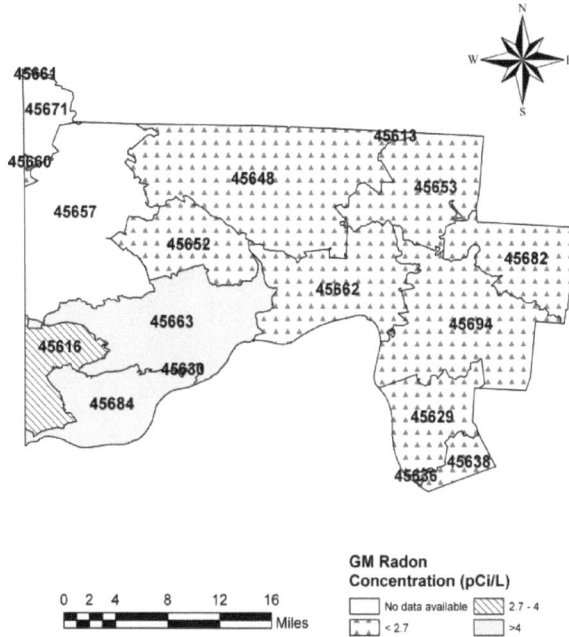

Figure 73a: GM ^{222}Rn concentrations in zip codes of Scioto county based on the WHO-USEPA classification.

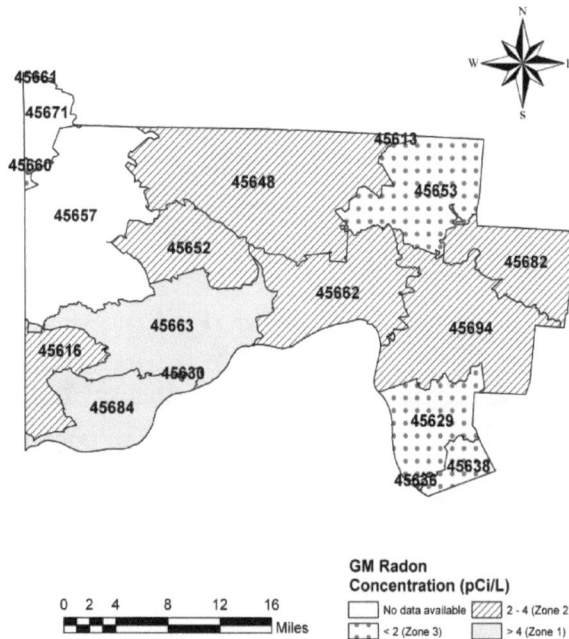

Figure 73b: GM ^{222}Rn concentrations in zip codes of Scioto county based on the USEPA classification.

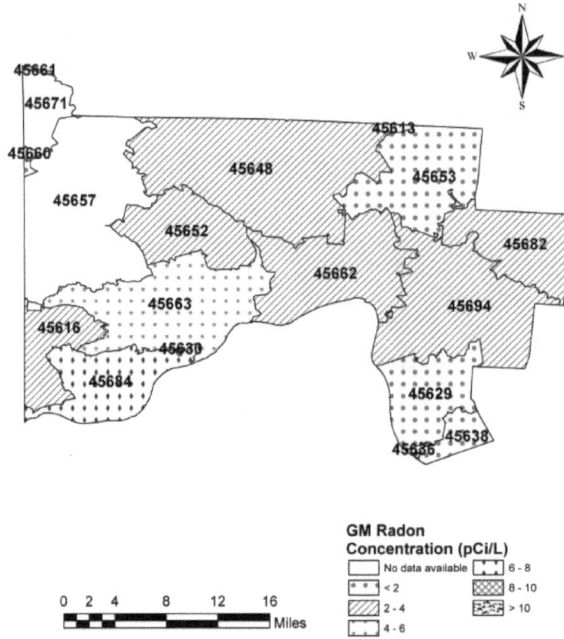

Figure 73c: GM ^{222}Rn concentrations in zip codes of Scioto county based on the 2 pCi/L breakdown classification.

Table 74: ^{222}Rn statistics for Seneca county

Zip code	No.	Min.	Max.	AM	GM	SD	CV
43316*	63	0.6	42.1	4.7	3.02	6.45	0.33
43410*	126	0.1	116.1	8.53	4.02	16.11	0.82
43457*	10	0.1	9.13	3.91	2.26	3.52	0.18
44802	19	0.1	120	11.21	4.05	26.65	1.36
44807	25	0.2	44.7	7.13	3.1	10.74	0.55
44809	2	7	14.9	10.95	10.21	5.59	0.29
44811*	262	0.1	159	15.21	6.34	24.33	1.25
44815	8	2.2	15.8	7.83	6.35	5.16	0.26
44818	21	0.1	16.4	4.3	2.72	3.99	0.2
44828	2	2.5	4.5	3.5	3.35	1.41	0.07
44830	272	0.1	86.2	6.32	3.21	8.51	0.44
44836	26	0.1	20.5	5.16	2.04	6.04	0.31
44841	10	0.1	5.2	1.69	1.01	1.59	0.08
44844*	13	0.2	26.1	7.32	3.32	8.3	0.42
44853	53	0.2	23	6.86	4.26	6.43	0.33
44861	2	2.2	3.7	2.95	2.85	1.06	0.05
44867	34	0.4	10.5	3.7	2.96	2.39	0.12
44882*	32	0.5	9.1	4.19	2.91	3	0.15
44883	1352	0.1	187.9	8.3	4.5	12.43	0.64

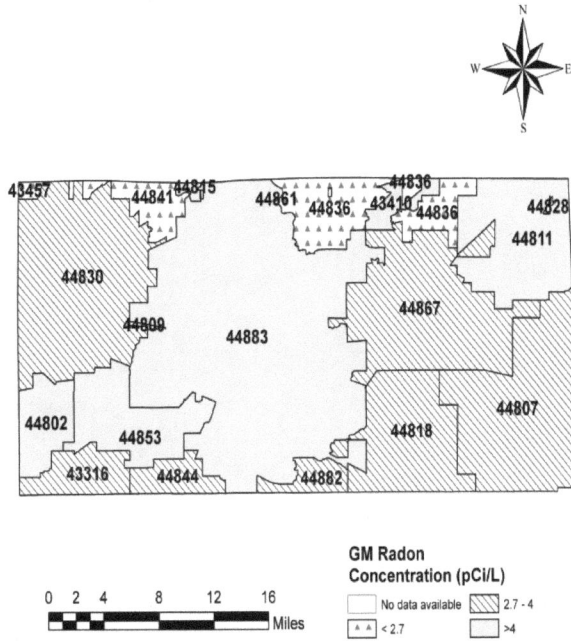

Figure 74a: GM ^{222}Rn concentrations in zip codes of Seneca county based on the WHO-USEPA classification.

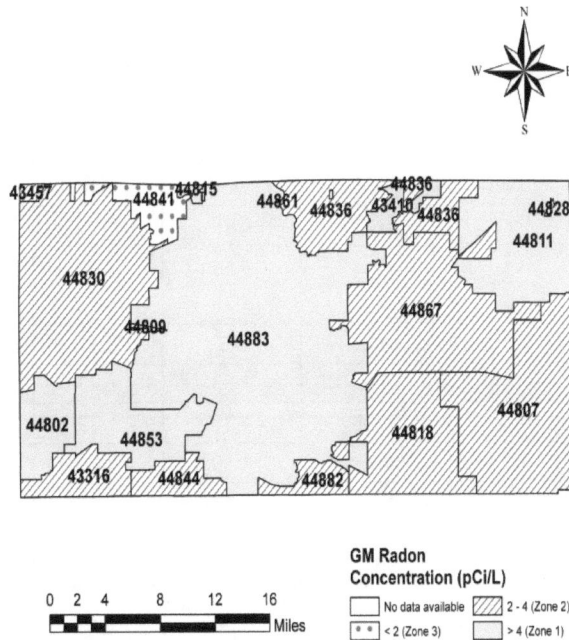

Figure 74b: GM ^{222}Rn concentrations in zip codes of Seneca county based on the USEPA classification.

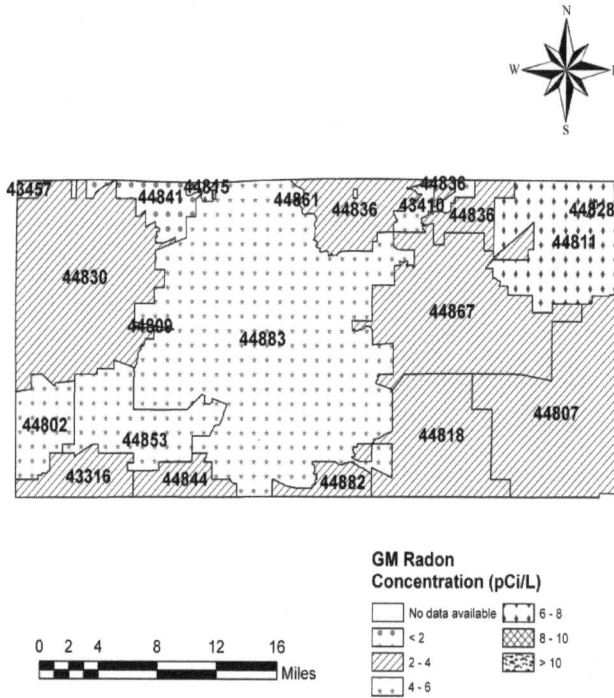

Figure 74c: GM ^{222}Rn concentrations in zip codes of Seneca county based on the 2 pCi/L breakdown classification.

Table 75: ^{222}Rn statistics for Shelby county

Zip code	No.	Min.	Max.	AM	GM	SD	CV
45302	65	0.2	61.3	10.37	5.11	12.75	0.65
45306	32	0.8	18.3	6.79	5.05	5.07	0.26
45317[*]	16	0.2	19.6	8.84	6.2	5.7	0.29
45333	19	1.3	36	8.16	5.12	8.8	0.45
45334	19	0.7	17	3.7	2.54	4.05	0.21
45336	4	0.6	5.8	2.83	1.95	2.43	0.12
45340	14	0.1	13.1	4.27	1.85	4.68	0.24
45356[*]	454	0.1	160	8.59	5.09	12.29	0.63
45363	26	1.1	17.7	5.58	3.96	4.77	0.24
45365	567	0.1	111.1	7.67	4.63	8.64	0.44
45845	60	0.5	48.6	8.24	5.98	8.11	0.42
45865[*]	85	0.1	64.5	8.56	4.35	10.44	0.53
45869[*]	71	0.4	24.9	6.51	4.18	6.03	0.31
45871[*]	33	0.1	38.9	9.63	5.23	9.82	0.5

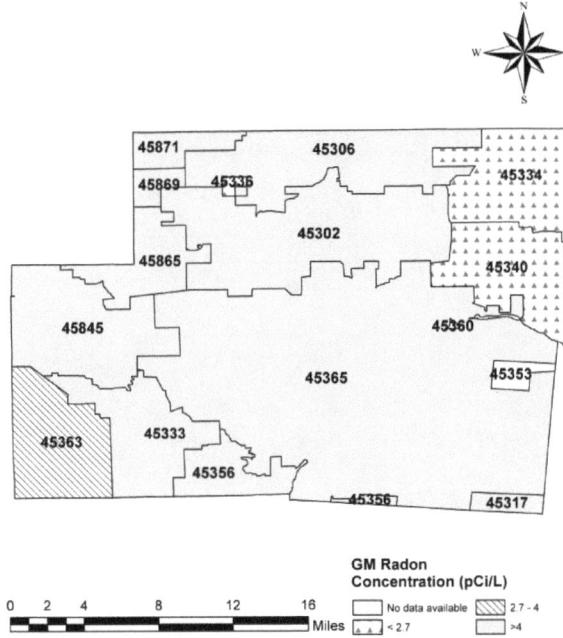

Figure 75a: GM ^{222}Rn concentrations in zip codes of Shelby county based on the WHO-USEPA classification.

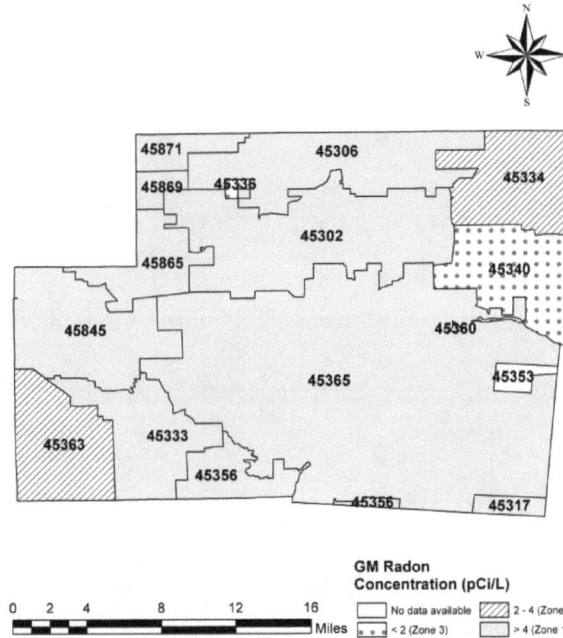

Figure 75b: GM ^{222}Rn concentrations in zip codes of Shelby county based on the USEPA classification.

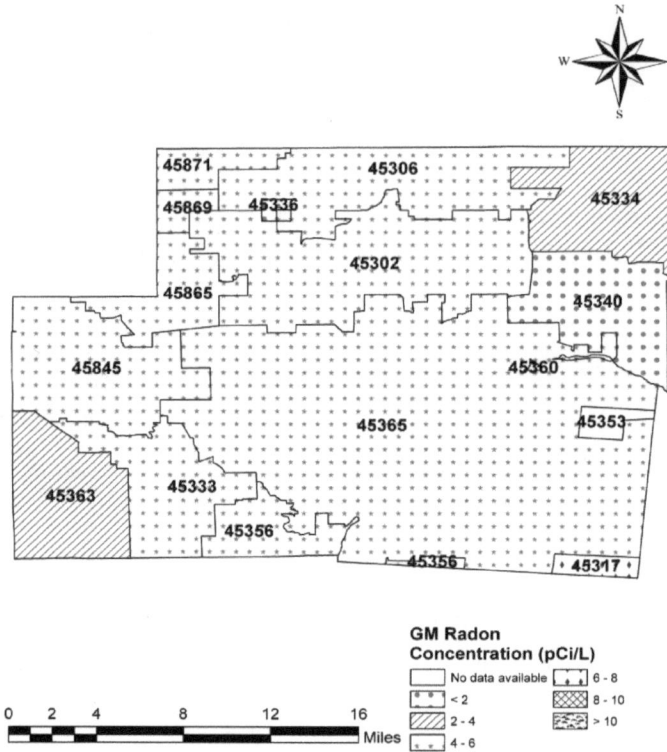

Figure 75c: GM ^{222}Rn concentrations in zip codes of Shelby county based on the 2 pCi/L breakdown classification.

Table 76: ^{222}Rn statistics for Stark county

Zip code	No.	Min.	Max.	AM	GM	SD	CV	Zip code	No.	Min.	Max.	AM	GM	SD	CV
44201[*]	38	0.2	8.9	3.11	2.25	2.41	0.12	44704	4	2	7.5	5.03	4.41	2.66	0.14
44216[*]	175	0.3	68.1	6.85	4.27	8.3	0.42	44705	99	0.1	31.6	5.83	3.32	5.79	0.3
44260[*]	197	0.1	44.9	5.71	3.38	6.09	0.31	44706	118	0.1	163.9	10.11	4.84	21.08	1.08
44601	214	0.1	16.3	2.62	1.89	2.21	0.11	44707	52	0.4	55.3	11.14	5.91	12.75	0.65
44608	10	0.1	19	7.39	3.32	7.26	0.37	44708	277	0.1	94.7	7.09	4.4	8.48	0.43
44612[*]	84	0.5	165.4	12.81	5.44	26.02	1.33	44709	242	0.2	49.9	6.07	3.99	6.29	0.32
44613	4	2.1	4.3	3.45	3.31	1.06	0.05	44710	38	0.1	31	8.54	5.1	7.44	0.38
44614	249	0.1	49.3	5.59	2.99	6.21	0.32	44714	120	0.1	34.9	5.91	3.6	6.35	0.33
44618[*]	47	0.3	17.9	5.59	3.43	5.31	0.27	44718	408	0.1	550.6	8.98	4.04	28.74	1.47
44626	26	0.1	245.1	33.31	7.82	60.41	3.09	44720	996	0.1	144.3	6.92	4	9.81	0.5
44632	128	0.1	23.6	4.67	2.92	4.71	0.24	44721	266	0.2	220	9.91	5.19	17.81	0.91

44634[*]	17	0.2	10.2	2.59	1.78	2.56	0.13	44730	47	0.1	63.6	6.88	3.51	11.13	0.57
44640	2	1.2	2.1	1.65	1.59	0.64	0.03	44630[**]	2	0.6	6.6	3.6	1.99	4.24	0.22
44641	167	0.1	65.2	5.56	3.53	6.69	0.34	44648[**]	2	0.1	3.3	1.7	0.57	2.26	0.12
44643	64	0.2	95.7	14.5	6.82	21.45	1.1	44701[**]	2	3.1	9.4	6.25	5.4	4.45	0.23
44646	603	0.1	125.2	7.17	3.69	12.4	0.63	44711[**]	4	1.7	10.9	5.6	4.48	4.03	0.21
44647	73	0.3	33.5	4.43	2.93	5.1	0.26	44735[**]	2	8.3	13.6	10.95	10.62	3.75	0.19
44657	111	0.1	580	17.61	6.09	56.16	2.88	44799[**]	3	3.1	8	6.07	5.6	2.61	0.13
44662	93	0.1	110	12.29	5.59	19.36	0.99								
44666	32	0.4	16.2	5.98	4.66	3.76	0.19								
44669	13	0.6	17.1	7.8	5.07	6.05	0.31								
44685	724	0.1	120	8.46	4.27	12.74	0.65								
44688	34	0.6	55.7	6.47	3.95	9.53	0.49								
44702	5	1.7	5.6	4.06	3.78	1.46	0.07								
44703	35	0.3	10.7	4.48	3.66	2.67	0.14								

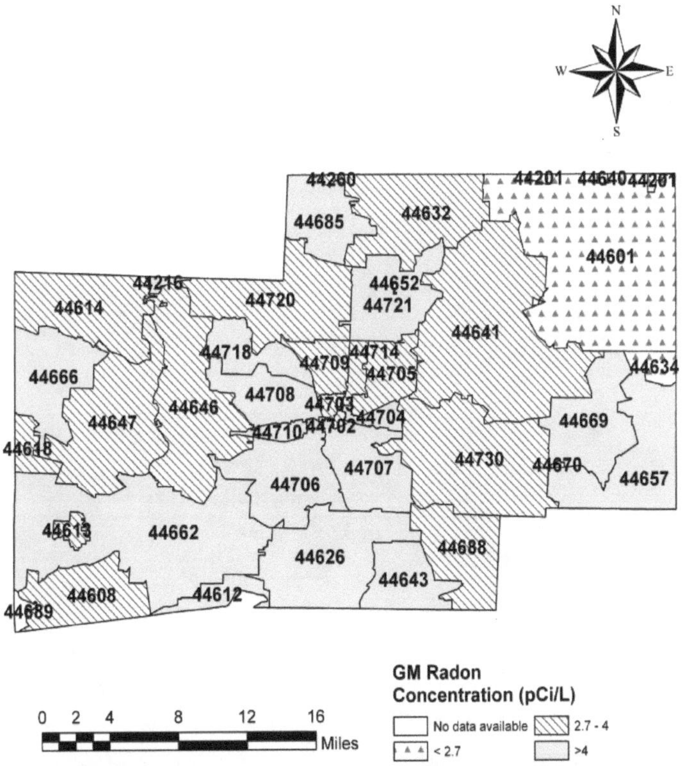

Figure 76a: GM ^{222}Rn concentrations in zip codes of Stark county based on the WHO-USEPA classification.

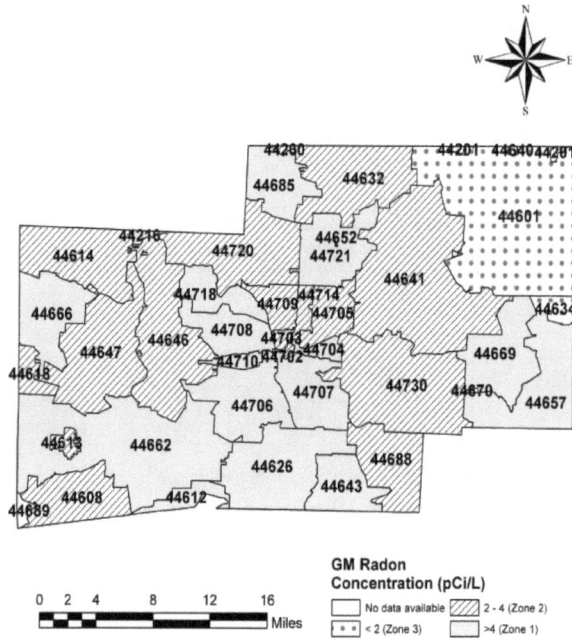

Figure 76b: GM ^{222}Rn concentrations in zip codes of Stark county based on the USEPA classification.

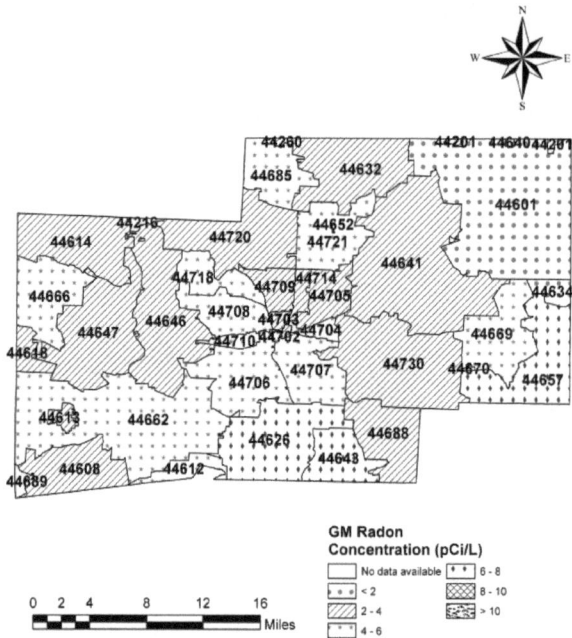

Figure 76c: GM ^{222}Rn concentrations in zip codes of Stark county based on the 2 pCi/L breakdown classification.

Table 77: ^{222}Rn statistics for Summit county

Zip code	No.	Min.	Max.	AM	GM	SD	CV	Zip code	No.	Min.	Max.	AM	GM	SD	CV
44056	274	0.1	32.3	3.49	2.27	3.91	0.2	44306	66	0.1	44.5	3.44	1.66	6.06	0.31
44067	471	0.1	47.5	4.02	2.3	5.85	0.3	44307	18	0.2	5.4	2.48	1.98	1.52	0.08
44087	610	0.1	908.1	35.11	3.48	129.15	6.61	44308	4	1.2	7.5	4.93	3.92	3.12	0.16
44141*	524	0.1	72.9	3.53	2.23	4.67	0.24	44310	127	0.1	20.8	3.52	2.33	3.18	0.16
44202*	465	0.1	22.4	2.43	1.65	2.35	0.12	44311	19	0.1	10.6	2.12	1.04	2.7	0.14
44203	482	0.1	30	4.21	2.74	4.16	0.21	44312	374	0.1	48.2	5.09	3.1	5.91	0.3
44216	175	0.3	68.1	6.85	4.27	8.3	0.42	44313	708	0.1	46.7	3.87	2.54	4.02	0.21
44221	346	0.1	22.5	3.01	1.93	3.09	0.16	44314	118	0.1	14.7	2.68	1.75	2.63	0.13
44223	418	0.1	20.9	3.45	2.23	3.01	0.15	44319	402	0.1	77.3	4.59	2.97	5.62	0.29
44224	1231	0.1	58.5	4.05	2.47	4.97	0.25	44320	163	0.1	131.9	4.57	1.51	14.93	0.76
44236	1270	0.1	319	3.76	2.21	10.53	0.54	44321	554	0.1	591.6	7.19	3.26	26.23	1.34
44240*	503	0.1	70.1	4.33	2.56	5.53	0.28	44333	621	0.1	50.8	6.67	3.75	8.28	0.42
44250	8	0.3	5.3	2.61	2.07	1.48	0.08	44614*	249	0.1	49.3	5.59	2.99	6.21	0.32
44256*	1446	0.1	60.3	3.93	2.73	4.35	0.22	44685*	724	0.1	120	8.46	4.27	12.74	0.65
44260*	197	0.1	44.9	5.71	3.38	6.09	0.31	44720*	996	0.1	144.3	6.92	4	9.81	0.5
44262	124	0.1	27	4.17	2.64	4.37	0.22	44210**	44	0.1	12.1	3.5	2.72	2.25	0.12
44264	49	0.1	13.2	2.21	1.43	2.32	0.12	44222**	3	0.7	2.7	1.97	1.68	1.1	0.06
44278	293	0.1	55.3	3.44	2.3	4.08	0.21	44232**	14	0.6	18.7	4.07	2.81	4.52	0.23
44281*	688	0.1	52.5	4.71	2.86	5.69	0.29	44237**	3	0.6	5.9	2.5	1.52	2.95	0.15
44286	195	0.1	27.6	3.39	2.5	3.12	0.16	44309**	1	4.8	4.8	4.8	4.8	0	0
44301	122	0.1	14.2	2.37	1.62	2.09	0.11	44315**	1	3.5	3.5	3.5	3.5	0	0
44302	34	0.1	11.5	2.23	1.67	2.2	0.11	44316**	2	1.5	3.2	2.35	2.19	1.2	0.06
44303	133	0.1	7.4	2.67	2.09	1.74	0.09	44326**	7	0.7	7.5	3.1	2.13	2.76	0.14
44304	11	0.4	6.8	2.62	1.56	2.53	0.13								
44305	122	0.1	24.2	3.82	2.52	3.61	0.18								

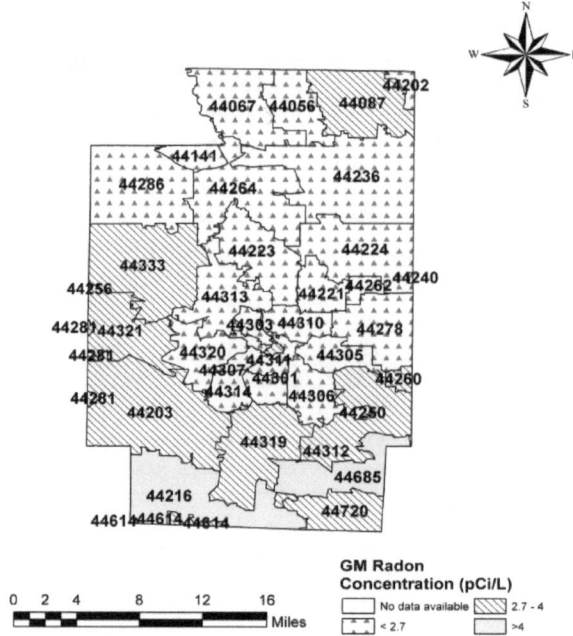

Figure 77a: GM ^{222}Rn concentrations in zip codes of Summit county based on the WHO-USEPA classification.

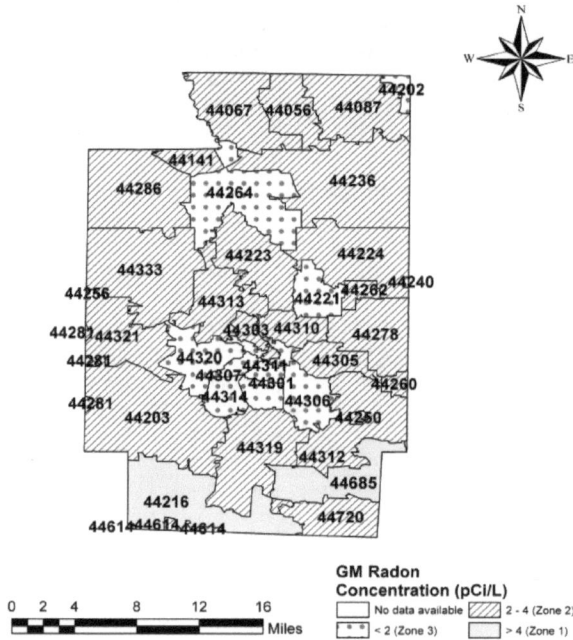

Figure 77b: GM ^{222}Rn concentrations in zip codes of Summit county based on the USEPA classification.

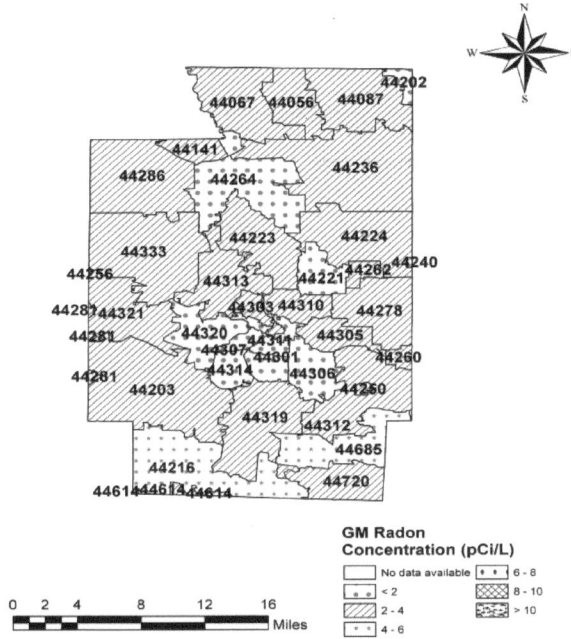

Figure 77c: GM ^{222}Rn concentrations in zip codes of Summit county based on the 2 pCi/L breakdown classification.

Table 78: ^{222}Rn statistics for Trumbull county

Zip code	No.	Min.	Max.	AM	GM	SD	CV
44062*	38	0.1	5.8	1.83	1.1	1.72	0.09
44076*	39	0.1	9.3	1.56	1.08	1.62	0.08
44402	24	0.1	11.6	3.26	2.22	3.17	0.16
44403	35	0.5	11	2.87	2.22	2.27	0.12
44404	12	0.2	8.5	2.2	1.28	2.29	0.12
44410	253	0.1	53	3.69	2.41	4.83	0.25
44417	17	0.3	3.6	2.06	1.63	1.22	0.06
44418	10	0.4	3.3	1.81	1.48	1.03	0.05
44420	100	0.1	17.1	2.47	1.54	2.44	0.12
44425	152	0.1	40.1	3.11	2.05	3.96	0.2
44428	21	0.3	10.7	2.73	1.74	2.79	0.14
44430	15	0.1	94.9	13.53	1.59	32.17	1.65
44437	14	0.1	6.2	2.21	1.72	1.41	0.07
44438	31	0.1	5.5	2.13	1.63	1.32	0.07
44439	2	0.9	1.3	1.1	1.08	0.28	0.01
44440	32	0.1	57	6.63	2.44	12.56	0.64

Zip code	No.	Min.	Max.	AM	GM	SD	CV
44444	64	0.1	13.9	2.47	1.74	2.22	0.11
44446	72	0.1	12.8	2.12	1.35	2.06	0.11
44450	14	0.4	5.9	2.71	1.89	1.93	0.1
44470	14	0.1	3.2	1.39	1.11	0.74	0.04
44473	28	0.3	6.4	2.2	1.84	1.38	0.07
44481	87	0.1	11.1	2.63	1.53	2.48	0.13
44483	110	0.1	13.6	2.1	1.5	2.01	0.1
44484	192	0.1	71.7	3.17	2.06	5.48	0.28
44485	35	0.1	8.8	1.41	0.81	1.58	0.08
44491	14	0.1	5.1	1.59	1.16	1.27	0.07
44504[*]	16	0.5	6.8	1.56	1.26	1.46	0.07
44505[*]	113	0.1	9.6	2.62	2.02	1.75	0.09
44424[**]	2	0.5	1.2	0.85	0.77	0.49	0.03
44453[**]	1	2.3	2.3	2.3	2.3	0	0
44482[**]	2	1.5	1.6	1.55	1.55	0.07	0
44486[**]	1	2.5	2.5	2.5	2.5	0	0

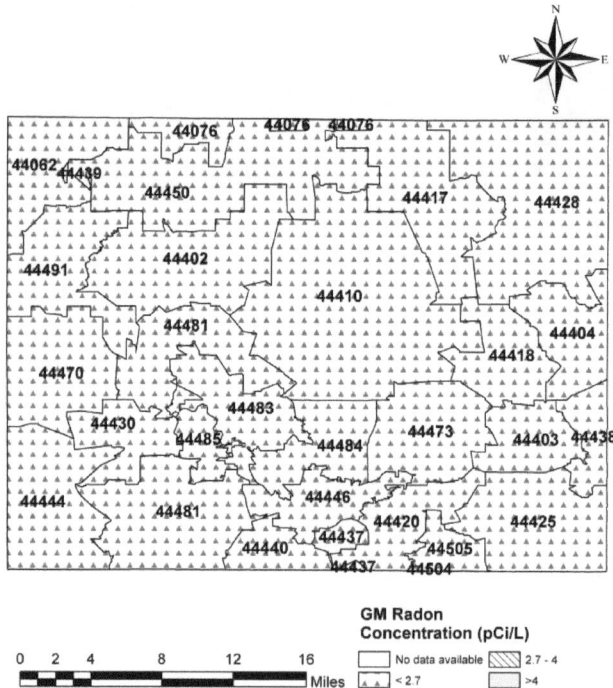

Figure 78a: GM ^{222}Rn concentrations in zip codes of Trumbull county based on the WHO-USEPA classification.

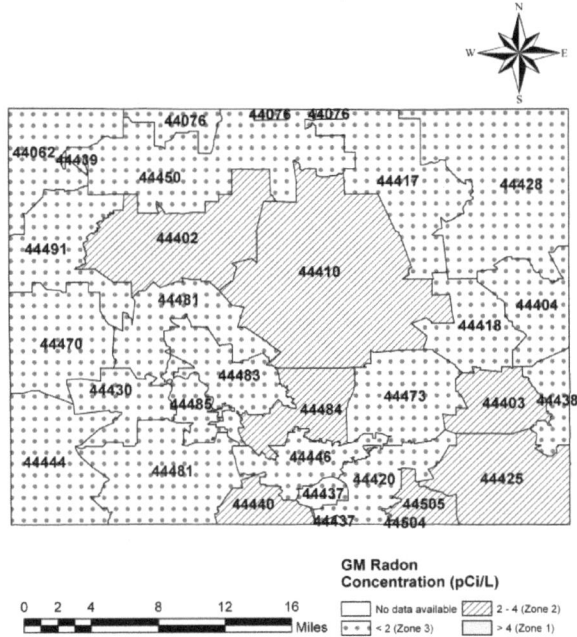

Figure 78b: GM ^{222}Rn concentrations in zip codes of Trumbull county based on the USEPA classification.

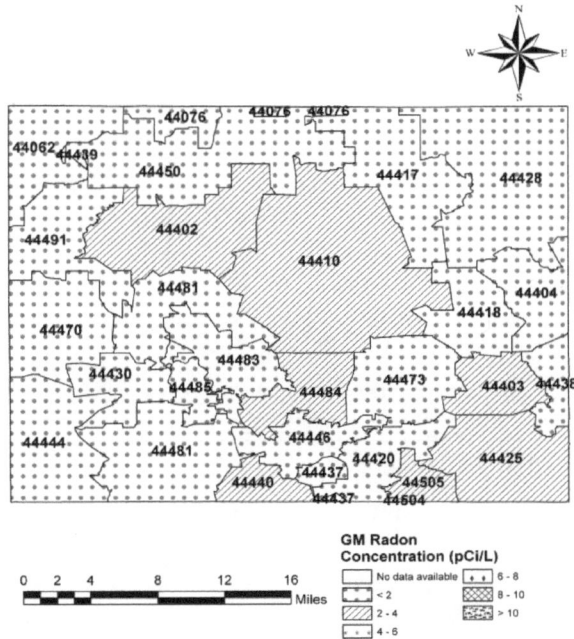

Figure 78c: GM ^{222}Rn concentrations in zip codes of Trumbull county based on the 2 pCi/L breakdown classification.

Table 79: ^{222}Rn statistics for Tuscarawas county

Zip code	No.	Min.	Max.	AM	GM	SD	CV
43749*	12	0.1	115.1	12.85	2.11	32.85	1.68
43804*	2	0.8	2.5	1.65	1.41	1.2	0.06
43824*	27	0.7	62.5	25.7	11.75	24.91	1.28
43832	26	0.2	20.4	6.77	3.22	7.41	0.38
43837	16	0.8	40.3	9.74	5.88	10.43	0.53
43840*	9	1.1	10.5	5.31	3.97	3.71	0.19
44608*	10	0.1	19	7.39	3.32	7.26	0.37
44612	84	0.5	165.4	12.81	5.44	26.02	1.33
44621	25	0.1	6.7	2.72	2.1	1.57	0.08
44622	341	0.1	56	6.7	4.18	7.08	0.36
44624	16	1.1	58	8.66	4.16	15.11	0.77
44626*	26	0.1	245.1	33.31	7.82	60.41	3.09
44629	22	0.1	68.1	9.64	4.95	13.84	0.71
44643*	64	0.2	95.7	14.5	6.82	21.45	1.1
44653	1	9.5	9.5	9.5	9.5	0	0
44656	36	0.2	47.5	11.71	5.44	12.81	0.66
44663	357	0.1	208.7	11.14	4.98	21.51	1.1
44671	4	1.7	43.9	24.1	12.26	22.96	1.18
44675*	12	0.5	55.6	10.6	4.68	16.79	0.86
44680	53	0.2	84.1	15	7.59	15.39	0.79
44681	29	0.1	88.2	8.7	3.76	16.84	0.86
44682	8	0.1	18.8	8.26	4.08	6.78	0.35
44683	38	0.2	46.6	8.02	3.45	11.44	0.59
44697	5	0.9	2.3	1.44	1.37	0.52	0.03
44699*	7	0.7	28.7	10.8	5.1	12.18	0.62
44679**	1	0.9	0.9	0.9	0.9	0	0

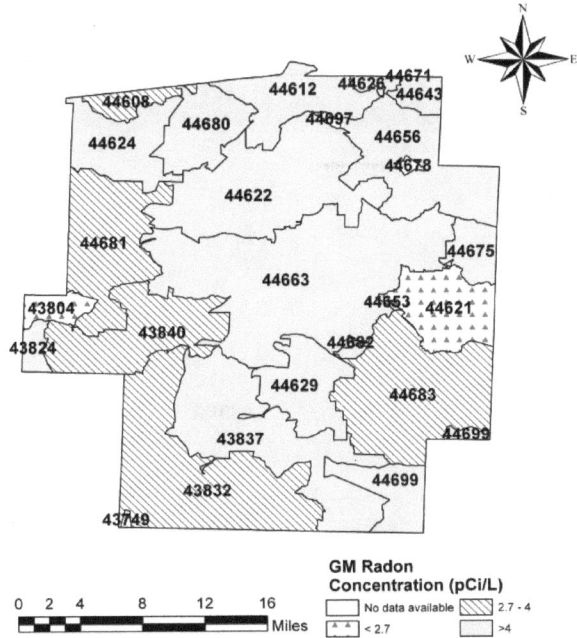

Figure 79a: GM ^{222}Rn concentrations in zip codes of Tuscarawas county based on the WHO-USEPA classification.

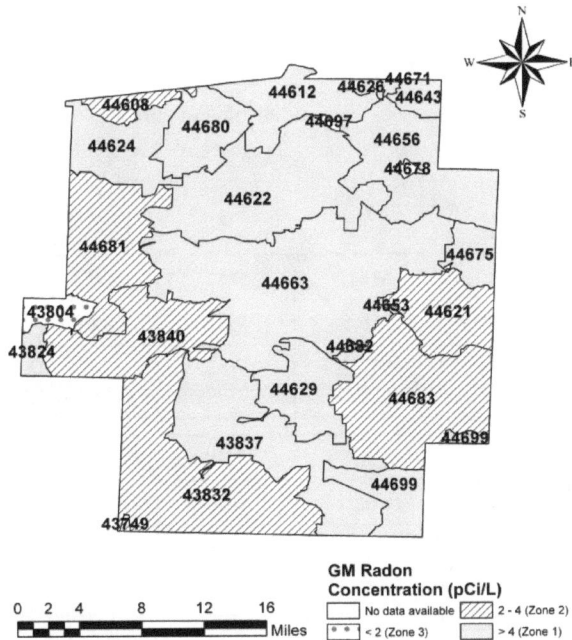

Figure 79b: GM ^{222}Rn concentrations in zip codes of Tuscarawas county based on the USEPA classification.

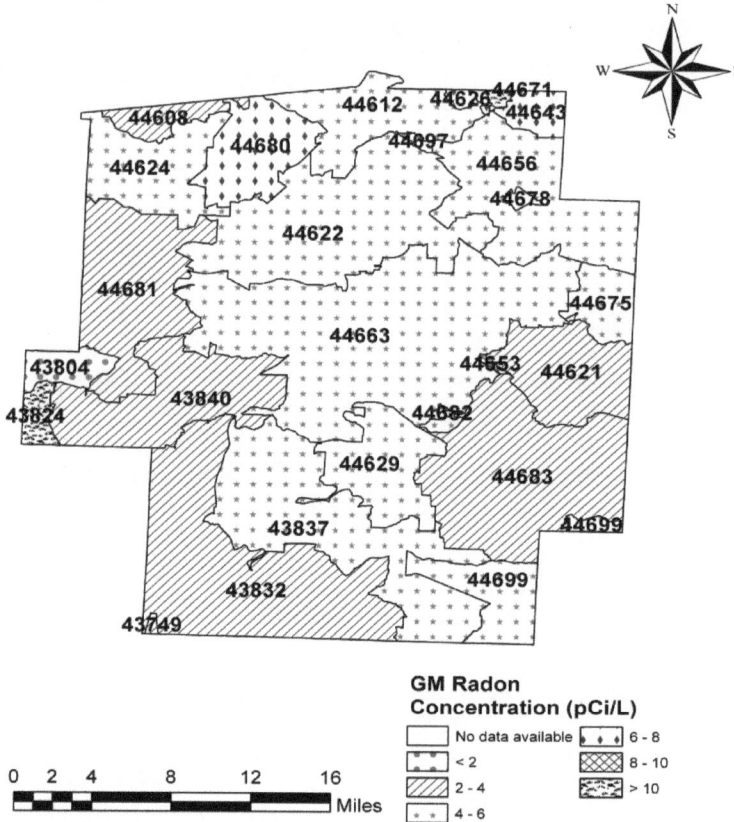

Figure 79c: GM ^{222}Rn concentrations in zip codes of Tuscarawas county based on the 2 pCi/L breakdown classification.

Table 80: ^{222}Rn statistics for Union county

Zip code	No.	Min.	Max.	AM	GM	SD	CV
43015*	1518	0.1	116.5	7.57	4.65	9.4	0.48
43016*	1225	0.1	137.2	7.83	4.57	8.55	0.44
43017*	2855	0.1	193.7	10.78	5.93	13.25	0.68
43029*	18	0.2	13.3	5.46	3.63	3.47	0.18
43036	6	1.4	10.2	6.83	5.86	2.98	0.15
43040	1063	0.1	63.5	5.28	3.28	6.13	0.31
43045	87	0.1	26.6	6.05	4	5.17	0.26
43060	51	0.2	62.7	8.24	4.51	11.03	0.56
43061*	133	0.1	29.8	6.1	3.97	5.38	0.28

Zip code	No.	Min.	Max.	AM	GM	SD	CV
43064[*]	329	0.5	70.4	8.28	5.12	10.09	0.52
43067	29	0.3	18.2	3.94	2.46	4.47	0.23
43077	4	0.1	5.4	2.23	1.12	2.25	0.12
43084[*]	12	0.1	18.6	5.54	2.53	5.78	0.3
43302[*]	911	0.1	67.1	6.25	3.47	7.2	0.37
43319[*]	20	0.6	20.4	3.97	2.81	4.37	0.22
43340[*]	9	1.1	13.8	3.52	2.41	4.02	0.21
43342[*]	58	0.6	21.3	9.15	6.79	5.92	0.3
43344	153	0.1	61.8	5.54	3.14	7.7	0.39
43358[*]	24	0.5	12.6	4.75	3.47	3.49	0.18
43007[**]	5	1.1	8.3	5.06	4.05	3.16	0.16

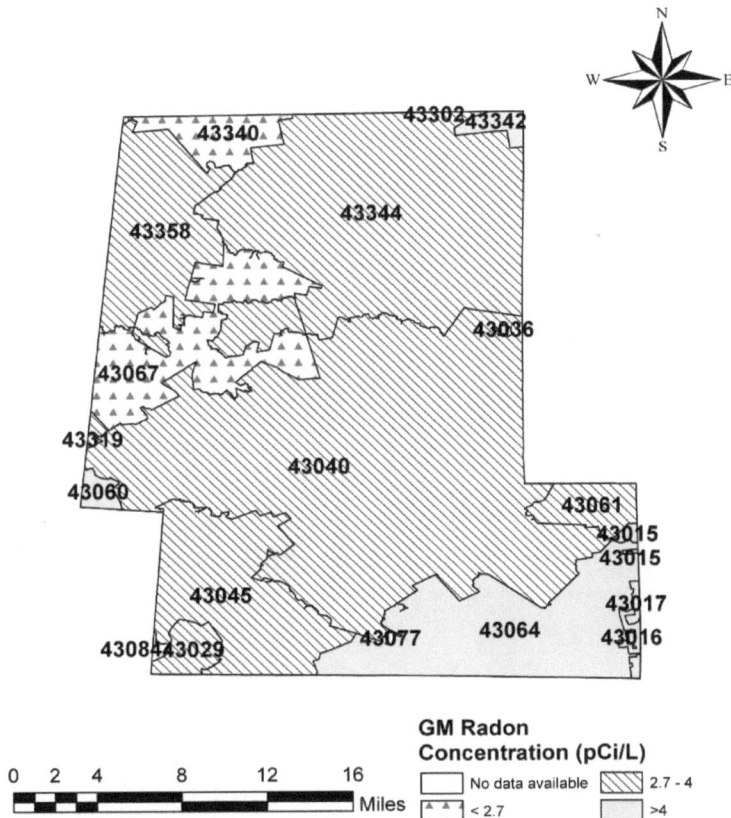

Figure 80a: GM ^{222}Rn concentrations in zip codes of Union county based on the WHO-USEPA classification.

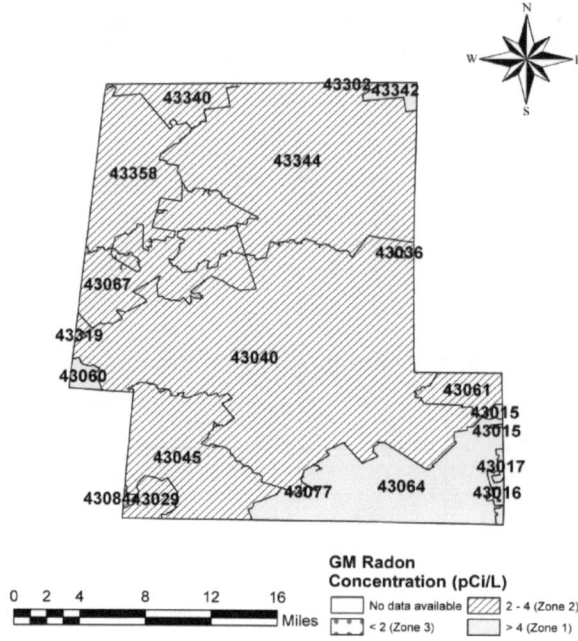

Figure 80b: GM [222]Rn concentrations in zip codes of Union county based on the USEPA classification.

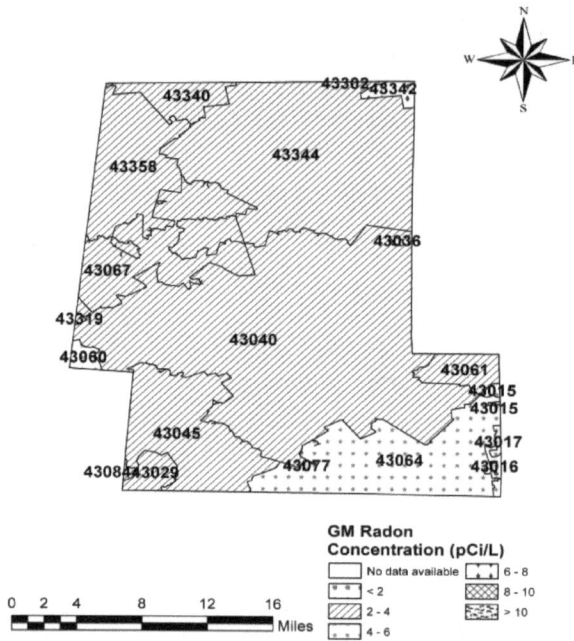

Figure 80c: GM [222]Rn concentrations in zip codes of Union county based on the 2 pCi/L breakdown classification.

Table 81: ^{222}Rn statistics for Van Wert county

Zip code	No.	Min.	Max.	AM	GM	SD	CV
45832	10	0.5	11.3	4.84	2.89	4.27	0.22
45833*	81	0.2	47.5	7.93	4.78	7.51	0.38
45838	1	1.5	1.5	1.5	1.5	0	0
45844*	36	1	28.7	7.38	4.99	6.97	0.36
45849*	4	0.3	0.8	0.5	0.47	0.22	0.01
45863	2	2.60	3.60	3.10	3.06	0.71	0.04
45874	10	0.10	7.30	4.73	3.01	2.80	0.14
45882*	17	0.1	13.3	3.48	1.83	4.14	0.21
45886	1	0.9	0.9	0.9	0.9	0	0
45887*	14	0.5	14.3	4.41	2.69	4.43	0.23
45891	70	0.50	24.50	6.17	4.07	5.65	0.29
45894	2	4.90	7.10	6.00	5.9	1.56	0.08
45898	8	1.8	7.2	4.61	4.08	2.26	0.12

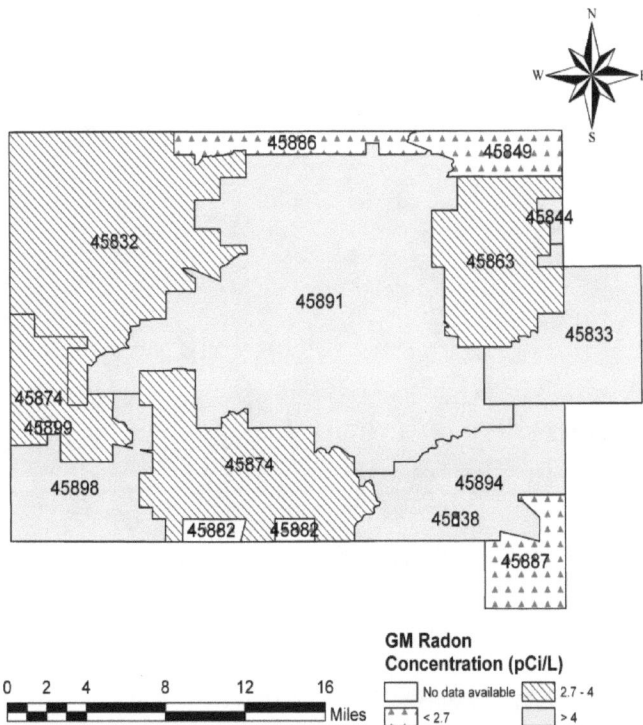

Figure 81a: GM ^{222}Rn concentrations in zip codes of Van Wert county based on the WHO-USEPA classification.

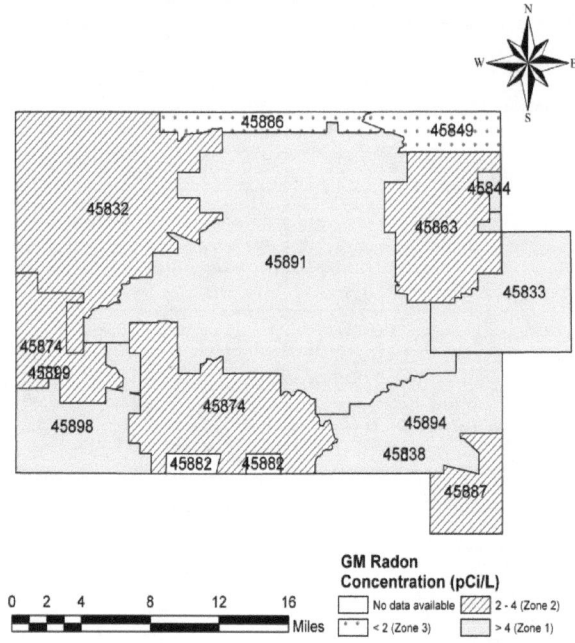

Figure 81b: GM ^{222}Rn concentrations in zip codes of Van Wert county based on the USEPA classification.

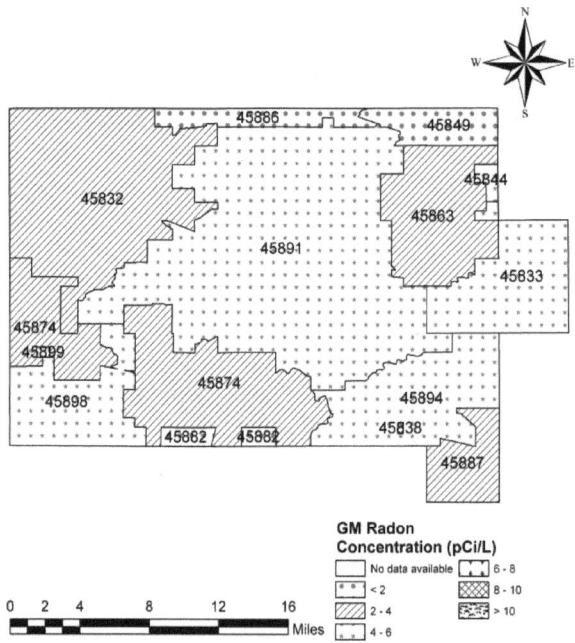

Figure 81c: GM ^{222}Rn concentrations in zip codes of Van Wert county based on the 2 pCi/L breakdown classification.

Table 82: ^{222}Rn statistics for Vinton county

Zip code	No.	Min.	Max.	AM	GM	SD	CV
43135*	26	0.8	93.2	11.76	6.68	18.43	0.94
43138*	169	0.1	99.5	6.96	3.63	9.68	0.5
43152*	10	0.1	10.2	3.48	2.01	3.02	0.15
45622	2	0.9	2.3	1.6	1.44	0.99	0.05
45634	8	0.7	16.4	4.98	2.77	5.45	0.28
45647*	25	0.7	21.4	6.92	6.09	3.59	0.18
45651	12	0.1	3.9	1.66	1.09	1.36	0.07
45654	4	1.1	5.8	2.63	2.13	2.14	0.11
45672	4	1.9	13.5	5	3.39	5.68	0.29
45686*	9	0.5	3.7	1.59	1.41	0.88	0.05
45695	1	1.7	1.7	1.7	1.7	0	0
45698	1	0.1	0.1	0.1	0.1	0	0
45710*	53	0.3	12.4	3.18	2.06	3.11	0.16
45766*	7	0.8	3.6	2.1	1.75	1.29	0.07

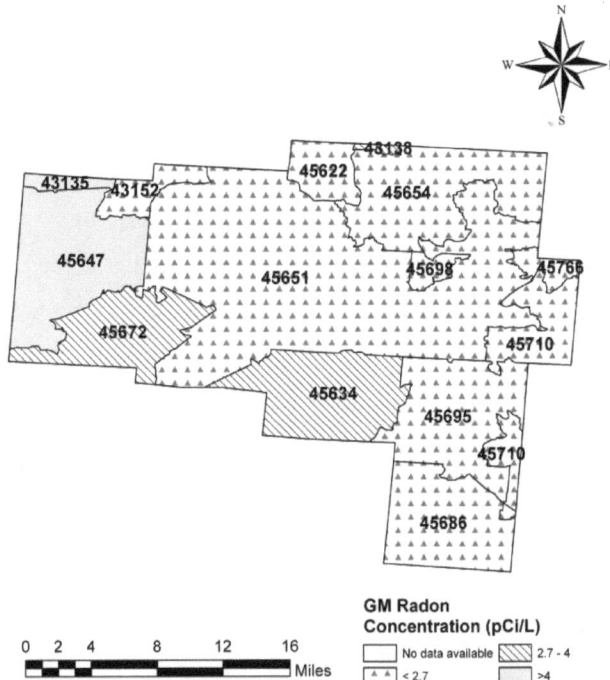

Figure 82a: GM ^{222}Rn concentrations in zip codes of Vinton county based on the WHO-USEPA classification.

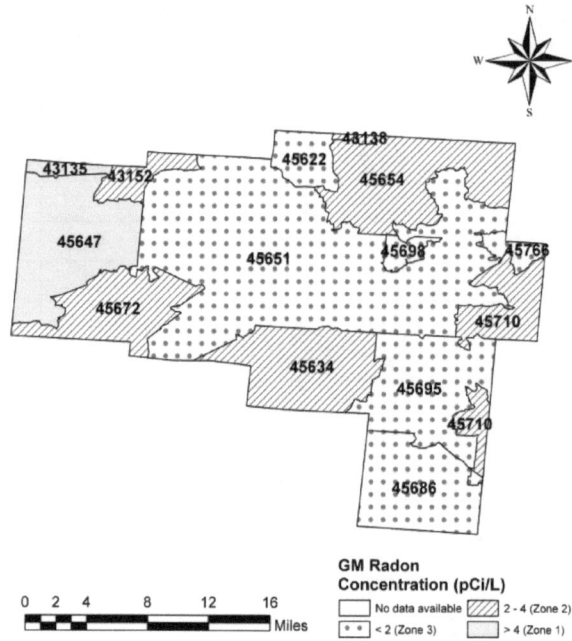

Figure 82b: GM ^{222}Rn concentrations in zip codes of Vinton county based on the USEPA classification.

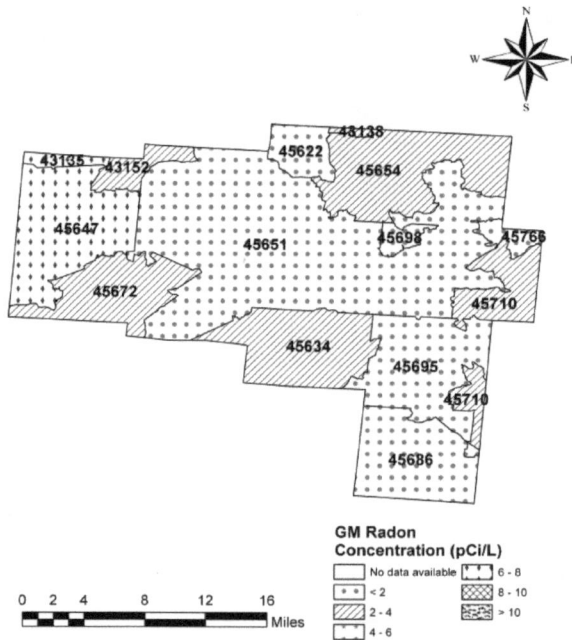

Figure 82c: GM ^{222}Rn concentrations in zip codes of Vinton county based on the 2 pCi/L breakdown classification.

Table 83: ^{222}Rn statistics for Warren county

Zip code	No.	Min.	Max.	AM	GM	SD	CV
45005	580	0.1	70.7	6.82	3.56	8.76	0.45
45034	44	0.1	27.4	3.52	1.95	4.65	0.24
45036	870	0.1	34.8	4.49	2.97	4.38	0.22
45039	875	0.1	219	4.87	3.08	9.13	0.47
45040	2066	0.1	44.1	4.43	2.89	4.44	0.23
45042***	640	0.1	143.8	4.51	1.96	7.76	0.4
45044*	963	0.1	50.2	4.83	3.14	4.94	0.25
45050*	153	0.1	30.6	4.75	2.78	5.09	0.26
45054	45	0.1	30.2	5.23	2.87	5.55	0.28
45065	118	0.2	34	5.4	3.79	4.82	0.25
45066	1044	0.1	58.1	4.5	2.89	4.82	0.25
45068	335	0.1	44.9	5.63	3.58	6	0.31
45107*	66	0.1	9.8	3.17	2.23	2.39	0.12
45113*	44	0.6	54.2	5.94	3.36	8.98	0.46
45122*	59	0.3	15.2	3.77	2.83	3.03	0.16
45140*	1852	0.1	470.8	4.95	2.99	12.09	0.62
45152	236	0.1	30.1	5.59	3.68	4.9	0.25
45162	35	0.2	6.4	2.05	1.51	1.61	0.08
45241*	1001	0.1	42.3	4.07	2.48	4.41	0.23
45249*	621	0.1	34.3	4.22	2.75	4.22	0.22
45342*	926	0.1	52.7	4.21	2.52	4.91	0.25
45458*	1266	0.1	35.5	3.82	2.27	3.94	0.2

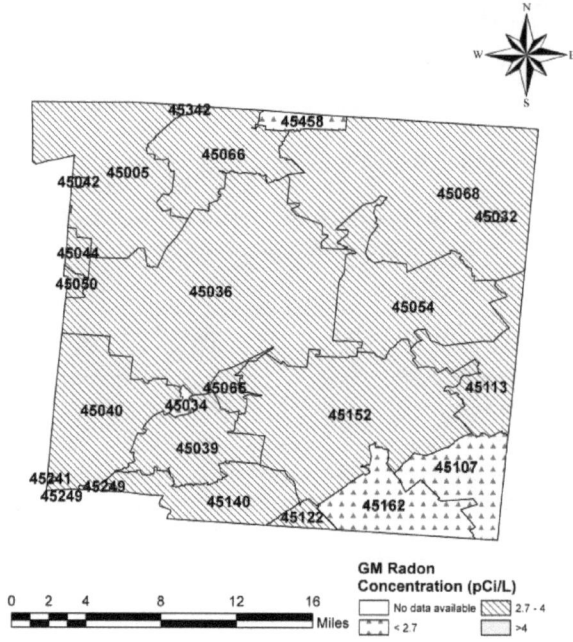

Figure 83a: GM ^{222}Rn concentrations in zip codes of Warren county based on the WHO-USEPA classification.

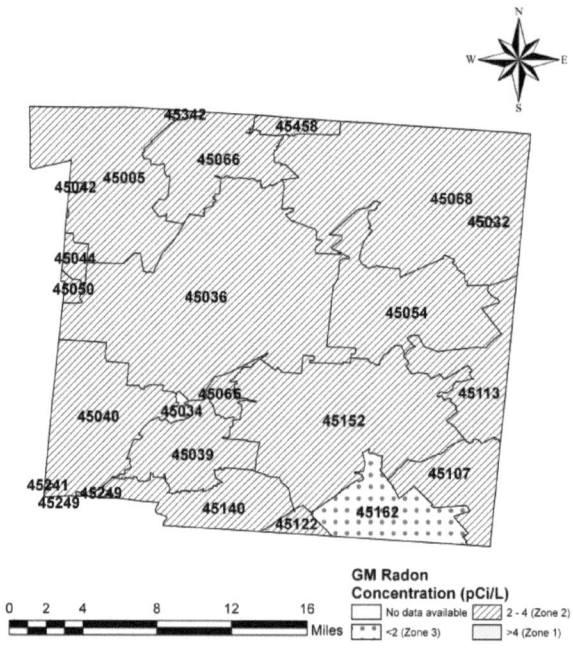

Figure 83b: GM ^{222}Rn concentrations in zip codes of Warren county based on the USEPA classification.

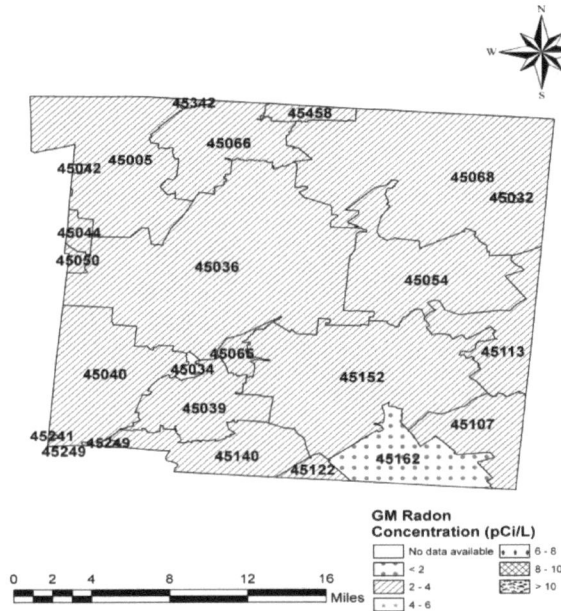

Figure 83c: GM ^{222}Rn concentrations in zip codes of Warren county based on the 2 pCi/L breakdown classification.

Table 84: ^{222}Rn statistics for Washington county

Zip code	No.	Min.	Max.	AM	GM	SD	CV
43787*	11	0.7	33.9	7.81	4.02	10.14	0.52
45711*	12	0.4	4.5	1.96	1.36	1.67	0.09
45714	81	0.1	23.1	2.98	1.93	3.27	0.17
45715	12	2.2	27	6.94	5.45	6.62	0.34
45723*	11	0.3	4.2	2.24	1.66	1.4	0.07
45724	11	0.2	5.8	2.91	2.19	1.83	0.09
45729	19	0.8	21	3.76	2.51	4.61	0.24
45742	20	0.3	18	2.9	1.65	3.97	0.2
45744	25	1	10.7	3.89	3	3.03	0.16
45745	5	1.1	3.9	2	1.74	1.23	0.06
45746	3	0.5	3.4	1.7	1.27	1.51	0.08
45750	496	0.1	91.9	5.45	2.91	8.67	0.44
45767*	10	0.3	9.35	2.19	1.38	2.67	0.14
45768	6	0.1	5.3	2.32	1.2	2.23	0.11
45773	9	0.7	5.4	3.36	2.76	1.8	0.09
45784	28	0.1	14.2	4.86	3.23	3.63	0.19
45786	17	0.4	19.6	3.84	2.48	4.71	0.24
45788	9	0.8	5.5	3.41	2.8	1.93	0.1
45789	1	2.9	2.9	2.9	2.9	0	0
45712**	2	1	3.7	2.35	1.92	1.91	0.1

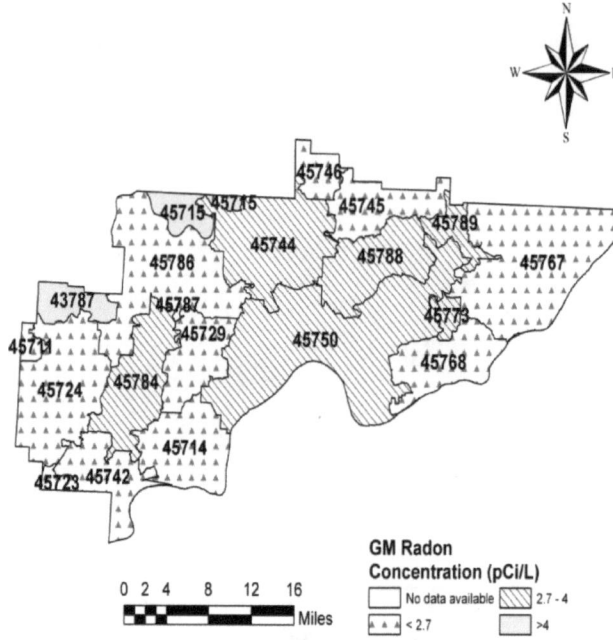

Figure 84a: GM ^{222}Rn concentrations in zip codes of Washington county based on the WHO-USEPA classification.

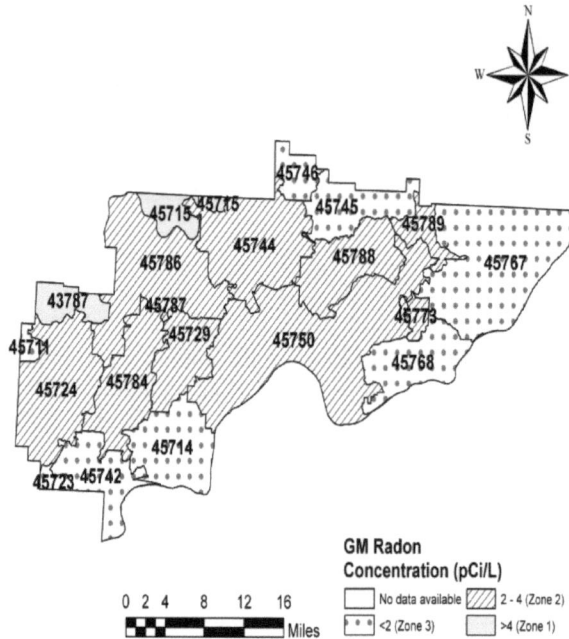

Figure 84b: GM ^{222}Rn concentrations in zip codes of Washington county based on the USEPA classification.

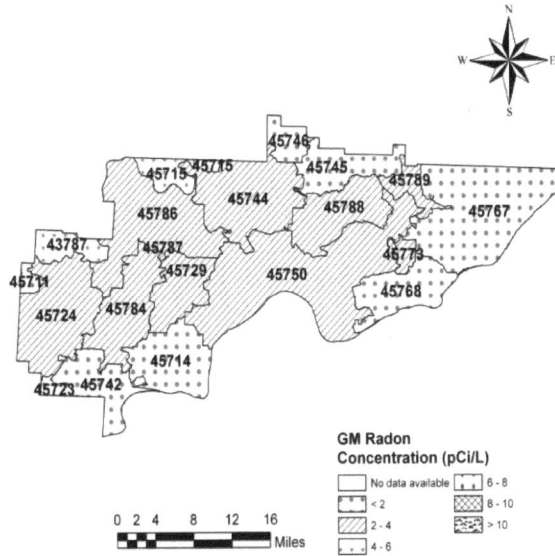

Figure 84c: GM ^{222}Rn concentrations in zip codes of Washington county based on the 2 pCi/L breakdown classification.

Table 85: ^{222}Rn statistics for Wayne county

Zip code	No.	Min.	Max.	AM	GM	SD	CV
44214	24	0.3	5.6	2.9	2.29	1.67	0.09
44217	28	0.1	90.4	9.45	2.88	18.33	0.94
44230	99	0.3	77.5	7.22	4.34	9.86	0.5
44270	43	0.1	14.7	3.9	2.6	3.37	0.17
44273*	93	0.4	48.4	6.49	4.08	8.96	0.46
44276	17	1.5	55	7.93	4.69	12.58	0.64
44287	39	0.1	17	4.78	2.85	4.49	0.23
44606	33	0.4	83	8.14	4.53	13.99	0.72
44611*	8	1.8	9.2	4.33	3.81	2.42	0.12
44618	47	0.3	17.9	5.59	3.43	5.31	0.27
44624*	16	1.1	58	8.66	4.16	15.11	0.77
44627	3	2.2	17.5	7.3	4.39	8.83	0.45
44638*	7	0.9	11.8	4.79	3.75	3.56	0.18
44645	34	0.1	24.2	4.13	2.33	4.92	0.25
44659	1	1.5	1.5	1.5	1.5	0	0
44662*	93	0.1	110	12.29	5.59	19.36	0.99
44666*	32	0.4	16.2	5.98	4.66	3.76	0.19
44667	224	0.1	368.7	17.3	4.51	49.8	2.55
44676	18	1.1	164.2	20.16	5.05	45.49	2.33
44677	43	0.8	259	14.89	5.11	40.29	2.06
44691	867	0.1	498.4	11.58	4.47	29.06	1.49
44840*	26	0.3	185.3	17.34	5.93	37.57	1.92
44636**	2	1	1	1	1	0	0

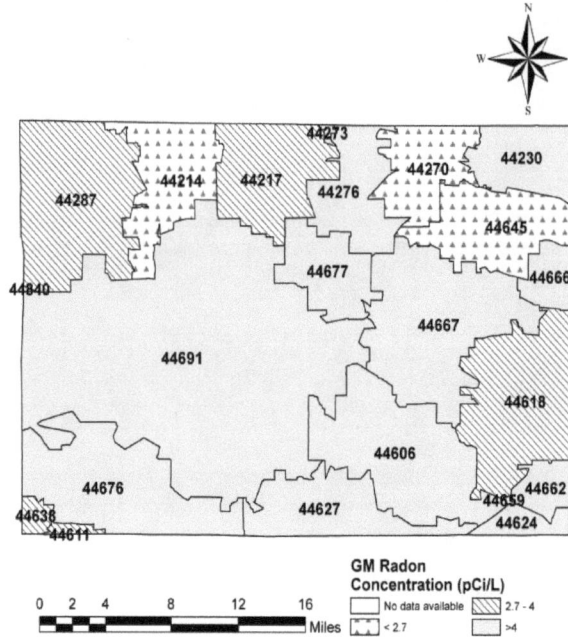

Figure 85a: GM ^{222}Rn concentrations in zip codes of Wayne county based on the WHO-USEPA classification.

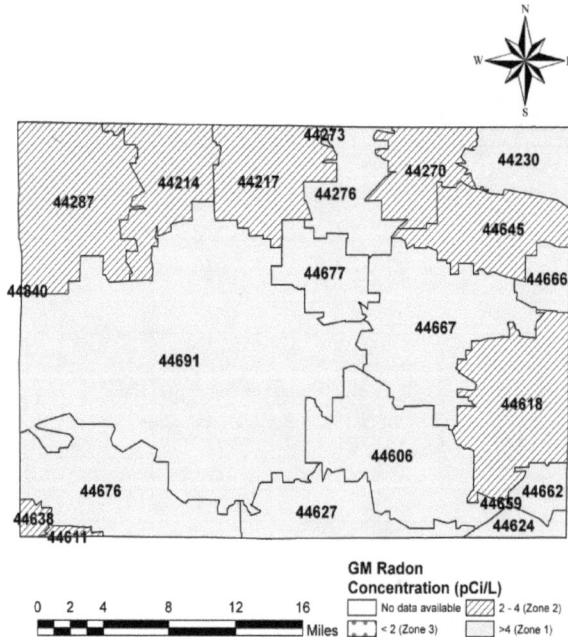

Figure 85b: GM ^{222}Rn concentrations in zip codes of Wayne county based on the USEPA classification.

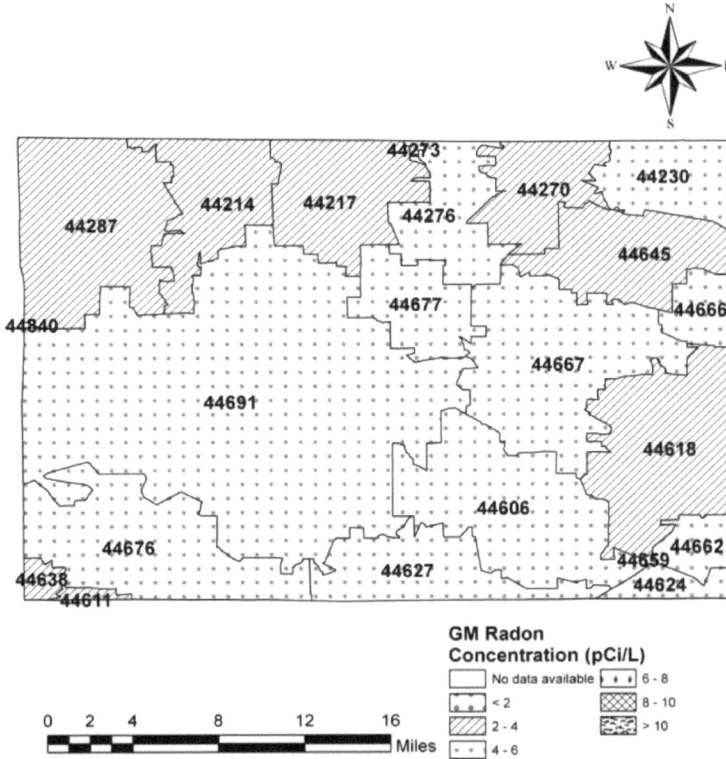

Figure 85c: GM ^{222}Rn concentrations in zip codes of Wayne county based on the 2 pCi/L breakdown classification.

Table 86: ^{222}Rn statistics for Williams county

Zip code	No.	Min.	Max.	AM	GM	SD	CV
43501	1	1.2	1.2	1.2	1.2	0	0
43502*	91	0.2	19.4	5.06	3.4	4.41	0.23
43505	1	1.4	1.4	1.4	1.4	0	0
43506	94	0.1	53.4	6.67	3.41	8.72	0.45
43517	11	0.4	67.8	8.99	2.51	19.89	1.02
43518	9	1	5.5	2.7	2.29	1.52	0.08
43521*	13	0.6	12.9	4.51	3.4	3.32	0.17
43531	1	1.6	1.6	1.6	1.6	0	0
43543	23	0.7	9.4	3.72	2.94	2.76	0.14
43554	10	0.5	2.5	1.26	1.12	0.6	0.03
43557	14	0.4	18.7	5.4	2.92	5.64	0.29
43570	16	0.7	22.9	6.53	3.81	6.7	0.34

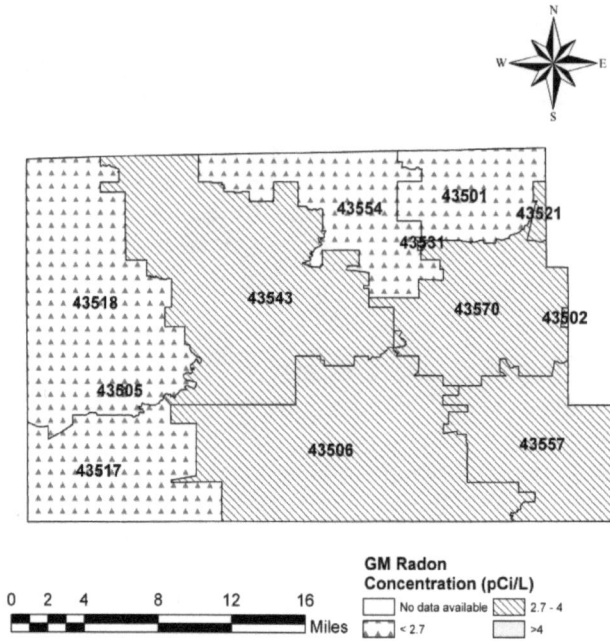

Figure 86a: GM ^{222}Rn concentrations in zip codes of Williams county based on the WHO-USEPA classification.

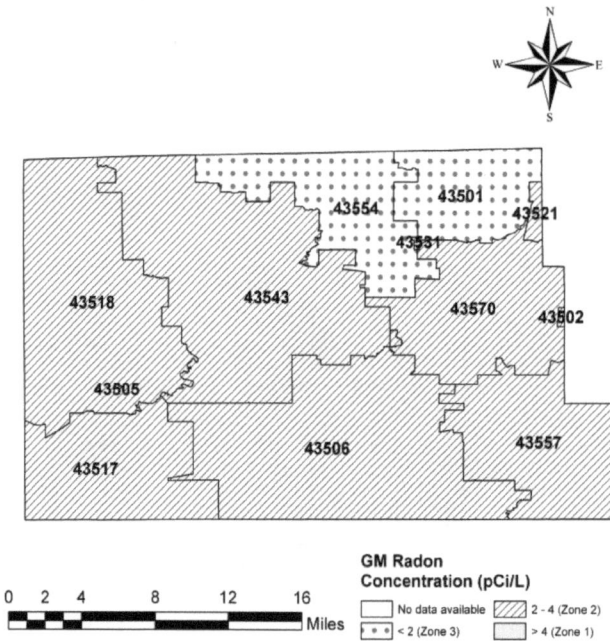

Figure 86b: GM ^{222}Rn concentrations in zip codes of Williams county based on the USEPA classification.

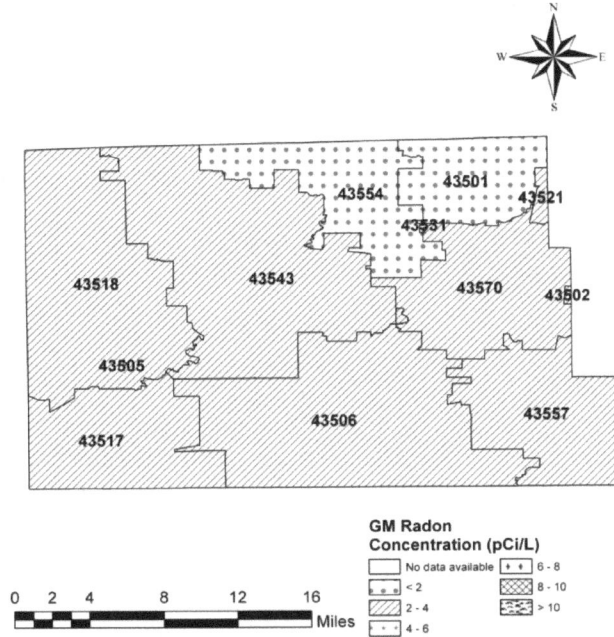

GM Radon Concentration (pCi/L)

No data available | 6 - 8
< 2 | 8 - 10
2 - 4 | > 10
4 - 6

Figure 86c: GM ^{222}Rn concentrations in zip codes of Williams county based on the 2 pCi/L breakdown classification.

Table 87: ^{222}Rn statistics for Wood county

Zip code	No.	Min.	Max.	AM	GM	SD	CV
43402	1188	0.1	100	3.72	1.96	5.07	0.26
43403	13	0.1	4.4	1.17	0.81	1.11	0.06
43406	11	0.1	5.9	3.1	2.12	2.11	0.11
43413	31	0.1	35	4.97	2.4	6.87	0.35
43430*	54	0.4	27.8	4.84	3.22	4.99	0.26
43437	6	0.1	31.2	6.32	1.07	12.32	0.63
43443	40	0.2	15.3	3.85	2.58	3.26	0.17
43447	42	0.2	17.2	4.98	2.96	4.8	0.25
43450	146	0.1	33	3.77	2.21	4.52	0.23
43451	14	0.1	7.8	2.1	1.2	2.04	0.1
43457	10	0.1	9.13	3.91	2.26	3.52	0.18
43460	118	0.1	25	4.24	2.51	4.49	0.23
43462	27	0.1	11.8	1.86	0.86	2.73	0.14
43465	66	0.1	18.8	4.54	2.44	4.7	0.24

Zip code	No.	Min.	Max.	AM	GM	SD	CV
43466	23	0.1	9	1.69	1.12	1.93	0.1
43511	15	0.1	24.1	5.45	2.12	7.84	0.4
43516[*]	13	0.7	6.1	2.35	2.04	1.39	0.07
43522	60	0.1	27.5	4.06	2.18	5.33	0.27
43525	35	0.1	14.1	3.52	1.84	3.94	0.2
43551	1875	0.1	145.5	4.68	2.62	6.87	0.35
43565	6	0.5	15.9	7.62	4.52	6.76	0.35
43569	44	0.1	18.1	2.57	1.31	3.77	0.19
43605[*]	58	0.1	12.1	3.29	2.09	2.8	0.14
43619	88	0.1	18.7	4.96	2.81	4.54	0.23
44817	30	0.1	18.4	4.31	2.59	3.91	0.2
44830[*]	272	0.1	86.2	6.32	3.21	8.51	0.44
45872	28	0.1	14.8	3.42	2.11	3.65	0.19
43414[**]	8	0.5	3	2.15	1.76	1.05	0.05
43441[**]	3	0.7	1.6	1.3	1.21	0.52	0.03
43552[**]	3	1	7.8	3.27	1.98	3.93	0.2

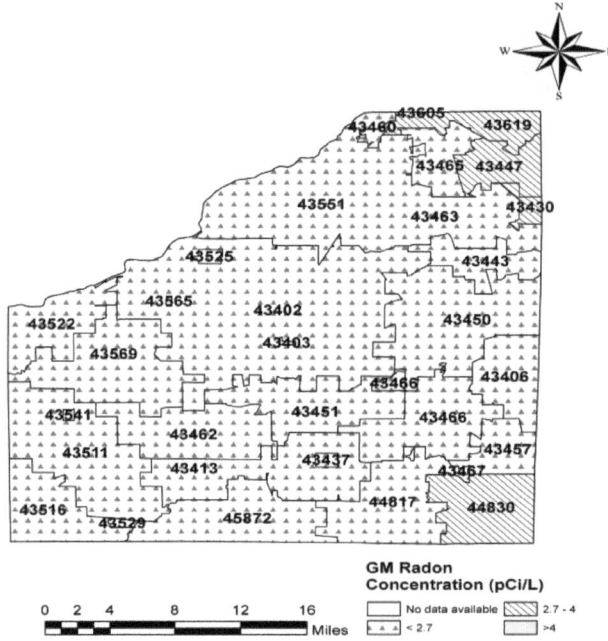

Figure 87a: GM ^{222}Rn concentrations in zip codes of Wood county based on the WHO-USEPA classification.

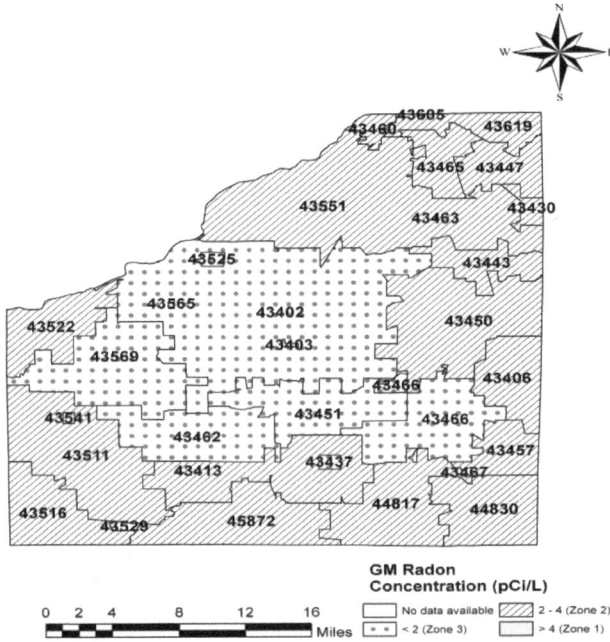

Figure 87b: GM ^{222}Rn concentrations in zip codes of Wood county based on the USEPA classification.

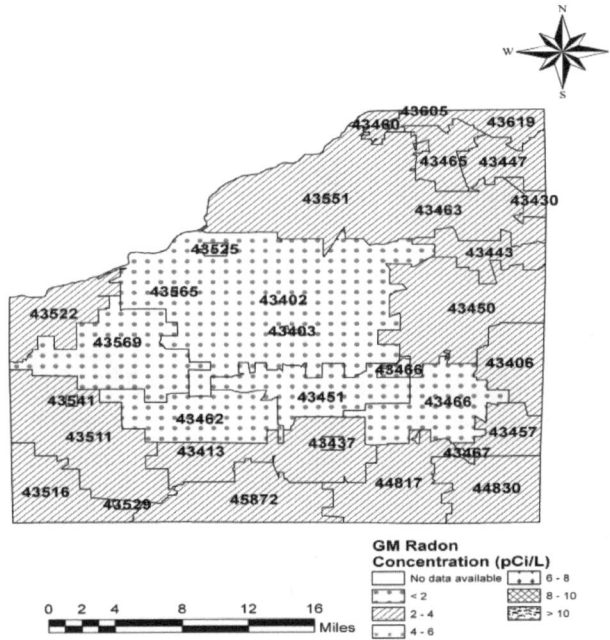

Figure 87c: GM ^{222}Rn concentrations in zip codes of Wood county based on the 2 pCi/L breakdown classification.

Table 88: ^{222}Rn statistics for Wyandot county

Zip code	No.	Min.	Max.	AM	GM	SD	CV
43316	63	0.6	42.1	4.7	3.02	6.45	0.33
43323	15	1.4	31.9	5.78	3.5	8.09	0.41
43330	2	0.9	3.1	2	1.67	1.56	0.08
43332*	18	2	30.3	12.28	8.78	8.91	0.46
43337*	7	0.8	8.6	4.06	3.05	2.96	0.15
43351	294	0.1	49.2	6.02	3.47	8.1	0.41
43359	9	0.5	10.8	3	2.04	3.16	0.16
44844	13	0.2	26.1	7.32	3.32	8.3	0.42
44849	52	0.2	20.9	4.15	2.88	4.36	0.22
44882	32	0.5	9.1	4.19	2.91	3	0.15
45843*	23	0.1	30.2	6.92	4.35	7.51	0.38
45867*	7	0.2	80.2	14.73	3.12	29.17	1.49

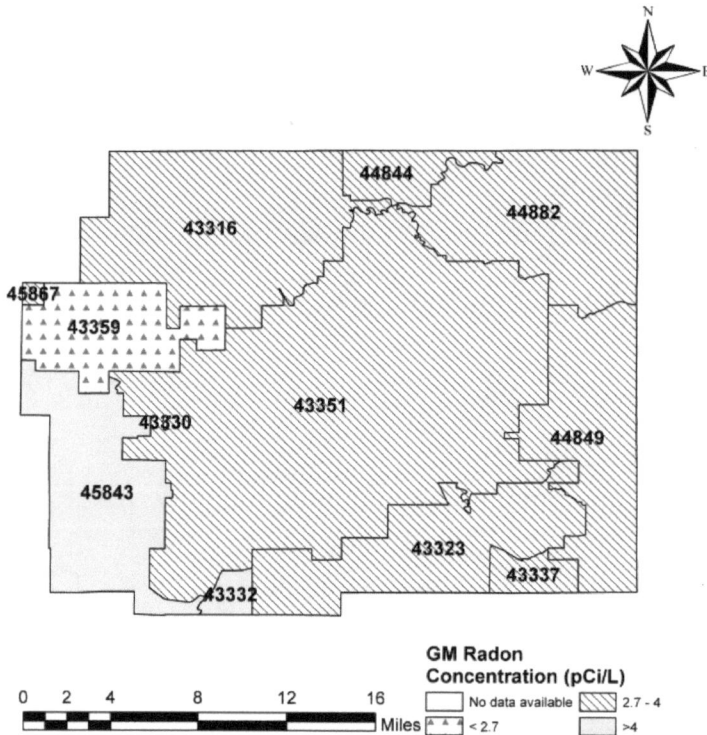

Figure 88a: GM ^{222}Rn concentrations in zip codes of Wyandot county based on the WHO-USEPA classification.

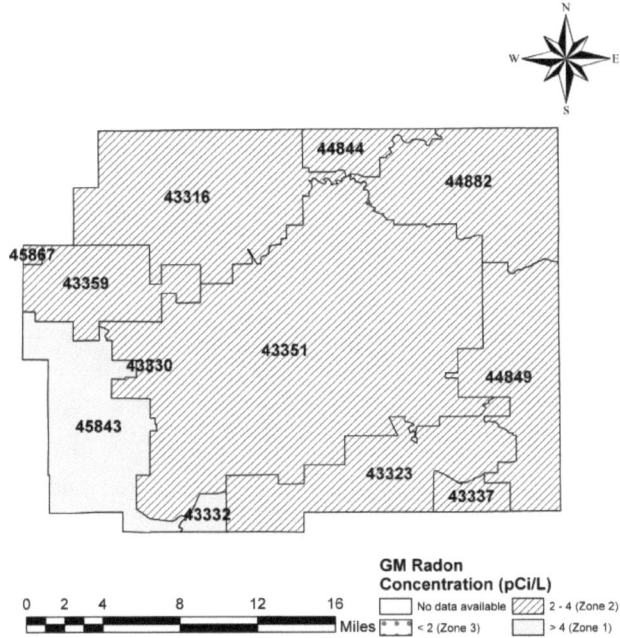

Figure 88b: GM ^{222}Rn concentrations in zip codes of Wyandot county based on the USEPA classification.

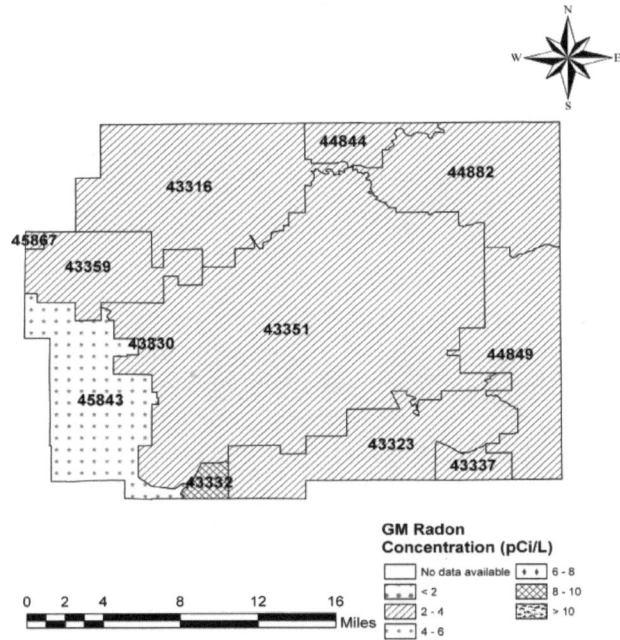

Figure 88c: GM ^{222}Rn concentrations in zip codes of Wyandot county based on the 2 pCi/L breakdown classification.

CHAPTER 4

Radon Mitigation in the State of Ohio, USA

Abstract: This chapter provides details on the mitigation of ^{222}Rn concentrations in Ohio, USA through development of associated geospatial maps and computation of statistical metrics. The geospatial and statistical analyses are examined at both the county level and the zip code level to have a thorough understanding of the ^{222}Rn mitigation that assists in accurately estimating the influence of adopted mitigation systems in controlling the in-house ^{222}Rn concentrations. Additionally, a novel approach to rank the performance of different mitigation systems has been developed. The developed methodology, incorporated in this chapter is used to identify the best performing mitigation system in Ohio and serves as a case study for radon mitigation specialists in ranking their respective mitigation systems.

Keywords: Radon, mitigation, pre-mitigation, post-mitigation, radon mitigation in Ohio, geospatial radon mitigation maps, radon mitigation statistical metrics, WHO-USEPA radon map for mitigation, USEPA radon map for mitigation, 2 pCi/L breakdown radon map for mitigation, minimum radon concentration statistics, maximum radon concentration statistics, arithmetic mean radon concentration statistics, geometric mean radon concentration statistics, standard deviation in radon measurements statistics, coefficient of variation in radon measurements statistics, median radon concentration statistics, first quartile radon concentration statistics, third quartile radon concentration statistics, mitigation system ranking methodology.

4.1. BACKGROUND ON MITIGATION OF INDOOR ^{222}RN CONCENTRA-TIONS IN OHIO

While ^{222}Rn monitoring in Ohio homes started as early as in 1988, mitigation data were only collected from 2002 onwards, after Ohio passed a law in 2001 that required the ^{222}Rn mitigation contractors and specialists to report the mitigation data to the ODH (ODH Chapter 2001). As a result of the law, mitigation data from homes were reported by the mitigation contractors and specialists to the ODH that were subsequently forwarded to the APRG group in

Ashok Kumar and Akhil Kadiyala

Civil Engineering Department at The University of Toledo, which maintained the ORIS database. The Ohio ^{222}Rn mitigation database includes details regarding the mitigation contractor (license number), test house contact (name, phone number), test house location (street address, city, county, zip code, state), and mitigation (type of the mitigation system installed, start date, completion date, pre-mitigation ^{222}Rn level, post-mitigation ^{222}Rn level). While the ^{222}Rn mitigation test end dates are used in assigning the year of mitigation test completion, the ^{222}Rn pre-mitigation level (Pre-ML) and post-mitigation levels (Post-ML) are used to compute the percentage (%) removal for each test house using equation 4.1. The accumulation of mitigation records with all the above mentioned details (provided by mitigation contractors) from 2002 to 2012 led to the development of an Ohio mitigation database with a total of 39,858 data points.

$$Percentage\ Removal = \frac{(Pre-ML - Post-ML)}{Pre-ML} \times 100 \qquad (4.1)$$

where,

Pre-ML is the pre-mitigation ^{222}Rn concentrations in the test house; and *Post-ML* is the post-mitigation ^{222}Rn concentrations in the test house.

This chapter provides detailed information regarding the performance of installed mitigation systems in controlling the ^{222}Rn problem in Ohio homes (using the GIS based ^{222}Rn GM concentration maps and the computation of different statistical metrics) in addition to the development of a statistical procedure to rank and identify the best performing mitigation system.

4.2. MITIGATION OF INDOOR ^{222}RN CONCENTRATIONS IN OHIO – A COUNTY BASED ANALYSIS

Table **1** presents a summary of the computed pre-mitigation ^{222}Rn statistics for about 82 counties in Ohio. Of the available 88 counties in Ohio, ^{222}Rn mitigation tests were performed in 82 counties. The six counties that were not tested for mitigation in Ohio are Adams, Gallia, Lawrence, Morgan, Noble, and Vinton. It may be noted from Table **1** that the 82 Ohio counties had pre-mitigation GM ^{222}Rn concentrations exceeding 4 pCi/L, thereby, indicating that 100% of the monitored Ohio counties exhibited the property of pre-mitigation GM ^{222}Rn concentrations exceeding both the WHO and the USEPA action levels. The relatively higher pre-mitigation GM ^{222}Rn concentrations observed in Table **1** when compared with the homes database county statistics (Table **1**) in Chapter 2 may be attributed to the reason that mitigation tests are generally performed only in homes with higher radon concentrations or those exceeding the USEPA action limit. Knox county had the highest pre-mitigation GM ^{222}Rn concentration (29.95 pCi/L), followed by Carroll (18.23 pCi/L), Logan (16.99 pCi/L), and Coshocton (16.53 pCi/L). The county with the maximum number of homes tested for ^{222}Rn mitigation in Ohio is Franklin (8,187 observations). Fig. **1** illustrates the graphical presentation of the number of Ohio homes tested for ^{222}Rn mitigation in each county. Figs. **2a**, **2b**, and **2c** illustrate the graphical representations of pre-mitigation ^{222}Rn GM concentrations in Ohio counties based on the WHO-USEPA, the USEPA, and the 2 pCi/L breakdown classifications, respectively.

Table **2** presents a summary of the post-mitigation ^{222}Rn statistics for the monitored 82 counties in Ohio. There are a total of 81 counties with post-mitigation GM ^{222}Rn concentrations less than 2.7 pCi/L and 1 county with post-mitigation GM ^{222}Rn concentration between 2.7 and 4 pCi/L. About 98.78% and 100% of the monitored Ohio counties had post-mitigation GM ^{222}Rn concentrations that did not exceed the WHO and the USEPA action levels, respectively. These statistics indicate that the installed mitigation systems in Ohio homes worked effectively in reducing the pre-mitigation GM ^{222}Rn concentrations. Meigs county had the highest post-mitigation GM ^{222}Rn concentration (3.8 pCi/L), followed by Coshocton (2.37 pCi/L), Fayette (2.3 pCi/L), and Highland (2.22 pCi/L). Figs. **3a**, **3b**, and **3c** illustrate the graphical representations of post-mitigation GM ^{222}Rn concentrations in Ohio counties based on the WHO-USEPA, the USEPA, and the 2 pCi/L breakdown classifications, respectively.

Table 1: Pre-mitigation ^{222}Rn statistics for Ohio counties

County Name	No.	Min.	Max.	AM	GM	SD	CV	Md	Q1	Q3
Allen	85	2.30	60.30	11.90	9.07	12.33	1.04	8.00	6.10	12.15
Ashland	162	2.30	141.00	15.02	10.83	18.25	1.22	9.35	6.60	15.00
Ashtabula	38	3.70	40.10	10.89	8.17	10.13	0.93	6.35	5.00	12.00
Athens	15	4.30	18.60	11.20	9.74	5.63	0.50	11.00	5.40	17.10
Auglaize	25	3.90	47.00	10.45	8.27	9.68	0.93	7.00	5.90	10.35
Belmont	80	3.90	87.20	13.51	9.86	14.65	1.08	8.75	5.60	12.00
Brown	10	2.70	15.00	6.61	5.99	3.39	0.51	5.75	4.60	8.10
Butler	1856	2.40	72.40	8.93	7.75	6.24	0.70	7.00	5.50	10.00
Carroll	84	4.40	201.00	33.03	18.23	45.75	1.39	14.90	8.40	32.00
Champaign	38	5.10	106.50	18.71	13.96	18.31	0.98	13.25	6.80	23.50
Clark	306	0.70	121.20	12.95	10.47	11.03	0.85	9.95	6.70	15.30
Clermont	1215	2.10	40.00	7.92	7.21	4.12	0.52	6.90	5.40	9.00
Clinton	32	3.40	20.90	9.37	8.39	4.50	0.48	7.95	5.30	12.10
Columbiana	85	4.00	136.50	19.73	11.96	26.04	1.32	9.20	6.20	19.00
Coshocton	9	6.20	33.60	19.16	16.53	10.43	0.54	17.30	11.00	31.00
Crawford	26	3.50	33.60	10.83	9.41	6.55	0.60	8.75	6.70	13.90
Cuyahoga	1948	1.20	116.40	8.14	6.94	6.41	0.79	6.10	4.90	8.60
Darke	35	3.30	30.70	11.84	10.28	6.43	0.54	9.60	7.50	16.00
Defiance	10	4.80	42.80	11.05	8.34	11.70	1.06	6.90	5.20	8.00
Delaware	2430	3.30	100.00	10.78	9.14	8.23	0.76	8.50	6.00	12.50
Erie	137	4.00	248.00	13.61	9.90	22.38	1.64	9.20	6.50	13.70
Fairfield	378	3.90	190.00	14.14	11.10	14.58	1.03	10.30	7.00	15.30
Fayette	2	5.20	12.70	8.95	8.13	5.30	0.59	8.95	5.20	12.70
Franklin	8187	0.90	232.00	12.95	10.45	11.65	0.90	9.60	6.60	15.00
Fulton	27	4.00	20.00	8.64	7.83	4.28	0.50	7.30	5.70	10.10
Geauga	208	2.30	45.00	7.25	6.37	5.09	0.70	5.80	4.70	7.70
Greene	1340	2.00	79.00	11.73	9.62	9.19	0.78	8.90	6.10	13.70
Guernsey	13	5.80	90.00	22.04	15.29	25.02	1.14	12.00	10.00	15.00
Hamilton	2227	1.80	1400.00	8.74	6.82	41.93	4.79	6.20	5.00	8.30
Hancock	515	1.00	538.00	10.78	8.28	24.53	2.27	7.40	5.50	10.80
Hardin	5	5.00	18.00	9.78	8.81	5.18	0.53	8.10	6.40	11.40
Harrison	12	4.60	57.70	17.53	11.82	18.23	1.04	9.00	5.90	16.70
Henry	11	2.60	15.00	8.59	7.41	4.52	0.53	7.90	4.50	13.00
Highland	6	4.50	11.90	7.42	6.95	2.91	0.39	7.30	4.70	8.80

County Name	No.	Min.	Max.	AM	GM	SD	CV	Md	Q1	Q3
Hocking	6	4.30	16.20	11.27	10.43	4.10	0.36	12.50	9.10	13.00
Holmes	41	4.20	72.00	14.77	11.34	12.98	0.88	10.30	6.20	20.90
Huron	72	4.00	68.80	14.44	11.07	12.48	0.86	9.90	6.00	18.00
Jackson	10	4.00	14.60	7.53	6.83	3.71	0.49	5.70	4.80	11.30
Jefferson	30	4.00	48.80	14.37	10.49	13.34	0.93	8.95	5.40	15.50
Knox	187	4.40	742.70	59.84	29.95	98.51	1.65	24.70	13.00	67.00
Lake	396	3.00	125.00	9.62	7.72	10.34	1.07	6.50	5.10	10.10
Licking	716	1.60	303.30	23.97	15.11	32.33	1.35	13.00	7.90	25.50
Logan	69	3.90	287.00	25.31	16.99	37.03	1.46	16.30	9.60	26.00
Lorain	1101	1.70	100.00	9.84	7.97	8.63	0.88	7.00	5.10	11.00
Lucas	1028	3.00	100.00	9.19	7.77	7.24	0.79	7.00	5.20	10.90
Madison	73	2.40	36.00	12.24	10.32	7.84	0.64	10.40	6.70	14.00
Mahoning	169	1.80	95.60	10.37	7.77	11.96	1.15	6.50	4.80	11.00
Marion	59	4.30	55.20	14.26	11.84	10.11	0.71	11.10	7.80	16.60
Medina	840	2.40	93.30	8.86	7.33	7.73	0.87	6.30	5.00	9.30
Meigs	1	11.60	11.60	11.60	11.60	0.00	0.00	11.60	11.60	11.60
Mercer	9	4.50	16.60	10.42	9.69	3.88	0.37	11.00	8.00	12.30
Miami	454.	2.50	3901.00	21.57	9.78	185.77	8.61	9.50	6.00	14.00
Monroe	2	10.00	20.30	15.15	14.25	7.28	0.48	15.15	10.00	20.30
Montgomery	2101	1.60	98.10	10.16	8.49	8.16	0.80	7.80	5.70	11.40
Morrow	26	4.20	23.70	12.29	10.78	6.32	0.51	10.65	6.50	18.00
Muskingum	97	4.00	129.70	16.54	11.84	18.52	1.12	10.50	6.60	18.60
Ottawa	67	4.00	46.00	10.15	8.43	7.74	0.76	7.60	5.00	12.80
Paulding	2	9.00	9.10	9.05	9.05	0.07	0.01	9.05	9.00	9.10
Perry	6	4.80	42.00	14.58	11.07	13.80	0.95	10.90	5.90	13.00
Pickaway	42	4.10	116.00	21.32	14.58	22.45	1.05	11.70	8.00	27.20
Pike	4	4.00	50.00	22.80	15.72	20.09	0.88	18.60	4.00	25.00
Portage	333	1.40	58.00	8.75	7.30	6.99	0.80	6.20	5.00	9.70
Preble	25	4.00	22.30	11.62	10.33	5.51	0.47	10.60	6.90	15.00
Putnam	24	4.15	26.90	11.25	9.85	6.31	0.56	8.95	6.80	13.30
Richland	315	3.10	316.50	18.19	11.60	30.81	1.69	10.00	6.40	18.00
Ross	63	4.00	200.00	14.52	9.83	25.33	1.75	8.10	5.90	14.60
Sandusky	84	4.00	804.00	24.37	11.75	88.17	3.62	10.95	6.50	16.70
Scioto	5	3.90	15.20	8.64	7.45	5.14	0.60	6.60	4.50	13.00
Seneca	89	2.30	70.00	15.36	11.78	13.89	0.90	10.00	7.60	16.90
Shelby	64	3.80	51.40	15.92	12.47	12.42	0.78	10.75	7.70	21.30

County Name	No.	Min.	Max.	AM	GM	SD	CV	Md	Q1	Q3
Stark	2571	1.60	150.10	11.91	9.55	10.91	0.92	8.80	6.00	13.80
Summit	2196	2.60	277.00	9.27	7.55	10.19	1.10	6.50	5.10	9.70
Trumbull	71	3.50	39.90	8.30	7.14	6.44	0.78	6.40	5.60	8.20
Tuscarawas	624	3.70	178.00	14.60	10.60	17.53	1.20	9.20	6.10	15.00
Union	130	4.10	49.00	10.66	9.12	7.12	0.67	8.95	6.00	12.50
Van Wert	1	6.00	6.00	6.00	6.00	0.00	0.00	6.00	6.00	6.00
Warren	2495	1.40	111.90	8.45	7.44	5.91	0.70	7.00	5.30	9.20
Washington	56	4.10	24.00	9.02	8.05	4.78	0.53	7.95	5.40	10.70
Wayne	390	3.50	539.00	16.84	10.07	36.04	2.14	8.00	5.70	14.10
Williams	6	4.20	17.90	7.77	6.70	5.30	0.68	5.10	4.90	9.40
Wood	718	1.00	51.90	9.26	7.99	6.06	0.65	7.30	5.30	11.20
Wyandot	11	4.20	66.60	16.20	10.33	19.96	1.23	7.60	6.05	11.90
Unknown	422	1.00	208.00	10.63	8.18	13.71	1.29	7.10	5.40	10.50
All of Ohio	39858	0.70	3901.00	11.52	8.85	26.71	2.32	8.00	5.70	12

Figure 1: Number of Ohio homes tested for [222]Rn mitigation in each county.

Figure 2a: GM ^{222}Rn Pre-mitigation concentrations in Ohio counties based on the WHO-USEPA classification.

Figure 2b: GM ^{222}Rn Pre-mitigation concentrations in Ohio counties based on the USEPA classification.

Figure 2c: GM ^{222}Rn Pre-mitigation concentrations in Ohio counties based on the 2 pCi/L breakdown classifications.

Table 2: Post-mitigation ^{222}Rn statistics for Ohio counties

County Name	No.	Min.	Max.	AM	GM	SD	CV	Md	Q1	Q3
Allen	85	0.20	3.85	1.26	0.96	0.93	0.74	1.00	0.60	1.60
Ashland	162	0.20	3.90	1.38	1.14	0.87	0.63	1.10	0.80	1.80
Ashtabula	38	0.30	3.60	1.16	0.95	0.76	0.66	0.95	0.60	1.40
Athens	15	0.80	7.60	2.12	1.66	1.81	0.86	1.30	0.90	2.60
Auglaize	25	0.40	3.70	1.35	1.12	0.91	0.67	1.10	0.70	1.60
Belmont	80	0.30	6.40	1.74	1.38	1.14	0.65	1.60	0.80	2.50
Brown	10	0.50	3.80	1.50	1.31	0.91	0.61	1.35	1.00	1.70
Butler	1856	0.10	7.50	1.20	1.01	0.74	0.61	1.00	0.70	1.50
Carroll	84	0.20	20.70	1.89	1.36	2.38	1.26	1.40	0.80	2.50
Champaign	38	0.20	4.20	1.73	1.41	1.04	0.60	1.45	1.00	2.40
Clark	306	0.10	8.00	1.63	1.31	1.06	0.65	1.30	0.80	2.30
Clermont	1215	0.00	12.30	1.21	1.03	0.80	0.66	1.00	0.70	1.50
Clinton	32	0.30	3.60	1.44	1.23	0.79	0.55	1.35	0.90	1.70
Columbiana	85	0.30	4.40	1.49	1.21	0.96	0.64	1.20	0.80	2.10
Coshocton	9	1.50	3.40	2.48	2.37	0.75	0.30	2.50	1.70	3.20
Crawford	26	0.30	3.80	1.48	1.27	0.81	0.55	1.25	1.00	1.80
Cuyahoga	1948	0.00	4.40	1.41	1.11	0.91	0.64	1.20	0.70	2.00
Darke	35	0.40	3.40	1.61	1.39	0.83	0.51	1.50	1.00	2.20
Defiance	10	0.30	2.90	0.91	0.73	0.75	0.83	0.68	0.50	1.00
Delaware	2430	0.06	7.70	1.36	1.10	0.88	0.65	1.10	0.70	1.80
Erie	137	0.30	21.10	1.66	1.17	2.34	1.41	1.10	0.70	1.70
Fairfield	378	0.10	8.20	1.41	1.11	0.97	0.69	1.10	0.70	1.90
Fayette	2	1.70	3.10	2.40	2.30	0.99	0.41	2.40	1.70	3.10
Franklin	8187	0.00	61.00	1.49	1.15	1.34	0.90	1.20	0.70	2.10
Fulton	27	0.20	3.10	1.14	0.93	0.78	0.69	0.80	0.60	1.60
Geauga	208	0.10	3.60	1.34	1.09	0.83	0.62	1.10	0.70	1.80
Greene	1340	0.10	11.50	1.29	1.04	0.89	0.69	1.00	0.70	1.60
Guernsey	13	0.80	3.80	1.86	1.71	0.81	0.43	1.60	1.40	2.10
Hamilton	2227	0.00	9.10	1.34	1.11	0.85	0.63	1.10	0.70	1.70
Hancock	515	0.10	5.40	1.20	0.94	0.84	0.70	0.90	0.60	1.60
Hardin	5	0.30	1.70	1.10	0.83	0.73	0.67	1.50	0.30	1.70
Harrison	12	0.40	3.90	2.54	2.15	1.22	0.48	2.75	1.50	3.60
Henry	11	0.30	2.30	1.04	0.83	0.66	0.64	1.10	0.30	1.30
Highland	6	0.80	2.94	2.41	2.22	0.81	0.34	2.65	2.50	2.90
Hocking	6	0.40	2.90	1.55	1.29	0.94	0.60	1.30	1.00	2.40
Holmes	41	0.60	3.80	1.96	1.72	0.97	0.49	1.90	1.10	2.70
Huron	72	0.30	3.70	1.41	1.20	0.78	0.56	1.25	0.90	1.80

County Name	No.	Min.	Max.	AM	GM	SD	CV	Md	Q1	Q3
Jackson	10	0.70	3.40	2.08	1.87	0.90	0.43	2.05	1.30	3.00
Jefferson	30	0.05	3.20	1.15	0.88	0.76	0.66	1.00	0.70	1.60
Knox	187	0.30	50.00	1.96	1.37	3.81	1.94	1.50	0.80	2.15
Lake	396	0.10	3.70	1.11	0.89	0.75	0.68	0.90	0.50	1.50
Licking	716	0.01	57.50	1.86	1.31	2.91	1.57	1.30	0.80	2.30
Logan	69	0.30	8.20	2.09	1.73	1.32	0.63	1.80	1.20	2.90
Lorain	1101	0.00	9.70	1.06	0.81	0.84	0.79	0.80	0.50	1.30
Lucas	1028	0.10	3.90	1.07	0.84	0.78	0.73	0.90	0.50	1.30
Madison	73	0.20	9.00	1.48	1.17	1.23	0.83	1.10	0.80	2.00
Mahoning	169	0.20	3.70	1.08	0.88	0.73	0.68	0.80	0.50	1.40
Marion	59	0.20	3.80	1.58	1.31	0.88	0.56	1.50	0.90	2.20
Medina	840	0.20	8.80	1.26	1.01	0.90	0.72	1.00	0.60	1.60
Meigs	1	3.80	3.80	3.80	3.80	0.00	0.00	3.80	3.80	3.80
Mercer	9	0.10	2.80	1.52	1.15	0.88	0.58	1.60	1.00	1.70
Miami	454	0.20	6.00	1.43	1.16	0.94	0.66	1.10	0.80	1.90
Monroe	2	1.20	2.60	1.90	1.77	0.99	0.52	1.90	1.20	2.60
Montgomery	2101	0.00	18.40	1.27	0.99	0.99	0.78	1.00	0.60	1.60
Morrow	26	0.20	3.80	1.43	1.18	0.84	0.59	1.20	0.80	2.00
Muskingum	97	0.20	14.50	1.89	1.41	2.01	1.07	1.40	0.90	2.60
Ottawa	67	0.30	3.90	1.60	1.24	1.07	0.67	1.20	0.70	2.70
Paulding	2	0.30	0.70	0.50	0.46	0.28	0.57	0.50	0.30	0.70
Perry	6	0.40	3.00	1.87	1.59	0.88	0.47	2.00	1.50	2.30
Pickaway	42	0.40	3.90	1.80	1.52	0.96	0.53	1.80	1.00	2.20
Pike	4	0.30	1.20	0.88	0.77	0.39	0.45	1.00	0.30	1.00
Portage	333	0.20	3.50	1.07	0.88	0.70	0.66	0.90	0.60	1.30
Preble	25	0.60	3.90	1.55	1.38	0.81	0.52	1.30	1.00	1.80
Putnam	24	0.30	3.90	1.13	0.89	0.88	0.78	0.85	0.60	1.15
Richland	315	0.20	5.20	1.49	1.28	0.82	0.55	1.30	0.90	2.00
Ross	63	0.40	3.60	1.64	1.42	0.87	0.53	1.40	1.00	2.20
Sandusky	84	0.20	4.10	1.29	0.99	0.94	0.73	1.05	0.60	1.80
Scioto	5	0.70	3.60	1.90	1.50	1.42	0.75	1.00	0.90	3.30
Seneca	89	0.16	6.10	1.66	1.28	1.13	0.68	1.45	0.70	2.40
Shelby	64	0.10	3.80	1.40	1.18	0.80	0.57	1.20	0.80	1.70
Stark	2571	0.00	7.20	1.10	0.89	0.78	0.71	0.90	0.60	1.40
Summit	2196	0.00	15.00	1.14	0.91	0.84	0.74	0.90	0.60	1.50
Trumbull	71	0.30	5.50	1.17	0.97	0.86	0.74	0.90	0.70	1.30
Tuscarawas	624	0.10	48.90	1.78	1.42	2.13	1.19	1.50	0.90	2.40
Union	130	0.10	4.05	1.44	1.13	0.94	0.65	1.15	0.70	2.10

County Name	No.	Min.	Max.	AM	GM	SD	CV	Md	Q1	Q3
Van Wert	1	0.60	0.60	0.60	0.60	0.00	0.00	0.60	0.60	0.60
Warren	2495	0.10	67.00	1.24	1.04	1.51	1.22	1.00	0.70	1.50
Washington	56	0.30	4.10	1.89	1.60	0.97	0.51	1.70	1.20	2.70
Wayne	390	0.10	5.10	1.49	1.21	0.93	0.63	1.20	0.80	2.20
Williams	6	0.30	3.80	1.08	0.70	1.35	1.24	0.60	0.40	0.80
Wood	718	0.10	16.00	1.21	0.93	1.07	0.89	0.90	0.60	1.50
Wyandot	11	0.30	3.50	1.09	0.81	0.95	0.87	0.80	0.30	1.10
Unknown	422	0.10	4.00	1.28	1.02	0.87	0.67	1.00	0.60	1.80
All of Ohio	39858	0.00	67.00	1.34	1.06	1.18	0.88	1.00	0.70	1.88

Figure 3a: GM ^{222}Rn Post-mitigation concentrations in Ohio counties based on the WHO-USEPA classification.

Figure 3b: GM ^{222}Rn Post-mitigation concentrations in Ohio counties based on the USEPA classification.

Figure 3c: GM ²²²Rn Post-mitigation concentrations in Ohio counties based on the 2 pCi/L breakdown classifications.

4.3. MITIGATION OF INDOOR ^{222}RN CONCENTRATIONS IN OHIO – A ZIP CODE BASED ANALYSIS

^{222}Rn mitigation tests were performed in 920 Ohio zip codes. Table **3** presents a summary of the computed pre-mitigation ^{222}Rn statistics for the 920 zip codes in Ohio. There are a total of 6 zip codes with pre-mitigation GM ^{222}Rn concentrations between 2.7 and 4 pCi/L, and 914 zip codes with pre-mitigation GM ^{222}Rn concentrations greater than 4 pCi/L; thereby, indicating that 100% and 99.34% of the monitored Ohio zip codes had the pre-mitigation GM ^{222}Rn concentrations exceeding the WHO and the USEPA action levels, respectively. Of the 914 zip codes that had the pre-mitigation GM ^{222}Rn concentrations greater than 4 pCi/L, 516 zip codes had the pre-mitigation GM ^{222}Rn concentrations exceeding 8 pCi/L and 54 zip codes had the pre-mitigation GM ^{222}Rn concentrations exceeding 20 pCi/L. Of the monitored zip codes, zip code 44432 had the highest pre-mitigation GM ^{222}Rn concentration (45.07 pCi/L), followed by the zip codes 43022 (42.27 pCi/L), 43028 (31.9 pCi/L), and 43050 (31.13 pCi/L). Figs. **4a**, **4b**, and **4c** illustrate the graphical representations of pre-mitigation GM ^{222}Rn concentrations in Ohio zip codes based on the WHO-USEPA, the USEPA, and the 2 pCi/L breakdown classifications, respectively.

Table **4** presents a summary of the post-mitigation ^{222}Rn statistics for the monitored 920 zip codes in Ohio. There are a total of 879 zip codes with post-mitigation GM ^{222}Rn concentrations less than 2.7 pCi/L, 37 zip codes with post-mitigation GM ^{222}Rn concentration between 2.7 and 4 pCi/L, and 3 zip codes with post-mitigation GM ^{222}Rn concentration exceeding 4 pCi/L. About 95.54% and 99.57% of the monitored Ohio zip codes had post-mitigation GM ^{222}Rn concentrations that did not exceed the WHO and the USEPA action levels, respectively. These statistics indicate that the installed mitigation systems in Ohio homes were effective in reducing the pre-mitigation GM ^{222}Rn concentrations. The post-mitigation GM ^{222}Rn concentrations were the highest in zip code 43053 (5.92 pCi/L), followed by the zip codes 43076 (4.48 pCi/L), and 44619 (4.34 pCi/L). Figs. **5a**, **5b**, and **5c** illustrate the graphical representations of post-mitigation GM ^{222}Rn concentrations in Ohio zip codes based on the WHO-USEPA, the USEPA, and the 2 pCi/L breakdown classifications, respectively.

Table 3: Pre-mitigation ^{222}Rn statistics for Ohio zip codes

Zip Code	No.	Min.	Max.	AM	GM	SD	CV	Md	Q1	Q3
43001	8	7.00	82.00	34.64	24.67	28.29	0.82	23.50	9.00	50.00
43003	4	9.20	30.60	22.53	20.30	10.26	0.46	25.15	9.20	30.60
43004	205	4.00	40.00	11.70	10.05	7.09	0.61	9.20	6.80	14.60
43005	1	18.20	18.20	18.20	18.20	0.00	0.00	18.20	18.20	18.20
43009	6	6.00	106.50	37.12	25.94	35.73	0.96	30.00	15.60	34.60
43011	13	5.70	58.30	18.05	14.07	14.94	0.83	11.80	8.00	23.00
43012	3	5.20	22.30	11.67	9.55	9.28	0.80	7.50	5.20	22.30
43013	3	8.20	16.80	12.00	11.49	4.39	0.37	11.00	8.20	16.80
43014	3	4.70	16.60	8.80	7.36	6.76	0.77	5.10	4.70	16.60
43015	379	4.00	100.00	12.05	10.05	10.05	0.83	9.20	6.80	13.50
43016	422	1.40	50.00	10.68	9.04	7.18	0.67	8.55	6.00	13.00
43017	1155	0.90	232.00	13.82	10.79	13.87	1.00	9.90	6.50	16.00
43018	1	6.40	6.40	6.40	6.40	0.00	0.00	6.40	6.40	6.40
43019	9	4.40	33.00	16.90	14.43	9.08	0.54	18.20	10.70	23.00
43020	1	46.70	46.70	46.70	46.70	0.00	0.00	46.70	46.70	46.70
43021	173	3.60	30.00	10.09	8.92	5.36	0.53	8.40	6.00	13.10
43022	10	8.10	242.00	66.21	42.27	72.28	1.09	34.45	26.00	80.00
43023	190	4.60	279.00	31.32	19.03	41.35	1.32	16.00	9.00	39.20
43024	3	8.00	32.00	16.20	13.01	13.69	0.84	8.60	8.00	32.00
43025	11	4.00	67.60	24.74	17.10	21.53	0.87	17.20	8.60	43.90
43026	707	4.00	158.00	13.10	10.77	11.40	0.87	10.00	7.10	14.80
43027	1	4.90	4.90	4.90	4.90	0.00	0.00	4.90	4.90	4.90
43028	114	4.60	317.00	57.99	31.90	70.38	1.21	25.20	14.00	80.00
43029	2	7.30	10.30	8.80	8.67	2.12	0.24	8.80	7.30	10.30
43030	1	8.20	8.20	8.20	8.20	0.00	0.00	8.20	8.20	8.20
43031	38	5.60	28.00	11.42	10.48	5.16	0.45	10.50	7.70	14.10
43033	3	9.00	145.00	57.67	29.16	75.80	1.31	19.00	9.00	145.00
43035	413	3.30	70.00	9.84	8.61	6.24	0.63	8.00	6.00	11.40
43036	1	15.00	15.00	15.00	15.00	0.00	0.00	15.00	15.00	15.00
43039	1	6.10	6.10	6.10	6.10	0.00	0.00	6.10	6.10	6.10
43040	80	4.20	38.60	8.89	7.88	5.40	0.61	7.55	5.30	10.60
43041	1	9.10	9.10	9.10	9.10	0.00	0.00	9.10	9.10	9.10
43042	1	32.00	32.00	32.00	32.00	0.00	0.00	32.00	32.00	32.00
43043	1	14.00	14.00	14.00	14.00	0.00	0.00	14.00	14.00	14.00

Zip Code	No.	Min.	Max.	AM	GM	SD	CV	Md	Q1	Q3
43044	4	6.90	48.00	22.23	16.52	18.99	0.85	17.00	6.90	25.00
43045	12	4.40	28.10	10.90	9.65	6.28	0.58	9.50	7.10	11.40
43046	3	5.50	12.10	8.13	7.68	3.50	0.43	6.80	5.50	12.10
43047	1	3.70	3.70	3.70	3.70	0.00	0.00	3.70	3.70	3.70
43050	44	6.20	742.70	80.77	31.13	163.09	2.02	21.65	12.80	53.20
43052	2	9.00	11.00	10.00	9.95	1.41	0.14	10.00	9.00	11.00
43053	2	5.00	32.30	18.65	12.71	19.30	1.04	18.65	5.00	32.30
43054	583	3.90	47.00	8.87	7.94	4.85	0.55	7.40	5.60	10.30
43055	235	3.50	303.30	28.16	17.96	34.50	1.22	17.90	8.10	34.00
43056	34	4.00	108.50	20.62	14.23	21.72	1.05	11.70	7.80	27.00
43058	2	31.70	65.00	48.35	45.39	23.55	0.49	48.35	31.70	65.00
43060	3	5.10	29.60	15.13	11.73	12.84	0.85	10.70	5.10	29.60
43061	20	4.40	29.80	10.95	9.58	6.17	0.56	9.50	6.00	14.40
43062	124	1.60	140.00	14.02	10.72	14.91	1.06	9.50	6.80	15.60
43064	52	4.10	49.00	12.96	10.49	9.52	0.73	10.10	5.80	14.30
43065	886	3.60	80.00	10.51	8.92	7.54	0.72	8.10	6.00	12.00
43066	4	5.00	21.30	12.18	10.65	6.98	0.57	11.20	5.00	13.40
43067	1	16.00	16.00	16.00	16.00	0.00	0.00	16.00	16.00	16.00
43068	170	4.00	60.00	12.38	10.04	9.72	0.78	8.65	6.40	14.10
43071	1	14.90	14.90	14.90	14.90	0.00	0.00	14.90	14.90	14.90
43072	1	6.80	6.80	6.80	6.80	0.00	0.00	6.80	6.80	6.80
43074	54	4.70	64.00	11.98	10.29	8.94	0.75	9.80	7.00	14.70
43075	1	5.50	5.50	5.50	5.50	0.00	0.00	5.50	5.50	5.50
43076	3	10.80	100.00	40.60	22.82	51.44	1.27	11.00	10.80	100.00
43078	22	5.90	33.00	13.51	11.55	8.18	0.61	11.60	6.50	16.80
43080	4	15.00	90.00	36.25	27.11	35.91	0.99	20.00	15.00	20.00
43081	499	1.00	93.00	12.47	10.18	10.11	0.81	9.30	6.40	14.80
43082	696	4.00	87.70	10.20	8.86	7.04	0.69	8.30	6.00	12.00
43084	1	5.00	5.00	5.00	5.00	0.00	0.00	5.00	5.00	5.00
43085	474	1.90	224.00	15.69	12.16	16.08	1.03	11.00	7.60	18.00
43086	1	4.90	4.90	4.90	4.90	0.00	0.00	4.90	4.90	4.90
43089	1	11.30	11.30	11.30	11.30	0.00	0.00	11.30	11.30	11.30
43103	12	6.40	82.00	24.44	17.16	22.30	0.91	20.50	6.80	28.00
43105	18	4.90	39.00	13.47	11.17	9.37	0.70	10.10	6.40	16.20
43106	1	9.00	9.00	9.00	9.00	0.00	0.00	9.00	9.00	9.00
43107	3	10.20	13.70	12.30	12.20	1.85	0.15	13.00	10.20	13.70

Zip Code	No.	Min.	Max.	AM	GM	SD	CV	Md	Q1	Q3
43110	116	4.00	82.00	13.42	10.78	11.14	0.83	9.95	6.80	15.00
43112	8	7.50	104.00	31.15	20.42	34.38	1.10	14.80	10.00	23.00
43113	22	4.10	116.00	22.55	14.51	25.83	1.15	12.50	8.00	27.20
43115	1	5.60	5.60	5.60	5.60	0.00	0.00	5.60	5.60	5.60
43116	2	6.00	11.30	8.65	8.23	3.75	0.43	8.65	6.00	11.30
43117	1	7.10	7.10	7.10	7.10	0.00	0.00	7.10	7.10	7.10
43119	148	4.30	50.50	12.44	10.60	8.02	0.64	9.60	7.00	16.00
43123	141	4.30	77.00	12.63	10.51	9.94	0.79	10.00	6.60	14.70
43124	1	17.70	17.70	17.70	17.70	0.00	0.00	17.70	17.70	17.70
43125	22	4.20	117.00	16.60	11.83	23.08	1.39	9.75	8.70	18.00
43130	56	4.10	70.00	12.35	10.30	10.50	0.85	9.80	7.20	12.40
43132	1	10.00	10.00	10.00	10.00	0.00	0.00	10.00	10.00	10.00
43135	2	4.30	6.50	5.40	5.29	1.56	0.29	5.40	4.30	6.50
43136	3	8.00	19.40	13.23	12.41	5.76	0.44	12.30	8.00	19.40
43138	3	9.10	16.20	12.43	12.09	3.57	0.29	12.00	9.10	16.20
43140	44	2.40	34.10	12.45	10.49	7.59	0.61	10.95	6.70	16.00
43142	2	12.70	13.00	12.85	12.85	0.21	0.02	12.85	12.70	13.00
43143	4	7.20	11.90	9.08	8.92	2.02	0.22	8.60	7.20	9.00
43146	7	5.80	35.00	20.47	17.35	11.25	0.55	20.70	11.00	31.00
43147	381	3.90	190.00	13.40	10.78	13.06	0.97	10.00	6.90	15.30
43148	4	4.00	24.00	10.23	7.94	9.26	0.91	6.45	4.00	6.90
43149	2	8.60	13.00	10.80	10.57	3.11	0.29	10.80	8.60	13.00
43150	1	65.90	65.90	65.90	65.90	0.00	0.00	65.90	65.90	65.90
43154	2	5.90	23.00	14.45	11.65	12.09	0.84	14.45	5.90	23.00
43155	2	9.60	60.00	34.80	24.00	35.64	1.02	34.80	9.60	60.00
43160	1	5.20	5.20	5.20	5.20	0.00	0.00	5.20	5.20	5.20
43162	9	8.00	20.00	12.74	12.35	3.46	0.27	12.00	11.00	14.00
43164	1	7.90	7.90	7.90	7.90	0.00	0.00	7.90	7.90	7.90
43174	1	18.00	18.00	18.00	18.00	0.00	0.00	18.00	18.00	18.00
43201	76	4.00	69.60	14.94	11.28	13.92	0.93	9.70	6.20	17.00
43202	60	4.30	65.00	12.02	10.00	9.69	0.81	8.70	7.00	14.00
43203	12	5.00	150.00	26.14	15.76	39.78	1.52	14.20	7.50	21.20
43204	54	4.40	182.20	19.56	12.28	30.12	1.54	9.65	6.90	16.40
43205	2	8.90	20.50	14.70	13.51	8.20	0.56	14.70	8.90	20.50
43206	72	4.00	50.40	12.25	9.87	9.60	0.78	8.80	5.90	13.50
43207	18	5.90	71.00	18.34	14.31	15.74	0.86	13.35	7.70	24.20

Zip Code	No.	Min.	Max.	AM	GM	SD	CV	Md	Q1	Q3
43208	1	4.20	4.20	4.20	4.20	0.00	0.00	4.20	4.20	4.20
43209	266	2.00	90.00	10.06	8.88	7.02	0.70	8.90	6.50	11.40
43210	1	16.40	16.40	16.40	16.40	0.00	0.00	16.40	16.40	16.40
43211	3	9.80	17.80	13.20	12.79	4.13	0.31	12.00	9.80	17.80
43212	93	3.30	31.20	10.84	9.53	5.93	0.55	9.00	6.40	13.00
43213	55	3.40	34.00	12.15	10.35	7.27	0.60	10.00	7.30	14.20
43214	229	4.00	140.00	14.46	11.57	13.12	0.91	10.60	7.30	16.20
43215	25	4.30	32.50	11.14	9.45	7.31	0.66	8.00	7.00	13.70
43216	1	9.00	9.00	9.00	9.00	0.00	0.00	9.00	9.00	9.00
43217	2	4.00	11.30	7.65	6.72	5.16	0.67	7.65	4.00	11.30
43219	19	3.80	40.00	18.76	14.76	12.44	0.66	14.00	8.00	26.50
43220	304	4.00	130.00	13.87	11.34	11.72	0.84	10.70	7.40	15.50
43221	615	3.60	121.00	13.36	11.04	10.62	0.79	10.70	7.10	16.00
43222	1	9.40	9.40	9.40	9.40	0.00	0.00	9.40	9.40	9.40
43223	22	5.60	139.20	37.25	26.05	32.99	0.89	27.65	18.00	50.90
43224	36	4.20	48.80	12.74	10.82	8.73	0.69	10.30	7.00	14.00
43225	1	24.00	24.00	24.00	24.00	0.00	0.00	24.00	24.00	24.00
43226	4	4.50	13.10	9.35	8.48	4.42	0.47	9.90	4.50	13.10
43227	7	7.00	20.40	13.39	12.43	5.42	0.41	12.90	9.00	20.00
43228	152	4.00	42.00	11.99	10.25	7.29	0.61	10.00	6.40	15.80
43229	73	4.30	35.00	12.61	11.07	6.96	0.55	11.30	7.40	16.00
43230	451	1.90	86.10	14.13	11.04	12.38	0.88	10.00	6.80	17.00
43231	60	4.60	60.00	14.31	11.63	10.82	0.76	10.60	7.30	19.00
43232	33	4.00	67.00	20.64	15.66	16.95	0.82	13.00	9.20	24.70
43235	446	4.00	158.50	14.45	11.26	13.64	0.94	10.00	6.90	17.00
43240	16	4.80	19.60	7.44	6.81	3.87	0.52	5.55	5.00	8.40
43255	1	6.10	6.10	6.10	6.10	0.00	0.00	6.10	6.10	6.10
43258	1	11.00	11.00	11.00	11.00	0.00	0.00	11.00	11.00	11.00
43290	1	10.50	10.50	10.50	10.50	0.00	0.00	10.50	10.50	10.50
43302	53	4.30	55.20	14.64	12.02	10.55	0.72	11.10	8.00	17.30
43303	2	10.40	13.10	11.75	11.67	1.91	0.16	11.75	10.40	13.10
43311	37	4.30	42.00	14.96	12.55	9.38	0.63	13.00	8.50	17.00
43313	1	9.80	9.80	9.80	9.80	0.00	0.00	9.80	9.80	9.80
43315	3	4.50	8.00	5.87	5.68	1.87	0.32	5.10	4.50	8.00
43316	4	4.80	11.90	7.46	7.04	3.10	0.42	6.58	4.80	7.10
43319	2	3.90	8.30	6.10	5.69	3.11	0.51	6.10	3.90	8.30

Zip Code	No.	Min.	Max.	AM	GM	SD	CV	Md	Q1	Q3
43320	2	6.30	18.10	12.20	10.68	8.34	0.68	12.20	6.30	18.10
43324	3	5.20	13.50	8.23	7.50	4.58	0.56	6.00	5.20	13.50
43326	2	5.00	8.10	6.55	6.36	2.19	0.33	6.55	5.00	8.10
43331	1	24.00	24.00	24.00	24.00	0.00	0.00	24.00	24.00	24.00
43334	7	5.90	18.00	11.57	11.03	3.74	0.32	11.20	10.00	13.80
43338	13	4.20	23.70	14.70	12.81	7.21	0.49	13.20	8.60	21.70
43342	3	6.70	16.20	11.37	10.67	4.75	0.42	11.20	6.70	16.20
43344	2	10.70	27.60	19.15	17.18	11.95	0.62	19.15	10.70	27.60
43345	1	15.50	15.50	15.50	15.50	0.00	0.00	15.50	15.50	15.50
43351	4	7.60	42.90	17.88	13.57	16.86	0.94	10.50	7.60	13.00
43356	2	7.00	11.00	9.00	8.77	2.83	0.31	9.00	7.00	11.00
43357	25	7.00	287.00	42.60	28.42	56.42	1.32	25.00	20.00	39.00
43360	5	6.10	60.00	22.14	15.44	22.14	1.00	16.00	6.90	21.70
43402	105	3.80	27.50	8.82	7.71	5.15	0.58	7.00	5.20	10.40
43403	2	4.60	7.70	6.15	5.95	2.19	0.36	6.15	4.60	7.70
43406	1	7.10	7.10	7.10	7.10	0.00	0.00	7.10	7.10	7.10
43407	1	17.40	17.40	17.40	17.40	0.00	0.00	17.40	17.40	17.40
43410	15	4.00	804.00	63.56	12.12	204.96	3.22	9.00	5.50	15.30
43412	7	4.40	12.00	8.07	7.56	3.09	0.38	7.00	5.40	11.30
43413	1	34.00	34.00	34.00	34.00	0.00	0.00	34.00	34.00	34.00
43416	2	5.90	16.00	10.95	9.72	7.14	0.65	10.95	5.90	16.00
43420	48	4.30	46.10	13.11	10.82	9.21	0.70	10.85	6.20	14.70
43421	1	6.90	6.90	6.90	6.90	0.00	0.00	6.90	6.90	6.90
43430	7	4.00	27.00	12.51	10.18	8.43	0.67	9.20	5.30	18.00
43431	3	7.90	13.00	10.40	10.19	2.55	0.25	10.30	7.90	13.00
43432	2	6.50	9.70	8.10	7.94	2.26	0.28	8.10	6.50	9.70
43435	1	4.50	4.50	4.50	4.50	0.00	0.00	4.50	4.50	4.50
43440	13	4.00	23.50	10.48	9.34	5.19	0.50	10.20	7.20	13.10
43443	3	4.00	8.60	5.87	5.56	2.42	0.41	5.00	4.00	8.60
43445	1	9.00	9.00	9.00	9.00	0.00	0.00	9.00	9.00	9.00
43446	1	13.00	13.00	13.00	13.00	0.00	0.00	13.00	13.00	13.00
43447	6	5.80	14.70	7.65	7.18	3.49	0.46	6.25	5.80	7.10
43449	9	4.20	13.00	7.84	6.95	4.04	0.51	4.80	4.50	12.00
43450	4	5.30	24.80	13.00	11.00	8.65	0.67	10.95	5.30	13.90
43452	36	4.00	46.00	10.21	8.21	9.19	0.90	7.00	5.10	10.30
43460	17	4.00	29.00	9.59	8.10	6.41	0.67	7.40	5.00	13.90

Zip Code	No.	Min.	Max.	AM	GM	SD	CV	Md	Q1	Q3
43462	1	5.30	5.30	5.30	5.30	0.00	0.00	5.30	5.30	5.30
43465	10	3.90	30.00	10.61	8.74	7.76	0.73	8.60	5.00	13.00
43466	1	4.00	4.00	4.00	4.00	0.00	0.00	4.00	4.00	4.00
43501	1	4.20	4.20	4.20	4.20	0.00	0.00	4.20	4.20	4.20
43502	6	6.00	18.00	11.27	10.27	5.21	0.46	10.05	7.00	16.50
43504	1	8.00	8.00	8.00	8.00	0.00	0.00	8.00	8.00	8.00
43506	4	4.90	9.40	6.13	5.88	2.19	0.36	5.10	4.90	5.10
43511	3	11.20	17.10	14.07	13.86	2.95	0.21	13.90	11.20	17.10
43512	6	5.00	8.00	6.78	6.69	1.17	0.17	6.90	6.00	7.90
43515	9	4.80	20.00	10.44	9.55	4.76	0.46	9.40	6.80	12.90
43520	1	5.20	5.20	5.20	5.20	0.00	0.00	5.20	5.20	5.20
43522	6	4.00	20.00	11.85	10.31	6.18	0.52	12.15	6.80	16.00
43523	1	4.60	4.60	4.60	4.60	0.00	0.00	4.60	4.60	4.60
43525	21	4.00	14.10	7.03	6.60	2.72	0.39	6.30	5.10	8.10
43526	2	4.80	17.00	10.90	9.03	8.63	0.79	10.90	4.80	17.00
43528	52	3.00	18.00	7.12	6.55	3.21	0.45	6.30	5.00	7.80
43530	1	27.00	27.00	27.00	27.00	0.00	0.00	27.00	27.00	27.00
43531	1	7.20	7.20	7.20	7.20	0.00	0.00	7.20	7.20	7.20
43532	3	4.50	15.00	9.10	8.07	5.37	0.59	7.80	4.50	15.00
43533	1	7.30	7.30	7.30	7.30	0.00	0.00	7.30	7.30	7.30
43534	1	13.00	13.00	13.00	13.00	0.00	0.00	13.00	13.00	13.00
43535	1	7.90	7.90	7.90	7.90	0.00	0.00	7.90	7.90	7.90
43537	291	3.10	100.00	10.47	8.65	8.79	0.84	8.00	5.60	12.20
43540	4	6.20	14.90	9.53	8.96	3.97	0.42	8.50	6.20	10.10
43542	52	3.70	42.00	9.52	7.91	7.18	0.75	7.00	4.80	10.00
43543	1	17.90	17.90	17.90	17.90	0.00	0.00	17.90	17.90	17.90
43545	2	11.50	14.60	13.05	12.96	2.19	0.17	13.05	11.50	14.60
43551	529	1.00	51.90	9.44	8.08	6.52	0.69	7.50	5.40	11.40
43552	1	6.80	6.80	6.80	6.80	0.00	0.00	6.80	6.80	6.80
43556	2	6.60	21.10	13.85	11.80	10.25	0.74	13.85	6.60	21.10
43557	2	12.10	20.10	16.10	15.60	5.66	0.35	16.10	12.10	20.10
43558	11	4.00	11.20	6.25	5.98	2.10	0.34	5.70	4.70	7.20
43560	185	3.00	79.00	8.12	7.00	6.74	0.83	6.20	5.00	9.00
43566	87	4.00	71.00	11.87	9.52	10.15	0.86	8.50	6.00	13.90
43567	2	7.50	9.20	8.35	8.31	1.20	0.14	8.35	7.50	9.20
43569	2	4.30	5.40	4.85	4.82	0.78	0.16	4.85	4.30	5.40

Zip Code	No.	Min.	Max.	AM	GM	SD	CV	Md	Q1	Q3
43571	28	4.00	22.00	8.35	7.30	4.85	0.58	6.05	4.60	10.00
43577	1	5.00	5.00	5.00	5.00	0.00	0.00	5.00	5.00	5.00
43601	2	6.30	30.00	18.15	13.75	16.76	0.92	18.15	6.30	30.00
43604	2	6.40	16.10	11.25	10.15	6.86	0.61	11.25	6.40	16.10
43605	5	5.10	10.90	6.56	6.27	2.47	0.38	5.30	5.20	6.30
43606	31	3.80	12.10	5.68	5.47	1.76	0.31	5.50	4.40	6.40
43607	4	7.30	14.50	10.38	10.03	3.14	0.30	9.85	7.30	11.00
43608	2	15.00	50.00	32.50	27.39	24.75	0.76	32.50	15.00	50.00
43609	3	7.00	9.40	8.43	8.37	1.27	0.15	8.90	7.00	9.40
43610	1	6.00	6.00	6.00	6.00	0.00	0.00	6.00	6.00	6.00
43611	17	4.30	17.50	8.93	8.07	4.28	0.48	7.40	6.00	12.00
43612	16	4.30	20.00	8.46	7.68	4.22	0.50	7.15	5.30	10.00
43613	8	4.00	24.10	10.55	8.71	7.47	0.71	6.75	5.60	11.20
43614	95	4.00	25.10	8.04	7.22	4.19	0.52	6.70	5.00	9.40
43615	48	3.40	22.10	8.34	7.41	4.42	0.53	6.85	4.90	10.00
43616	39	3.70	23.10	8.59	7.76	4.40	0.51	7.20	5.70	10.40
43617	9	4.00	9.00	5.29	5.13	1.55	0.29	4.90	4.20	5.80
43618	3	6.80	13.70	11.23	10.71	3.85	0.34	13.20	6.80	13.70
43619	17	4.00	18.00	8.61	7.62	4.57	0.53	6.90	5.20	12.00
43623	6	4.10	7.30	5.45	5.36	1.10	0.20	5.40	4.70	5.80
43635	1	7.00	7.00	7.00	7.00	0.00	0.00	7.00	7.00	7.00
43661	1	5.20	5.20	5.20	5.20	0.00	0.00	5.20	5.20	5.20
43701	71	4.10	73.00	15.29	11.94	12.62	0.83	11.70	6.90	18.60
43713	4	7.10	25.00	12.88	11.37	8.18	0.64	9.70	7.10	10.00
43716	5	4.70	87.20	24.92	12.59	35.39	1.42	6.80	5.60	20.30
43718	8	9.10	60.00	20.80	15.84	18.91	0.91	10.75	10.00	15.80
43719	2	11.20	25.00	18.10	16.73	9.76	0.54	18.10	11.20	25.00
43720	1	12.20	12.20	12.20	12.20	0.00	0.00	12.20	12.20	12.20
43723	3	10.00	14.50	11.50	11.32	2.60	0.23	10.00	10.00	14.50
43725	8	5.80	25.00	12.51	11.52	5.79	0.46	11.85	8.60	12.80
43739	3	4.80	13.00	8.60	7.93	4.13	0.48	8.00	4.80	13.00
43749	2	61.90	90.00	75.95	74.64	19.87	0.26	75.95	61.90	90.00
43760	1	5.90	5.90	5.90	5.90	0.00	0.00	5.90	5.90	5.90
43762	7	5.20	96.60	28.27	17.97	32.01	1.13	16.80	7.50	35.90
43767	1	6.10	6.10	6.10	6.10	0.00	0.00	6.10	6.10	6.10
43777	1	5.10	5.10	5.10	5.10	0.00	0.00	5.10	5.10	5.10

Zip Code	No.	Min.	Max.	AM	GM	SD	CV	Md	Q1	Q3
43783	1	42.00	42.00	42.00	42.00	0.00	0.00	42.00	42.00	42.00
43812	9	5.80	33.60	13.88	11.95	8.75	0.63	11.00	9.00	17.30
43821	2	8.40	10.60	9.50	9.44	1.56	0.16	9.50	8.40	10.60
43822	3	9.00	31.00	21.33	18.85	11.24	0.53	24.00	9.00	31.00
43824	1	30.00	30.00	30.00	30.00	0.00	0.00	30.00	30.00	30.00
43830	12	4.00	129.70	19.93	9.85	35.47	1.78	6.40	4.70	12.50
43832	7	5.20	11.70	8.49	8.12	2.61	0.31	9.00	5.30	11.40
43837	4	8.60	40.00	18.20	14.98	14.63	0.80	12.10	8.60	12.20
43840	2	10.50	30.00	20.25	17.75	13.79	0.68	20.25	10.50	30.00
43842	1	6.90	6.90	6.90	6.90	0.00	0.00	6.90	6.90	6.90
43845	1	31.30	31.30	31.30	31.30	0.00	0.00	31.30	31.30	31.30
43867	1	6.50	6.50	6.50	6.50	0.00	0.00	6.50	6.50	6.50
43901	2	8.00	9.00	8.50	8.49	0.71	0.08	8.50	8.00	9.00
43903	1	14.20	14.20	14.20	14.20	0.00	0.00	14.20	14.20	14.20
43906	2	7.00	8.10	7.55	7.53	0.78	0.10	7.55	7.00	8.10
43907	2	4.00	10.40	7.20	6.45	4.53	0.63	7.20	4.00	10.40
43912	4	6.00	16.00	10.25	9.64	4.19	0.41	9.50	6.00	10.00
43913	1	4.30	4.30	4.30	4.30	0.00	0.00	4.30	4.30	4.30
43916	2	4.20	4.40	4.30	4.30	0.14	0.03	4.30	4.20	4.40
43920	27	4.10	40.20	14.85	11.35	11.54	0.78	10.00	6.70	27.00
43930	1	15.50	15.50	15.50	15.50	0.00	0.00	15.50	15.50	15.50
43935	3	4.00	30.00	14.30	10.22	13.82	0.97	8.90	4.00	30.00
43942	5	9.00	31.00	22.30	19.11	11.91	0.53	31.00	9.50	31.00
43943	2	14.00	40.00	27.00	23.66	18.38	0.68	27.00	14.00	40.00
43945	2	9.50	177.00	93.25	41.01	118.44	1.27	93.25	9.50	177.00
43947	4	5.90	37.60	20.73	16.88	13.55	0.65	19.70	5.90	24.40
43950	41	3.90	62.80	10.07	7.71	11.79	1.17	7.20	5.10	9.00
43952	14	4.90	17.00	8.85	8.03	4.20	0.47	7.05	5.40	13.70
43953	8	4.60	48.80	21.41	13.39	20.55	0.96	8.90	4.80	40.20
43964	2	14.10	16.00	15.05	15.02	1.34	0.09	15.05	14.10	16.00
43968	1	40.00	40.00	40.00	40.00	0.00	0.00	40.00	40.00	40.00
43977	1	12.00	12.00	12.00	12.00	0.00	0.00	12.00	12.00	12.00
43988	11	5.90	93.00	28.70	19.47	27.47	0.96	20.00	9.60	50.00
44001	49	3.20	30.00	8.56	7.45	5.25	0.61	6.70	5.10	10.70
44004	10	4.30	40.10	13.77	9.55	13.20	0.96	6.35	4.50	18.00
44010	4	4.00	9.00	6.60	6.32	2.12	0.32	6.70	4.00	7.40

Zip Code	No.	Min.	Max.	AM	GM	SD	CV	Md	Q1	Q3
44011	315	2.20	51.00	9.35	7.64	8.12	0.87	6.90	5.10	9.90
44012	396	1.70	100.00	10.14	8.14	9.17	0.90	7.20	5.20	11.30
44015	1	4.30	4.30	4.30	4.30	0.00	0.00	4.30	4.30	4.30
44017	49	3.80	30.00	9.42	7.91	6.77	0.72	7.10	5.30	10.00
44020	3	5.20	7.90	6.73	6.63	1.39	0.21	7.10	5.20	7.90
44021	7	4.00	13.30	6.61	6.18	3.04	0.46	5.70	5.20	6.20
44022	79	2.60	30.00	6.44	5.91	3.59	0.56	5.30	4.70	7.20
44023	111	2.30	45.00	7.59	6.54	5.89	0.78	6.00	4.70	8.10
44024	41	2.30	14.00	5.66	5.37	2.09	0.37	5.10	4.40	6.00
44026	24	4.00	14.20	6.53	6.07	2.84	0.43	5.70	4.20	7.70
44028	19	3.60	30.00	8.32	7.21	5.88	0.71	7.10	4.90	8.90
44030	8	4.20	16.40	8.18	7.09	5.10	0.62	6.25	4.40	6.80
44033	1	4.70	4.70	4.70	4.70	0.00	0.00	4.70	4.70	4.70
44034	2	5.50	6.00	5.75	5.74	0.35	0.06	5.75	5.50	6.00
44035	61	3.90	41.90	8.98	7.53	6.91	0.77	6.50	5.00	10.30
44037	1	11.00	11.00	11.00	11.00	0.00	0.00	11.00	11.00	11.00
44039	97	3.30	58.00	9.95	8.01	8.54	0.86	7.00	5.10	11.00
44040	23	3.30	44.00	8.45	6.95	8.17	0.97	6.80	5.00	8.30
44041	2	7.10	34.00	20.55	15.54	19.02	0.93	20.55	7.10	34.00
44044	24	4.10	43.00	9.88	7.89	8.97	0.91	6.70	5.00	11.00
44045	1	7.30	7.30	7.30	7.30	0.00	0.00	7.30	7.30	7.30
44046	2	5.00	10.00	7.50	7.07	3.54	0.47	7.50	5.00	10.00
44047	4	5.60	30.00	12.15	9.17	11.91	0.98	6.50	5.60	7.00
44048	2	35.00	35.50	35.25	35.25	0.35	0.01	35.25	35.00	35.50
44049	2	6.00	14.70	10.35	9.39	6.15	0.59	10.35	6.00	14.70
44050	14	4.00	18.00	9.19	8.14	4.74	0.52	8.20	4.80	11.20
44052	10	4.10	13.90	8.39	7.62	3.87	0.46	6.55	5.50	12.60
44053	20	2.50	30.00	8.99	7.48	6.46	0.72	6.90	6.00	9.10
44054	37	4.00	47.00	10.17	8.44	8.15	0.80	8.30	5.10	10.40
44055	4	7.00	27.00	12.03	9.84	9.98	0.83	7.05	7.00	7.10
44056	56	3.10	25.00	8.06	6.94	5.25	0.65	6.00	4.70	8.90
44057	24	3.90	25.00	10.00	8.40	6.49	0.65	7.00	5.30	11.40
44060	144	3.00	125.00	9.13	7.18	12.35	1.35	6.10	5.00	8.60
44062	6	4.50	20.00	9.65	8.47	5.71	0.59	8.30	5.50	11.30
44064	2	5.50	7.80	6.65	6.55	1.63	0.24	6.65	5.50	7.80
44065	7	6.00	14.60	8.77	8.44	2.84	0.32	8.00	6.70	9.70

Zip Code	No.	Min.	Max.	AM	GM	SD	CV	Md	Q1	Q3
44067	117	1.30	84.00	9.92	7.52	11.46	1.16	6.30	5.10	8.80
44068	2	3.70	3.70	3.70	3.70	0.00	0.00	3.70	3.70	3.70
44070	103	3.50	116.40	11.90	9.03	13.43	1.13	7.50	6.00	12.00
44072	6	4.60	7.90	5.58	5.49	1.21	0.22	5.35	4.70	5.60
44074	24	4.10	55.00	9.63	7.74	10.21	1.06	7.20	5.40	9.10
44076	1	6.70	6.70	6.70	6.70	0.00	0.00	6.70	6.70	6.70
44077	151	3.50	53.40	9.31	7.83	6.80	0.73	6.90	5.00	11.00
44080	1	12.50	12.50	12.50	12.50	0.00	0.00	12.50	12.50	12.50
44081	20	4.10	98.00	19.18	12.23	23.06	1.20	11.18	5.20	21.00
44084	8	4.00	9.00	5.89	5.65	1.90	0.32	5.30	4.10	5.80
44085	1	8.00	8.00	8.00	8.00	0.00	0.00	8.00	8.00	8.00
44086	2	4.40	5.80	5.10	5.05	0.99	0.19	5.10	4.40	5.80
44087	89	3.60	29.00	6.60	6.08	3.45	0.52	5.50	4.60	7.20
44088	1	12.00	12.00	12.00	12.00	0.00	0.00	12.00	12.00	12.00
44089	37	4.00	47.00	11.05	9.16	8.65	0.78	8.00	5.90	11.20
44090	12	4.10	30.00	8.66	6.96	7.51	0.87	5.15	4.40	7.90
44092	8	4.00	13.00	7.09	6.53	3.21	0.45	6.30	4.50	7.40
44094	57	3.90	33.00	8.92	7.61	6.47	0.73	7.00	5.30	9.00
44095	8	3.80	9.00	6.48	6.19	1.98	0.31	6.90	4.00	7.70
44101	4	8.20	27.00	17.53	15.18	10.11	0.58	17.45	8.20	25.50
44102	11	4.10	15.40	8.93	8.00	4.20	0.47	7.40	4.90	12.00
44103	12	4.00	9.40	6.59	6.40	1.63	0.25	6.25	5.70	8.00
44105	7	4.50	13.60	7.51	6.99	3.32	0.44	5.90	5.40	10.50
44106	76	4.00	19.80	6.81	6.33	3.08	0.45	5.85	4.90	7.50
44107	15	4.00	33.20	7.88	6.51	7.26	0.92	6.20	4.10	8.00
44109	10	4.00	11.50	6.10	5.76	2.38	0.39	5.30	4.40	7.40
44110	1	7.00	7.00	7.00	7.00	0.00	0.00	7.00	7.00	7.00
44111	20	4.00	20.00	8.78	7.69	4.93	0.56	7.55	5.00	10.20
44112	2	6.30	7.60	6.95	6.92	0.92	0.13	6.95	6.30	7.60
44113	12	4.00	10.30	7.18	6.81	2.30	0.32	7.60	4.50	8.40
44114	9	4.10	20.00	6.96	6.02	5.09	0.73	5.30	4.60	5.50
44116	50	3.80	43.00	6.96	5.98	6.22	0.89	5.20	4.40	7.00
44117	5	4.10	8.50	6.42	6.23	1.68	0.26	6.50	5.60	7.40
44118	19	3.80	30.00	7.94	6.87	5.84	0.74	6.60	4.40	8.30
44119	1	9.30	9.30	9.30	9.30	0.00	0.00	9.30	9.30	9.30
44120	6	4.10	8.50	6.50	6.24	1.93	0.30	6.95	4.50	8.00

Zip Code	No.	Min.	Max.	AM	GM	SD	CV	Md	Q1	Q3
44121	14	4.00	7.90	5.48	5.32	1.38	0.25	5.35	4.20	6.50
44122	45	3.80	12.00	6.05	5.82	1.80	0.30	5.90	4.70	6.50
44123	3	4.10	14.50	9.37	8.27	5.20	0.56	9.50	4.10	14.50
44124	69	2.80	30.00	6.23	5.72	3.59	0.58	5.40	4.40	6.50
44125	9	4.50	20.00	9.39	8.27	5.50	0.59	7.50	6.00	9.80
44126	32	4.00	52.00	10.24	8.19	9.64	0.94	7.00	5.00	10.40
44128	2	5.10	7.40	6.25	6.14	1.63	0.26	6.25	5.10	7.40
44129	7	3.00	30.00	8.44	6.12	9.58	1.14	5.00	4.10	6.80
44130	18	3.50	30.00	7.29	6.15	6.17	0.85	5.15	4.50	7.00
44131	31	3.00	13.00	6.38	6.00	2.34	0.37	6.00	4.90	8.00
44132	1	4.60	4.60	4.60	4.60	0.00	0.00	4.60	4.60	4.60
44133	62	3.60	31.00	7.15	6.31	5.17	0.72	5.90	4.80	7.60
44134	10	3.30	14.00	6.03	5.54	3.06	0.51	5.05	4.10	7.00
44135	6	4.30	30.00	10.87	8.50	9.70	0.89	7.75	5.00	10.40
44136	57	3.90	30.00	6.46	5.90	3.90	0.60	5.50	4.50	6.60
44137	3	4.10	5.30	4.50	4.47	0.69	0.15	4.10	4.10	5.30
44138	77	1.60	30.00	7.64	6.66	4.79	0.63	6.00	4.90	9.00
44139	98	2.90	24.50	5.76	5.37	2.95	0.51	5.10	4.30	6.00
44140	227	3.90	40.00	9.84	8.32	6.80	0.69	7.40	5.40	11.30
44141	68	3.00	30.00	7.39	6.61	4.59	0.62	6.50	4.80	7.80
44142	2	5.90	11.30	8.60	8.17	3.82	0.44	8.60	5.90	11.30
44143	62	4.00	35.00	7.65	6.58	6.30	0.82	6.00	5.00	7.10
44144	1	5.90	5.90	5.90	5.90	0.00	0.00	5.90	5.90	5.90
44145	334	2.80	61.00	9.46	7.87	7.23	0.76	7.00	5.00	11.20
44146	10	4.20	9.00	5.60	5.46	1.41	0.25	5.45	4.70	6.10
44147	83	1.20	30.00	6.86	6.11	4.21	0.61	5.50	4.60	8.00
44149	34	3.80	22.00	6.93	6.31	3.69	0.53	5.75	5.00	8.10
44154	2	4.00	11.00	7.50	6.63	4.95	0.66	7.50	4.00	11.00
44184	1	4.10	4.10	4.10	4.10	0.00	0.00	4.10	4.10	4.10
44188	1	12.00	12.00	12.00	12.00	0.00	0.00	12.00	12.00	12.00
44201	5	4.20	12.00	7.22	6.75	3.05	0.42	6.80	5.10	8.00
44202	103	3.40	23.00	6.65	6.17	3.05	0.46	5.80	4.70	8.00
44203	90	4.00	57.00	8.45	7.30	6.53	0.77	6.45	5.20	9.80
44210	3	4.50	10.00	6.83	6.46	2.84	0.42	6.00	4.50	10.00
44212	78	4.00	30.00	8.69	7.30	6.60	0.76	6.20	5.00	9.80
44214	5	5.10	22.00	10.16	8.40	7.42	0.73	5.50	5.20	13.00

Zip Code	No.	Min.	Max.	AM	GM	SD	CV	Md	Q1	Q3
44215	7	4.10	9.80	6.40	6.06	2.38	0.37	5.40	4.70	9.80
44216	33	4.10	144.00	15.46	10.51	24.20	1.57	9.70	6.50	14.10
44217	17	4.20	539.00	53.28	17.55	127.04	2.38	18.50	4.50	37.50
44219	1	6.20	6.20	6.20	6.20	0.00	0.00	6.20	6.20	6.20
44220	1	6.40	6.40	6.40	6.40	0.00	0.00	6.40	6.40	6.40
44221	60	3.50	32.50	8.06	7.10	5.34	0.66	6.10	5.20	8.90
44223	76	3.80	30.00	8.15	7.32	4.46	0.55	7.10	5.00	9.00
44224	250	3.30	46.90	8.55	7.34	6.06	0.71	6.40	5.10	9.60
44230	27	3.50	20.70	8.71	7.47	5.40	0.62	6.50	4.80	12.00
44231	6	5.30	33.00	14.90	11.81	11.22	0.75	10.15	6.80	24.00
44232	8	4.30	15.00	10.36	9.52	4.04	0.39	11.00	5.20	13.00
44233	23	4.00	30.00	8.11	6.79	6.45	0.80	5.60	4.70	7.70
44234	6	4.00	17.80	9.58	8.41	5.16	0.54	9.50	4.70	12.00
44235	7	5.00	29.90	13.43	10.60	9.96	0.74	9.70	5.10	24.00
44236	320	3.30	36.00	6.83	6.20	3.91	0.57	5.60	4.60	7.80
44237	2	6.00	11.20	8.60	8.20	3.68	0.43	8.60	6.00	11.20
44238	1	7.00	7.00	7.00	7.00	0.00	0.00	7.00	7.00	7.00
44239	1	7.00	7.00	7.00	7.00	0.00	0.00	7.00	7.00	7.00
44240	102	2.30	40.00	9.49	7.77	7.62	0.80	6.50	5.00	11.00
44241	62	1.40	58.00	9.92	7.43	10.24	1.03	6.00	4.80	9.00
44242	4	5.10	9.00	7.63	7.45	1.73	0.23	8.20	5.10	8.40
44243	1	12.50	12.50	12.50	12.50	0.00	0.00	12.50	12.50	12.50
44244	2	5.00	6.00	5.50	5.48	0.71	0.13	5.50	5.00	6.00
44251	6	4.70	11.80	7.48	7.14	2.58	0.34	6.95	5.50	9.00
44253	12	4.40	33.00	10.94	9.21	7.86	0.72	8.70	5.40	12.50
44254	12	5.10	34.00	11.68	9.18	10.10	0.87	6.85	5.80	9.30
44255	10	5.00	23.00	10.08	8.76	6.02	0.60	9.05	5.10	11.60
44256	472	2.40	93.30	8.98	7.20	8.85	0.99	6.10	4.90	9.00
44260	46	4.10	40.00	10.62	8.85	7.76	0.73	8.00	6.00	11.60
44261	2	5.00	12.00	8.50	7.75	4.95	0.58	8.50	5.00	12.00
44262	20	4.10	41.65	9.47	7.73	8.45	0.89	6.25	5.30	9.50
44264	10	4.40	11.60	6.75	6.48	2.14	0.32	6.50	5.40	7.80
44265	1	8.40	8.40	8.40	8.40	0.00	0.00	8.40	8.40	8.40
44266	46	4.10	31.00	9.51	8.05	6.79	0.71	7.80	5.40	9.90
44270	14	4.10	9.00	6.40	6.21	1.60	0.25	6.60	4.80	8.10
44272	12	4.60	20.90	9.32	8.21	5.52	0.59	7.25	5.60	9.30

Zip Code	No.	Min.	Max.	AM	GM	SD	CV	Md	Q1	Q3
44273	30	4.10	18.00	8.23	7.66	3.41	0.41	6.80	5.90	10.60
44274	1	6.00	6.00	6.00	6.00	0.00	0.00	6.00	6.00	6.00
44275	3	6.30	9.40	8.07	7.95	1.59	0.20	8.50	6.30	9.40
44278	48	4.50	18.30	8.09	7.48	3.53	0.44	7.35	5.30	9.10
44280	23	4.10	76.40	11.34	8.16	15.17	1.34	6.80	4.60	11.00
44281	177	4.00	70.00	9.23	7.85	7.40	0.80	7.00	5.30	10.00
44284	1	6.00	6.00	6.00	6.00	0.00	0.00	6.00	6.00	6.00
44286	37	3.40	19.80	7.27	6.72	3.24	0.45	6.00	5.20	9.00
44287	16	4.10	16.00	8.60	7.98	3.35	0.39	8.50	5.40	10.50
44288	1	5.40	5.40	5.40	5.40	0.00	0.00	5.40	5.40	5.40
44291	1	5.50	5.50	5.50	5.50	0.00	0.00	5.50	5.50	5.50
44296	1	30.00	30.00	30.00	30.00	0.00	0.00	30.00	30.00	30.00
44301	6	5.10	12.00	6.58	6.25	2.68	0.41	5.60	5.20	6.00
44302	3	4.50	6.60	5.60	5.53	1.05	0.19	5.70	4.50	6.60
44303	20	4.50	27.20	8.43	7.40	5.41	0.64	6.55	5.00	9.00
44304	1	8.00	8.00	8.00	8.00	0.00	0.00	8.00	8.00	8.00
44305	16	4.30	10.20	6.51	6.33	1.63	0.25	6.25	5.00	7.30
44306	7	4.00	50.00	14.31	9.66	16.35	1.14	9.00	4.50	17.00
44307	2	5.30	11.40	8.35	7.77	4.31	0.52	8.35	5.30	11.40
44309	2	7.00	10.00	8.50	8.37	2.12	0.25	8.50	7.00	10.00
44310	23	3.90	30.00	9.32	8.05	5.99	0.64	8.00	5.20	11.10
44312	86	4.00	48.10	9.19	7.91	6.74	0.73	7.25	5.30	10.40
44313	145	3.90	44.00	9.27	7.92	6.59	0.71	7.00	5.50	10.50
44314	9	4.80	14.70	7.37	6.85	3.36	0.46	5.90	5.60	7.40
44317	1	4.50	4.50	4.50	4.50	0.00	0.00	4.50	4.50	4.50
44319	67	3.50	68.10	8.16	6.87	8.39	1.03	6.00	5.00	8.00
44320	18	4.10	38.00	11.15	9.60	7.67	0.69	9.65	6.80	12.00
44321	196	4.00	80.00	11.51	8.65	12.12	1.05	7.00	5.20	12.00
44324	1	4.50	4.50	4.50	4.50	0.00	0.00	4.50	4.50	4.50
44333	211	3.00	85.10	12.71	9.53	12.33	0.97	8.00	5.70	13.80
44336	1	6.00	6.00	6.00	6.00	0.00	0.00	6.00	6.00	6.00
44401	1	6.10	6.10	6.10	6.10	0.00	0.00	6.10	6.10	6.10
44405	1	5.00	5.00	5.00	5.00	0.00	0.00	5.00	5.00	5.00
44406	44	3.90	50.00	9.31	7.43	8.65	0.93	6.60	4.40	9.00
44408	19	4.00	28.80	11.54	9.63	7.40	0.64	8.00	6.00	16.00
44410	25	3.50	39.90	7.93	6.78	6.97	0.88	6.40	5.20	8.20

Zip Code	No.	Min.	Max.	AM	GM	SD	CV	Md	Q1	Q3
44411	1	6.00	6.00	6.00	6.00	0.00	0.00	6.00	6.00	6.00
44412	1	5.20	5.20	5.20	5.20	0.00	0.00	5.20	5.20	5.20
44413	8	4.10	30.00	9.30	7.37	8.62	0.93	6.10	4.70	8.40
44420	11	4.00	32.00	11.05	8.34	10.16	0.92	6.30	4.60	12.70
44423	1	10.30	10.30	10.30	10.30	0.00	0.00	10.30	10.30	10.30
44425	6	5.80	36.00	12.78	10.07	11.66	0.91	7.60	6.70	13.00
44427	2	25.50	83.50	54.50	46.14	41.01	0.75	54.50	25.50	83.50
44429	4	3.40	6.20	4.65	4.53	1.23	0.26	4.50	3.40	5.00
44431	2	6.10	10.40	8.25	7.96	3.04	0.37	8.25	6.10	10.40
44432	11	5.00	136.50	65.94	45.07	43.24	0.66	70.00	22.50	90.00
44436	1	8.00	8.00	8.00	8.00	0.00	0.00	8.00	8.00	8.00
44437	2	4.40	13.50	8.95	7.71	6.43	0.72	8.95	4.40	13.50
44438	1	8.60	8.60	8.60	8.60	0.00	0.00	8.60	8.60	8.60
44440	3	6.00	18.00	10.20	8.93	6.76	0.66	6.60	6.00	18.00
44441	3	4.00	12.00	6.73	5.86	4.56	0.68	4.20	4.00	12.00
44442	4	4.70	12.20	7.68	7.18	3.31	0.43	6.90	4.70	8.00
44443	1	6.90	6.90	6.90	6.90	0.00	0.00	6.90	6.90	6.90
44444	1	5.70	5.70	5.70	5.70	0.00	0.00	5.70	5.70	5.70
44445	3	9.10	15.30	13.13	12.78	3.50	0.27	15.00	9.10	15.30
44451	1	4.90	4.90	4.90	4.90	0.00	0.00	4.90	4.90	4.90
44452	3	8.20	22.00	13.33	12.09	7.55	0.57	9.80	8.20	22.00
44460	30	4.00	95.60	14.49	8.96	22.26	1.54	6.60	5.20	10.20
44466	1	16.20	16.20	16.20	16.20	0.00	0.00	16.20	16.20	16.20
44471	1	4.10	4.10	4.10	4.10	0.00	0.00	4.10	4.10	4.10
44477	1	5.40	5.40	5.40	5.40	0.00	0.00	5.40	5.40	5.40
44481	5	4.40	9.50	6.22	5.96	2.10	0.34	5.60	4.60	7.00
44483	2	6.00	6.80	6.40	6.39	0.57	0.09	6.40	6.00	6.80
44484	13	5.70	18.00	7.77	7.38	3.22	0.41	6.80	6.30	7.80
44485	1	4.00	4.00	4.00	4.00	0.00	0.00	4.00	4.00	4.00
44491	1	6.00	6.00	6.00	6.00	0.00	0.00	6.00	6.00	6.00
44505	5	4.10	10.90	7.50	7.03	2.87	0.38	7.70	5.20	9.60
44510	9	1.50	41.90	9.30	5.44	12.65	1.36	4.60	2.80	8.00
44511	6	3.90	32.60	14.17	9.90	12.57	0.89	8.35	4.80	27.00
44512	25	4.10	30.00	8.17	7.14	5.50	0.67	6.50	5.00	9.00
44514	34	3.50	24.00	6.51	6.01	3.55	0.55	5.75	4.70	6.60
44515	14	4.20	27.00	8.80	7.67	5.83	0.66	7.30	5.00	11.00

Zip Code	No.	Min.	Max.	AM	GM	SD	CV	Md	Q1	Q3
44517	1	24.00	24.00	24.00	24.00	0.00	0.00	24.00	24.00	24.00
44535	1	25.50	25.50	25.50	25.50	0.00	0.00	25.50	25.50	25.50
44555	1	10.00	10.00	10.00	10.00	0.00	0.00	10.00	10.00	10.00
44572	2	4.70	4.70	4.70	4.70	0.00	0.00	4.70	4.70	4.70
44601	41	4.50	16.00	8.24	7.77	2.93	0.36	7.70	6.10	10.00
44606	9	4.20	43.00	12.43	9.60	11.96	0.96	8.80	5.90	12.10
44607	2	4.60	7.30	5.95	5.79	1.91	0.32	5.95	4.60	7.30
44608	6	10.00	25.00	17.87	17.23	4.94	0.28	18.00	16.20	20.00
44609	1	6.30	6.30	6.30	6.30	0.00	0.00	6.30	6.30	6.30
44610	2	13.00	73.50	43.25	30.91	42.78	0.99	43.25	13.00	73.50
44611	3	6.70	9.00	7.70	7.64	1.18	0.15	7.40	6.70	9.00
44612	51	4.00	165.40	19.63	11.80	30.97	1.58	10.20	6.10	17.00
44614	59	4.00	35.40	9.07	8.07	5.31	0.59	7.40	6.00	10.00
44615	14	5.60	201.00	47.26	23.73	66.77	1.41	18.50	10.00	47.00
44616	1	13.60	13.60	13.60	13.60	0.00	0.00	13.60	13.60	13.60
44618	11	5.00	36.40	17.18	13.89	11.03	0.64	17.00	7.10	30.00
44619	2	30.00	147.00	88.50	66.41	82.73	0.93	88.50	30.00	147.00
44620	5	4.40	28.00	15.64	12.70	9.72	0.62	18.60	7.20	20.00
44621	11	4.20	51.60	14.41	9.79	15.47	1.07	6.30	5.00	14.00
44622	198	4.00	78.00	11.07	9.08	8.86	0.80	8.15	5.80	12.70
44624	4	5.70	58.00	19.65	11.62	25.58	1.30	7.45	5.70	8.00
44626	11	6.00	69.00	16.76	12.78	17.89	1.07	11.10	9.00	15.00
44628	1	21.60	21.60	21.60	21.60	0.00	0.00	21.60	21.60	21.60
44629	11	4.20	68.10	17.58	12.40	18.31	1.04	11.10	5.60	23.10
44631	1	5.50	5.50	5.50	5.50	0.00	0.00	5.50	5.50	5.50
44632	42	3.50	20.00	8.14	7.35	4.09	0.50	7.00	5.00	9.10
44633	2	11.90	28.90	20.40	18.54	12.02	0.59	20.40	11.90	28.90
44634	1	6.00	6.00	6.00	6.00	0.00	0.00	6.00	6.00	6.00
44636	1	11.00	11.00	11.00	11.00	0.00	0.00	11.00	11.00	11.00
44637	2	16.30	72.00	44.15	34.26	39.39	0.89	44.15	16.30	72.00
44638	2	6.20	14.90	10.55	9.61	6.15	0.58	10.55	6.20	14.90
44641	89	4.00	65.00	9.98	8.38	8.04	0.81	7.30	5.20	11.50
44643	12	5.40	49.00	16.06	12.20	14.84	0.92	9.75	6.90	14.80
44644	44	4.70	190.60	33.79	18.89	43.49	1.29	15.85	8.40	29.30
44645	8	4.00	10.00	6.61	6.29	2.17	0.33	6.95	4.00	7.90
44646	408	3.10	111.70	11.71	9.18	11.66	1.00	8.00	5.70	13.00

Zip Code	No.	Min.	Max.	AM	GM	SD	CV	Md	Q1	Q3
44647	34	4.30	100.00	17.19	11.73	20.54	1.20	9.25	6.40	18.30
44648	1	5.40	5.40	5.40	5.40	0.00	0.00	5.40	5.40	5.40
44651	2	5.00	12.40	8.70	7.87	5.23	0.60	8.70	5.00	12.40
44653	3	5.10	9.50	7.20	6.97	2.21	0.31	7.00	5.10	9.50
44654	25	4.20	36.00	13.45	10.38	10.48	0.78	9.00	5.60	20.90
44656	7	4.40	108.30	28.84	13.02	40.35	1.40	6.30	4.70	60.00
44657	9	4.50	38.00	12.03	9.72	10.29	0.85	9.00	7.60	10.00
44661	1	11.10	11.10	11.10	11.10	0.00	0.00	11.10	11.10	11.10
44662	18	6.20	93.00	19.18	13.59	21.59	1.13	10.65	7.30	23.60
44663	236	4.00	178.00	14.37	10.66	16.77	1.17	9.50	6.00	14.50
44666	2	4.50	15.00	9.75	8.22	7.42	0.76	9.75	4.50	15.00
44667	19	4.30	37.90	13.75	11.43	8.79	0.64	11.00	6.90	18.00
44671	2	22.40	43.00	32.70	31.04	14.57	0.45	32.70	22.40	43.00
44672	1	5.10	5.10	5.10	5.10	0.00	0.00	5.10	5.10	5.10
44675	7	4.30	34.30	13.70	9.44	13.54	0.99	6.20	5.10	32.60
44676	6	6.20	12.50	8.87	8.60	2.42	0.27	8.25	7.00	11.00
44677	3	10.00	21.10	15.37	14.68	5.56	0.36	15.00	10.00	21.10
44680	40	3.70	84.10	21.25	15.74	18.76	0.88	15.25	8.10	22.00
44681	6	5.60	22.00	12.82	11.40	6.45	0.50	12.70	6.90	17.00
44682	6	4.30	23.00	11.82	10.24	6.80	0.58	9.95	7.70	16.00
44683	16	4.00	136.10	18.59	10.02	32.60	1.75	7.70	4.70	12.00
44685	332	2.00	277.00	13.23	10.17	17.67	1.34	9.00	6.40	15.30
44688	3	10.80	55.00	28.60	22.82	23.32	0.82	20.00	10.80	55.00
44689	1	7.10	7.10	7.10	7.10	0.00	0.00	7.10	7.10	7.10
44691	240	4.00	274.00	17.14	10.38	28.31	1.65	7.95	5.70	15.00
44695	6	4.60	36.00	13.95	10.74	11.68	0.84	11.90	4.70	14.60
44696	1	4.70	4.70	4.70	4.70	0.00	0.00	4.70	4.70	4.70
44697	3	11.00	18.00	14.60	14.31	3.50	0.24	14.80	11.00	18.00
44698	1	9.60	9.60	9.60	9.60	0.00	0.00	9.60	9.60	9.60
44699	1	6.70	6.70	6.70	6.70	0.00	0.00	6.70	6.70	6.70
44701	3	5.80	10.10	7.43	7.21	2.33	0.31	6.40	5.80	10.10
44703	19	4.10	17.50	8.17	7.47	3.72	0.46	7.20	5.50	10.20
44704	3	4.10	20.00	10.50	8.47	8.39	0.80	7.40	4.10	20.00
44705	37	4.00	35.70	11.03	9.29	7.40	0.67	9.10	6.00	13.40
44706	37	4.20	69.00	13.09	9.58	13.74	1.05	7.10	5.80	12.00
44707	20	4.20	30.00	11.36	9.53	7.39	0.65	8.30	6.00	14.80

Zip Code	No.	Min.	Max.	AM	GM	SD	CV	Md	Q1	Q3
44708	211	4.00	67.00	10.78	9.04	8.62	0.80	8.40	6.00	11.30
44709	110	3.90	53.00	10.59	8.83	8.01	0.76	7.80	6.00	11.10
44710	18	4.00	17.90	8.67	7.85	4.00	0.46	7.70	5.10	12.10
44711	1	6.30	6.30	6.30	6.30	0.00	0.00	6.30	6.30	6.30
44714	69	4.00	43.90	14.57	12.35	8.80	0.60	10.80	7.90	20.00
44718	180	3.40	47.00	10.36	8.91	6.71	0.65	8.80	6.00	11.10
44719	1	4.50	4.50	4.50	4.50	0.00	0.00	4.50	4.50	4.50
44720	660	1.60	80.00	12.14	9.85	9.68	0.80	9.10	6.10	14.50
44721	195	1.80	150.10	14.09	10.63	16.89	1.20	9.70	6.70	16.00
44730	15	4.50	95.20	21.18	12.34	30.07	1.42	9.00	6.90	15.00
44746	2	4.50	51.30	27.90	15.19	33.09	1.19	27.90	4.50	51.30
44770	1	4.80	4.80	4.80	4.80	0.00	0.00	4.80	4.80	4.80
44774	1	7.80	7.80	7.80	7.80	0.00	0.00	7.80	7.80	7.80
44802	1	8.40	8.40	8.40	8.40	0.00	0.00	8.40	8.40	8.40
44803	2	14.00	17.00	15.50	15.43	2.12	0.14	15.50	14.00	17.00
44804	1	4.70	4.70	4.70	4.70	0.00	0.00	4.70	4.70	4.70
44805	114	2.30	141.00	13.54	9.88	16.58	1.22	8.40	6.20	12.50
44811	21	4.10	166.60	25.35	15.24	35.54	1.40	12.00	8.00	32.20
44813	20	4.10	316.50	58.14	26.81	80.29	1.38	29.30	6.70	55.00
44814	9	7.50	248.00	44.17	20.25	77.95	1.76	15.90	9.50	22.00
44817	1	17.90	17.90	17.90	17.90	0.00	0.00	17.90	17.90	17.90
44818	1	7.80	7.80	7.80	7.80	0.00	0.00	7.80	7.80	7.80
44820	7	4.40	19.00	9.07	8.25	4.74	0.52	8.00	6.30	10.10
44822	5	4.40	131.00	35.20	16.79	53.75	1.53	15.10	9.70	15.80
44824	4	7.20	18.00	13.38	12.64	4.75	0.35	14.15	7.20	16.00
44825	1	15.80	15.80	15.80	15.80	0.00	0.00	15.80	15.80	15.80
44827	3	6.70	22.70	13.83	12.25	8.14	0.59	12.10	6.70	22.70
44830	19	4.00	53.00	17.84	13.49	14.33	0.80	13.20	7.00	24.00
44833	12	3.50	14.30	8.73	8.18	3.13	0.36	8.40	6.10	9.70
44836	3	2.30	10.30	7.37	6.08	4.41	0.60	9.50	2.30	10.30
44837	1	8.90	8.90	8.90	8.90	0.00	0.00	8.90	8.90	8.90
44839	56	4.00	59.00	12.43	9.63	11.03	0.89	9.10	5.20	13.50
44840	7	4.10	24.50	12.44	10.11	8.29	0.67	8.40	5.20	21.60
44841	1	7.40	7.40	7.40	7.40	0.00	0.00	7.40	7.40	7.40
44842	20	5.80	32.00	17.05	15.10	8.39	0.49	15.00	10.50	22.80
44843	8	6.70	34.00	16.38	14.48	9.16	0.56	14.05	10.30	14.80

Zip Code	No.	Min.	Max.	AM	GM	SD	CV	Md	Q1	Q3
44846	16	5.10	41.90	13.86	11.04	11.22	0.81	8.30	6.60	14.10
44847	6	4.10	28.00	11.37	9.08	8.98	0.79	8.00	5.60	14.50
44851	6	4.50	17.90	9.20	8.12	5.09	0.55	8.80	4.50	10.70
44853	1	12.00	12.00	12.00	12.00	0.00	0.00	12.00	12.00	12.00
44856	1	4.60	4.60	4.60	4.60	0.00	0.00	4.60	4.60	4.60
44857	31	4.00	68.80	18.28	14.07	14.95	0.82	13.10	7.10	23.80
44859	3	6.60	20.00	12.20	10.97	6.97	0.57	10.00	6.60	20.00
44862	5	5.10	24.00	10.34	8.49	8.04	0.78	6.20	5.20	11.20
44864	10	4.00	138.00	33.35	17.19	42.97	1.29	16.30	7.10	41.00
44866	1	11.90	11.90	11.90	11.90	0.00	0.00	11.90	11.90	11.90
44870	48	4.20	30.00	10.75	9.36	6.16	0.57	9.40	5.70	13.00
44875	13	6.20	32.60	13.05	11.21	8.38	0.64	10.30	7.40	14.20
44878	2	4.50	13.30	8.90	7.74	6.22	0.70	8.90	4.50	13.30
44882	2	4.20	66.60	35.40	16.72	44.12	1.25	35.40	4.20	66.60
44883	65	4.20	70.00	13.84	11.33	11.47	0.83	10.00	8.00	15.00
44886	1	70.00	70.00	70.00	70.00	0.00	0.00	70.00	70.00	70.00
44888	1	5.10	5.10	5.10	5.10	0.00	0.00	5.10	5.10	5.10
44889	4	4.60	11.60	8.55	8.08	3.02	0.35	9.00	4.60	10.00
44890	4	4.80	14.70	10.50	9.68	4.30	0.41	11.25	4.80	12.70
44901	2	20.00	31.60	25.80	25.14	8.20	0.32	25.80	20.00	31.60
44903	75	4.10	232.00	16.07	10.34	28.65	1.78	9.00	5.60	14.90
44904	102	4.30	180.20	15.73	11.62	19.87	1.26	10.45	6.80	19.50
44905	13	5.30	55.50	13.89	10.80	13.68	0.98	9.80	7.00	11.20
44906	35	4.70	36.40	13.42	11.23	8.69	0.65	9.40	6.80	19.00
44907	32	3.10	74.20	11.51	8.88	12.92	1.12	7.90	6.40	10.80
44920	1	4.00	4.00	4.00	4.00	0.00	0.00	4.00	4.00	4.00
44961	3	7.90	27.40	15.43	13.35	10.48	0.68	11.00	7.90	27.40
45001	1	17.10	17.10	17.10	17.10	0.00	0.00	17.10	17.10	17.10
45002	23	3.90	12.00	7.43	6.96	2.73	0.37	6.90	4.90	10.50
45005	66	4.40	70.70	14.65	11.15	13.38	0.91	9.00	6.30	17.50
45006	2	6.10	11.70	8.90	8.45	3.96	0.44	8.90	6.10	11.70
45010	1	4.30	4.30	4.30	4.30	0.00	0.00	4.30	4.30	4.30
45011	478	2.40	55.00	7.99	7.16	5.11	0.64	6.80	5.40	8.90
45012	1	7.20	7.20	7.20	7.20	0.00	0.00	7.20	7.20	7.20
45013	79	3.80	50.40	9.79	8.31	7.16	0.73	7.80	5.30	12.10
45014	165	3.90	52.00	10.76	9.08	7.57	0.70	8.70	6.00	12.40

Zip Code	No.	Min.	Max.	AM	GM	SD	CV	Md	Q1	Q3
45015	11	4.30	20.60	8.43	7.77	4.30	0.51	7.20	6.90	8.40
45018	1	4.90	4.90	4.90	4.90	0.00	0.00	4.90	4.90	4.90
45019	1	6.20	6.20	6.20	6.20	0.00	0.00	6.20	6.20	6.20
45028	1	6.90	6.90	6.90	6.90	0.00	0.00	6.90	6.90	6.90
45029	1	8.00	8.00	8.00	8.00	0.00	0.00	8.00	8.00	8.00
45030	50	4.00	19.50	8.65	7.71	4.51	0.52	7.00	5.60	11.00
45032	1	4.50	4.50	4.50	4.50	0.00	0.00	4.50	4.50	4.50
45034	15	4.40	15.00	8.69	7.95	3.90	0.45	7.00	5.50	14.00
45036	250	3.80	42.00	8.58	7.65	4.96	0.58	7.00	5.50	10.20
45038	1	8.20	8.20	8.20	8.20	0.00	0.00	8.20	8.20	8.20
45039	363	2.00	82.00	8.14	7.26	5.77	0.71	7.00	5.60	8.80
45040	1034	1.40	30.00	7.60	6.99	3.64	0.48	6.60	5.20	8.60
45041	1	6.00	6.00	6.00	6.00	0.00	0.00	6.00	6.00	6.00
45042	39	4.00	37.50	11.34	9.55	7.69	0.68	8.90	6.00	13.10
45044	260	2.60	36.10	8.17	7.26	4.81	0.59	6.70	5.20	9.00
45049	1	6.10	6.10	6.10	6.10	0.00	0.00	6.10	6.10	6.10
45050	57	3.90	25.80	9.39	8.20	5.46	0.58	7.40	6.00	11.20
45051	1	5.00	5.00	5.00	5.00	0.00	0.00	5.00	5.00	5.00
45052	3	3.60	19.50	9.50	7.24	8.71	0.92	5.40	3.60	19.50
45053	6	4.00	12.00	6.62	6.15	2.98	0.45	5.65	4.70	7.70
45054	13	4.80	17.00	10.52	9.77	4.09	0.39	10.40	7.00	13.60
45056	90	3.70	22.90	9.17	8.22	4.55	0.50	7.95	5.80	10.40
45058	1	8.00	8.00	8.00	8.00	0.00	0.00	8.00	8.00	8.00
45061	1	4.10	4.10	4.10	4.10	0.00	0.00	4.10	4.10	4.10
45062	1	6.20	6.20	6.20	6.20	0.00	0.00	6.20	6.20	6.20
45064	5	4.50	6.70	5.74	5.69	0.87	0.15	5.70	5.40	6.40
45065	59	4.00	22.10	8.84	7.94	4.58	0.52	7.20	5.70	10.10
45066	405	3.10	111.90	8.95	7.61	7.85	0.88	6.80	5.30	9.70
45067	41	4.70	52.20	18.97	15.15	12.99	0.68	14.00	11.10	23.40
45068	87	4.10	39.50	10.42	8.84	6.50	0.62	7.70	5.40	14.60
45069	659	2.70	72.40	8.64	7.61	5.94	0.69	7.00	5.60	9.00
45102	74	2.70	20.40	7.04	6.56	2.98	0.42	6.00	5.10	8.00
45103	148	3.90	30.00	7.85	7.14	4.03	0.51	6.00	5.40	8.60
45104	2	9.40	10.10	9.75	9.74	0.49	0.05	9.75	9.40	10.10
45106	7	4.80	10.20	7.27	7.04	1.96	0.27	7.40	5.40	9.00
45107	6	3.40	14.00	8.57	7.68	4.07	0.48	8.25	5.20	12.30

Zip Code	No.	Min.	Max.	AM	GM	SD	CV	Md	Q1	Q3
45111	1	17.90	17.90	17.90	17.90	0.00	0.00	17.90	17.90	17.90
45113	8	5.20	12.00	8.30	7.98	2.53	0.30	7.60	6.00	9.20
45118	1	8.10	8.10	8.10	8.10	0.00	0.00	8.10	8.10	8.10
45120	2	4.50	13.50	9.00	7.79	6.36	0.71	9.00	4.50	13.50
45121	1	2.70	2.70	2.70	2.70	0.00	0.00	2.70	2.70	2.70
45122	13	4.20	20.00	9.28	8.41	4.62	0.50	7.30	6.10	10.00
45133	4	5.80	11.90	8.83	8.55	2.49	0.28	8.80	5.80	8.80
45135	1	4.50	4.50	4.50	4.50	0.00	0.00	4.50	4.50	4.50
45140	764	1.70	42.60	7.99	7.26	4.09	0.51	7.00	5.30	9.00
45146	2	6.20	6.80	6.50	6.49	0.42	0.07	6.50	6.20	6.80
45147	1	8.00	8.00	8.00	8.00	0.00	0.00	8.00	8.00	8.00
45148	2	6.40	9.90	8.15	7.96	2.47	0.30	8.15	6.40	9.90
45150	194	3.80	35.60	8.25	7.47	4.46	0.54	6.85	5.70	9.00
45152	101	3.70	74.70	10.12	8.40	8.81	0.87	7.80	5.70	11.00
45154	5	4.20	8.30	5.72	5.56	1.61	0.28	5.50	4.60	6.00
45157	13	4.20	25.20	7.60	6.54	5.69	0.75	5.80	4.70	7.00
45159	2	4.10	7.40	5.75	5.51	2.33	0.41	5.75	4.10	7.40
45171	3	5.10	15.00	8.80	7.84	5.40	0.61	6.30	5.10	15.00
45174	77	3.20	44.00	8.31	7.19	5.80	0.70	6.70	5.00	9.00
45176	4	6.00	21.80	11.80	10.21	7.40	0.63	9.70	6.00	13.00
45177	22	4.30	20.90	10.20	9.09	4.91	0.48	10.10	5.40	13.00
45202	11	4.20	16.00	8.23	7.41	4.25	0.52	6.40	4.80	11.00
45203	3	6.00	7.10	6.37	6.35	0.64	0.10	6.00	6.00	7.10
45205	3	4.00	11.30	7.27	6.65	3.71	0.51	6.50	4.00	11.30
45206	5	4.10	6.00	4.76	4.71	0.79	0.17	4.60	4.10	5.00
45208	120	3.10	28.00	5.95	5.48	3.34	0.56	5.00	4.30	6.30
45209	24	3.90	9.00	5.48	5.31	1.47	0.27	5.05	4.30	6.20
45211	20	3.40	15.00	6.70	6.29	2.65	0.40	6.30	4.70	8.10
45212	14	4.10	26.40	7.87	6.82	5.70	0.72	6.75	4.70	8.00
45213	22	4.00	25.00	8.03	7.04	5.04	0.63	6.00	5.00	8.50
45214	4	8.30	8.60	8.38	8.37	0.15	0.02	8.30	8.30	8.30
45215	78	4.00	20.00	6.95	6.56	2.88	0.41	6.00	5.10	7.80
45216	1	6.30	6.30	6.30	6.30	0.00	0.00	6.30	6.30	6.30
45217	10	4.30	12.00	7.07	6.64	2.72	0.38	6.30	5.00	10.00
45218	5	4.50	8.50	6.20	6.01	1.75	0.28	6.00	4.60	7.40
45219	1	13.10	13.10	13.10	13.10	0.00	0.00	13.10	13.10	13.10

Zip Code	No.	Min.	Max.	AM	GM	SD	CV	Md	Q1	Q3
45220	12	3.60	14.00	7.31	6.86	2.83	0.39	7.00	6.00	7.50
45221	1	12.00	12.00	12.00	12.00	0.00	0.00	12.00	12.00	12.00
45223	17	4.00	17.50	7.19	6.69	3.18	0.44	6.70	5.90	8.10
45224	21	4.00	21.00	7.14	6.34	4.31	0.60	5.40	4.60	7.20
45225	3	5.50	10.00	7.95	7.71	2.28	0.29	8.34	5.50	10.00
45226	22	4.00	9.40	5.93	5.73	1.65	0.28	5.65	4.80	6.40
45227	86	3.50	23.00	7.05	6.54	3.20	0.45	6.05	5.00	8.00
45229	6	6.00	8.80	6.82	6.73	1.25	0.18	6.05	6.00	8.00
45230	140	3.60	18.00	6.70	6.34	2.53	0.38	6.00	5.00	7.90
45231	42	4.00	17.00	6.90	6.49	2.71	0.39	6.05	5.00	8.00
45232	2	8.40	20.00	14.20	12.96	8.20	0.58	14.20	8.40	20.00
45233	16	4.10	14.90	6.99	6.47	3.18	0.46	5.95	4.90	8.00
45234	1	9.90	9.90	9.90	9.90	0.00	0.00	9.90	9.90	9.90
45235	3	6.00	16.00	9.33	8.32	5.77	0.62	6.00	6.00	16.00
45236	66	1.90	18.30	6.97	6.43	3.02	0.43	6.45	4.50	8.10
45237	20	3.90	14.20	7.27	6.70	3.17	0.44	6.40	4.70	8.60
45238	11	4.10	13.20	7.54	7.14	2.67	0.35	6.90	5.80	9.10
45239	21	4.20	23.50	7.68	6.87	4.51	0.59	7.00	4.70	8.40
45240	19	4.00	14.00	6.61	6.17	2.82	0.43	6.00	4.60	7.90
45241	173	3.70	42.30	8.51	7.26	5.96	0.70	6.30	4.90	9.70
45242	241	1.80	41.00	7.67	6.87	4.60	0.60	6.50	5.00	8.30
45243	120	4.00	22.00	6.43	6.02	2.79	0.43	5.50	4.70	7.00
45244	254	3.60	45.90	7.96	7.12	4.75	0.60	6.25	5.00	9.00
45245	90	2.60	16.20	6.98	6.57	2.65	0.38	6.70	5.00	8.00
45246	29	4.20	18.00	8.04	7.45	3.47	0.43	7.00	5.90	9.10
45247	29	4.00	17.50	7.43	7.07	2.61	0.35	6.90	6.00	8.50
45248	32	4.20	14.30	7.71	7.30	2.72	0.35	6.85	6.00	8.60
45249	157	1.80	52.70	8.37	7.22	6.50	0.78	6.40	5.10	9.00
45250	1	32.00	32.00	32.00	32.00	0.00	0.00	32.00	32.00	32.00
45251	14	5.10	41.00	9.64	8.01	9.16	0.95	7.00	6.00	9.60
45252	6	7.00	10.00	8.62	8.53	1.31	0.15	8.75	7.20	10.00
45255	161	3.10	1400.00	25.01	7.66	154.73	6.19	7.00	5.60	9.00
45273	1	9.00	9.00	9.00	9.00	0.00	0.00	9.00	9.00	9.00
45277	2	7.00	8.00	7.50	7.48	0.71	0.09	7.50	7.00	8.00
45302	8	7.70	26.00	12.50	11.41	6.39	0.51	9.85	7.70	11.70
45304	6	3.30	14.00	7.53	6.64	4.11	0.55	6.60	4.00	10.70

Zip Code	No.	Min.	Max.	AM	GM	SD	CV	Md	Q1	Q3
45305	186	3.90	66.00	11.51	9.67	8.64	0.75	9.20	6.40	13.00
45306	1	16.00	16.00	16.00	16.00	0.00	0.00	16.00	16.00	16.00
45307	1	6.50	6.50	6.50	6.50	0.00	0.00	6.50	6.50	6.50
45308	4	4.40	16.00	9.30	8.34	4.98	0.54	8.40	4.40	9.80
45309	19	6.00	20.00	11.33	10.65	4.21	0.37	10.40	8.40	14.40
45312	4	9.10	43.90	18.33	14.22	17.07	0.93	10.15	9.10	11.00
45314	9	8.00	79.00	24.98	19.80	21.55	0.86	18.00	13.00	26.00
45315	33	4.10	25.00	9.60	8.53	5.12	0.53	8.20	6.00	11.30
45316	1	7.50	7.50	7.50	7.50	0.00	0.00	7.50	7.50	7.50
45317	1	28.60	28.60	28.60	28.60	0.00	0.00	28.60	28.60	28.60
45318	5	6.90	16.90	11.74	10.89	4.85	0.41	11.70	6.90	16.30
45319	2	5.60	19.10	12.35	10.34	9.55	0.77	12.35	5.60	19.10
45320	16	4.00	22.30	10.93	9.55	5.75	0.53	9.70	6.00	15.00
45321	1	6.00	6.00	6.00	6.00	0.00	0.00	6.00	6.00	6.00
45322	72	3.30	60.00	10.62	8.94	7.93	0.75	8.00	6.00	13.50
45323	27	4.40	36.00	13.59	11.80	7.85	0.58	12.50	7.20	16.80
45324	203	3.70	54.70	12.08	9.67	9.63	0.80	8.50	5.90	14.50
45325	4	5.40	13.70	9.93	9.30	3.86	0.39	10.30	5.40	12.50
45326	2	6.00	6.00	6.00	6.00	0.00	0.00	6.00	6.00	6.00
45327	26	4.00	41.80	11.73	9.93	7.90	0.67	11.20	5.90	13.60
45331	16	4.50	30.70	11.60	10.05	7.16	0.62	9.10	7.50	10.60
45334	1	12.00	12.00	12.00	12.00	0.00	0.00	12.00	12.00	12.00
45335	12	4.10	20.00	7.81	7.12	4.17	0.53	7.05	5.00	8.40
45337	2	4.00	4.80	4.40	4.38	0.57	0.13	4.40	4.00	4.80
45338	6	5.70	22.00	11.83	10.65	6.01	0.51	10.90	6.60	14.90
45339	4	5.50	27.40	11.35	8.73	10.72	0.94	6.25	5.50	7.00
45340	2	5.20	13.30	9.25	8.32	5.73	0.62	9.25	5.20	13.30
45341	11	8.00	41.40	18.03	16.30	9.21	0.51	15.60	11.10	20.80
45342	126	3.80	26.80	8.87	7.90	4.78	0.54	7.50	5.60	10.00
45343	3	4.70	15.20	8.87	7.82	5.58	0.63	6.70	4.70	15.20
45344	38	4.10	59.00	18.89	14.26	14.58	0.77	14.95	7.40	24.50
45345	4	7.00	22.00	15.93	14.51	7.01	0.44	17.35	7.00	21.00
45349	2	6.60	13.50	10.05	9.44	4.88	0.49	10.05	6.60	13.50

Zip Code	No.	Min.	Max.	AM	GM	SD	CV	Md	Q1	Q3
45354	1	9.60	9.60	9.60	9.60	0.00	0.00	9.60	9.60	9.60
45356	33	3.80	41.20	13.39	11.54	7.64	0.57	12.40	8.70	17.10
45357	1	30.00	30.00	30.00	30.00	0.00	0.00	30.00	30.00	30.00
45358	2	8.40	12.80	10.60	10.37	3.11	0.29	10.60	8.40	12.80
45359	2	7.10	8.70	7.90	7.86	1.13	0.14	7.90	7.10	8.70
45360	1	5.70	5.70	5.70	5.70	0.00	0.00	5.70	5.70	5.70
45363	1	8.20	8.20	8.20	8.20	0.00	0.00	8.20	8.20	8.20
45365	52	3.80	51.40	17.38	13.25	13.53	0.78	11.40	7.00	25.00
45368	4	5.90	7.30	6.65	6.63	0.58	0.09	6.70	5.90	6.80
45369	5	5.90	14.00	9.68	9.30	3.00	0.31	10.00	8.10	10.40
45370	24	3.70	27.00	8.33	7.36	4.98	0.60	6.80	4.60	9.20
45371	144	3.50	3901.00	37.87	9.48	324.29	8.56	9.00	6.00	11.90
45372	2	20.00	20.50	20.25	20.25	0.35	0.02	20.25	20.00	20.50
45373	237	2.50	734.00	14.14	9.61	47.43	3.35	9.70	6.00	14.00
45375	2	5.30	25.00	15.15	11.51	13.93	0.92	15.15	5.30	25.00
45377	49	4.20	25.10	9.33	8.08	5.64	0.61	7.60	5.10	11.10
45380	9	4.20	21.50	11.47	10.01	6.06	0.53	9.50	9.00	12.90
45381	5	3.90	18.00	10.06	8.80	5.59	0.56	8.20	6.90	13.30
45382	1	22.00	22.00	22.00	22.00	0.00	0.00	22.00	22.00	22.00
45383	10	4.70	25.50	10.58	9.33	6.16	0.58	8.55	6.30	12.00
45385	235	4.00	74.00	16.06	12.68	12.48	0.78	12.20	7.50	21.00
45387	73	4.10	43.00	12.44	10.65	7.71	0.62	10.20	6.70	15.00
45389	2	5.50	11.30	8.40	7.88	4.10	0.49	8.40	5.50	11.30
45402	14	4.10	84.80	22.21	14.84	21.92	0.99	12.20	6.80	32.30
45403	12	4.20	20.00	7.94	7.15	4.45	0.56	6.25	5.20	8.30
45404	8	4.60	15.30	9.46	8.80	3.77	0.40	8.75	6.00	11.00
45405	15	4.00	30.00	11.12	9.05	8.09	0.73	8.50	5.00	13.20
45406	10	4.30	8.70	6.22	6.04	1.59	0.26	6.05	4.60	8.00
45407	30	1.60	98.10	25.50	14.72	23.16	0.91	19.85	5.10	39.40
45408	5	4.00	8.80	6.12	5.88	1.94	0.32	5.40	5.00	7.40
45409	47	2.30	62.00	9.18	7.62	8.76	0.95	6.80	5.40	10.30
45410	27	3.60	16.60	7.17	6.71	2.89	0.40	6.40	5.00	8.80
45413	1	8.80	8.80	8.80	8.80	0.00	0.00	8.80	8.80	8.80

Zip Code	No.	Min.	Max.	AM	GM	SD	CV	Md	Q1	Q3
45414	56	4.00	23.30	9.51	8.43	5.15	0.54	7.90	5.70	11.30
45415	50	3.20	38.60	12.14	9.76	8.67	0.71	9.35	5.40	15.70
45416	2	4.90	4.90	4.90	4.90	0.00	0.00	4.90	4.90	4.90
45417	20	4.10	49.90	20.45	15.55	14.10	0.69	17.15	6.90	29.30
45418	6	2.40	18.30	8.43	7.12	5.35	0.63	7.20	6.20	9.30
45419	186	1.70	42.00	8.13	7.41	4.25	0.52	7.10	5.70	9.40
45420	75	4.00	55.00	9.43	8.14	7.10	0.75	7.40	5.70	10.00
45423	1	12.10	12.10	12.10	12.10	0.00	0.00	12.10	12.10	12.10
45424	160	3.90	83.00	11.40	9.29	10.16	0.89	8.30	6.00	12.40
45426	10	4.20	18.00	8.14	7.19	4.66	0.57	6.10	5.30	11.00
45428	1	4.50	4.50	4.50	4.50	0.00	0.00	4.50	4.50	4.50
45429	226	4.00	37.00	9.79	8.58	5.99	0.61	7.95	6.00	11.00
45430	93	3.80	27.00	9.92	8.93	4.90	0.49	8.50	6.80	12.10
45431	135	2.90	63.00	9.20	7.83	6.97	0.76	7.00	5.30	10.90
45432	107	3.30	43.60	9.44	8.18	6.00	0.63	8.00	5.50	11.20
45433	2	4.40	4.70	4.55	4.55	0.21	0.05	4.55	4.40	4.70
45434	219	2.00	75.00	10.69	8.95	8.32	0.78	8.30	6.00	12.20
45439	22	4.00	27.40	11.57	10.26	6.17	0.53	9.80	7.20	14.70
45440	182	1.70	40.00	9.63	8.49	5.41	0.56	8.00	5.60	12.00
45449	40	3.90	83.70	12.70	9.82	13.37	1.05	8.75	6.60	13.50
45453	1	13.60	13.60	13.60	13.60	0.00	0.00	13.60	13.60	13.60
45458	292	3.90	58.50	8.75	7.60	5.99	0.68	7.00	5.10	10.20
45459	276	3.80	64.00	10.14	8.72	7.10	0.70	8.00	6.00	11.80
45480	1	6.80	6.80	6.80	6.80	0.00	0.00	6.80	6.80	6.80
45485	1	5.10	5.10	5.10	5.10	0.00	0.00	5.10	5.10	5.10
45499	1	64.70	64.70	64.70	64.70	0.00	0.00	64.70	64.70	64.70
45501	1	17.60	17.60	17.60	17.60	0.00	0.00	17.60	17.60	17.60
45502	79	4.20	40.00	11.21	9.56	7.20	0.64	9.60	6.00	13.00
45503	79	0.70	68.00	12.04	9.77	9.32	0.77	9.80	6.80	14.20
45504	37	5.00	121.20	15.92	11.44	20.54	1.29	9.00	7.00	16.00
45505	9	4.50	24.00	10.39	9.23	5.93	0.57	9.30	6.80	10.10
45506	12	4.00	20.00	11.32	10.24	4.95	0.44	11.65	6.70	15.00
45560	1	6.30	6.30	6.30	6.30	0.00	0.00	6.30	6.30	6.30
45585	1	7.40	7.40	7.40	7.40	0.00	0.00	7.40	7.40	7.40

Zip Code	No.	Min.	Max.	AM	GM	SD	CV	Md	Q1	Q3
45601	57	4.00	200.00	14.79	9.74	26.60	1.80	8.00	5.90	14.20
45612	4	4.00	12.90	7.25	6.57	3.94	0.54	6.05	4.00	6.80
45628	1	14.60	14.60	14.60	14.60	0.00	0.00	14.60	14.60	14.60
45634	1	11.70	11.70	11.70	11.70	0.00	0.00	11.70	11.70	11.70
45640	10	4.00	14.60	7.53	6.83	3.71	0.49	5.70	4.80	11.30
45644	1	14.00	14.00	14.00	14.00	0.00	0.00	14.00	14.00	14.00
45647	1	36.00	36.00	36.00	36.00	0.00	0.00	36.00	36.00	36.00
45648	1	6.60	6.60	6.60	6.60	0.00	0.00	6.60	6.60	6.60
45656	1	8.00	8.00	8.00	8.00	0.00	0.00	8.00	8.00	8.00
45662	2	13.00	15.20	14.10	14.06	1.56	0.11	14.10	13.00	15.20
45690	4	4.00	50.00	22.80	15.72	20.09	0.88	18.60	4.00	25.00
45694	2	3.90	4.50	4.20	4.19	0.42	0.10	4.20	3.90	4.50
45697	1	15.00	15.00	15.00	15.00	0.00	0.00	15.00	15.00	15.00
45701	7	4.30	15.20	7.21	6.50	3.97	0.55	5.40	4.50	9.50
45714	12	4.20	10.10	6.06	5.80	2.02	0.33	5.10	4.80	6.00
45716	1	6.00	6.00	6.00	6.00	0.00	0.00	6.00	6.00	6.00
45723	1	6.00	6.00	6.00	6.00	0.00	0.00	6.00	6.00	6.00
45732	1	11.00	11.00	11.00	11.00	0.00	0.00	11.00	11.00	11.00
45742	3	5.30	14.10	9.83	9.10	4.41	0.45	10.10	5.30	14.10
45743	1	11.60	11.60	11.60	11.60	0.00	0.00	11.60	11.60	11.60
45750	31	4.20	24.00	11.00	9.89	5.32	0.48	9.30	7.00	15.00
45766	1	13.20	13.20	13.20	13.20	0.00	0.00	13.20	13.20	13.20
45780	4	14.70	18.60	17.55	17.47	1.91	0.11	18.45	14.70	18.60
45784	4	4.20	8.90	6.48	6.24	1.98	0.31	6.40	4.20	7.00
45786	1	4.10	4.10	4.10	4.10	0.00	0.00	4.10	4.10	4.10
45801	3	3.20	14.00	7.77	6.49	5.59	0.72	6.10	3.20	14.00
45802	1	9.80	9.80	9.80	9.80	0.00	0.00	9.80	9.80	9.80
45804	2	5.25	8.60	6.93	6.72	2.37	0.34	6.93	5.25	8.60
45805	34	2.30	60.30	12.75	8.68	15.11	1.18	6.85	5.40	9.00
45806	9	4.30	60.00	19.39	12.02	21.90	1.13	8.60	6.50	17.60
45807	15	4.20	15.60	8.41	7.90	3.17	0.38	7.70	5.80	10.85
45809	1	7.70	7.70	7.70	7.70	0.00	0.00	7.70	7.70	7.70
45810	2	6.40	11.40	8.90	8.54	3.54	0.40	8.90	6.40	11.40

Zip Code	No.	Min.	Max.	AM	GM	SD	CV	Md	Q1	Q3
45813	1	20.00	20.00	20.00	20.00	0.00	0.00	20.00	20.00	20.00
45814	6	3.40	20.10	10.08	8.84	5.58	0.55	9.15	7.70	11.00
45816	1	21.60	21.60	21.60	21.60	0.00	0.00	21.60	21.60	21.60
45817	17	5.00	18.00	10.42	9.59	4.19	0.40	11.80	6.30	13.20
45822	2	8.00	10.00	9.00	8.94	1.41	0.16	9.00	8.00	10.00
45826	1	11.60	11.60	11.60	11.60	0.00	0.00	11.60	11.60	11.60
45828	3	11.00	14.10	12.47	12.40	1.56	0.12	12.30	11.00	14.10
45830	5	8.90	21.90	13.56	12.75	5.49	0.40	12.00	9.00	16.00
45833	5	6.00	21.00	10.74	9.73	5.94	0.55	8.40	8.00	10.30
45840	472	1.00	538.00	10.77	8.20	25.55	2.37	7.20	5.50	10.50
45841	1	10.00	10.00	10.00	10.00	0.00	0.00	10.00	10.00	10.00
45843	2	7.90	29.00	18.45	15.14	14.92	0.81	18.45	7.90	29.00
45844	1	13.30	13.30	13.30	13.30	0.00	0.00	13.30	13.30	13.30
45845	2	7.20	8.00	7.60	7.59	0.57	0.07	7.60	7.20	8.00
45846	1	5.70	5.70	5.70	5.70	0.00	0.00	5.70	5.70	5.70
45850	1	10.40	10.40	10.40	10.40	0.00	0.00	10.40	10.40	10.40
45856	3	6.00	25.00	13.93	11.74	9.88	0.71	10.80	6.00	25.00
45858	10	5.00	28.50	14.30	12.15	8.38	0.59	13.75	6.90	17.00
45865	4	7.00	47.00	24.40	19.60	17.14	0.70	21.80	7.00	27.00
45868	1	6.70	6.70	6.70	6.70	0.00	0.00	6.70	6.70	6.70
45869	12	3.90	27.00	8.33	7.04	6.35	0.76	6.15	4.40	9.30
45872	3	4.75	19.00	12.92	11.06	7.35	0.57	15.00	4.75	19.00
45873	1	5.70	5.70	5.70	5.70	0.00	0.00	5.70	5.70	5.70
45875	15	4.15	18.10	7.80	7.24	3.46	0.44	7.40	5.00	8.90
45877	2	10.80	15.05	12.93	12.75	3.01	0.23	12.93	10.80	15.05
45879	1	9.00	9.00	9.00	9.00	0.00	0.00	9.00	9.00	9.00
45881	3	11.60	17.40	14.87	14.66	2.97	0.20	15.60	11.60	17.40
45883	2	4.50	5.80	5.15	5.11	0.92	0.18	5.15	4.50	5.80
45884	1	13.80	13.80	13.80	13.80	0.00	0.00	13.80	13.80	13.80
45885	1	11.00	11.00	11.00	11.00	0.00	0.00	11.00	11.00	11.00
45889	4	10.00	22.50	15.58	14.66	6.15	0.40	14.90	10.00	19.00
45890	1	12.70	12.70	12.70	12.70	0.00	0.00	12.70	12.70	12.70
45895	9	5.80	10.00	7.13	7.01	1.46	0.20	6.38	6.10	8.00
Unknown	488	1.00	208.00	10.86	8.37	13.97	1.29	7.35	5.40	11.00

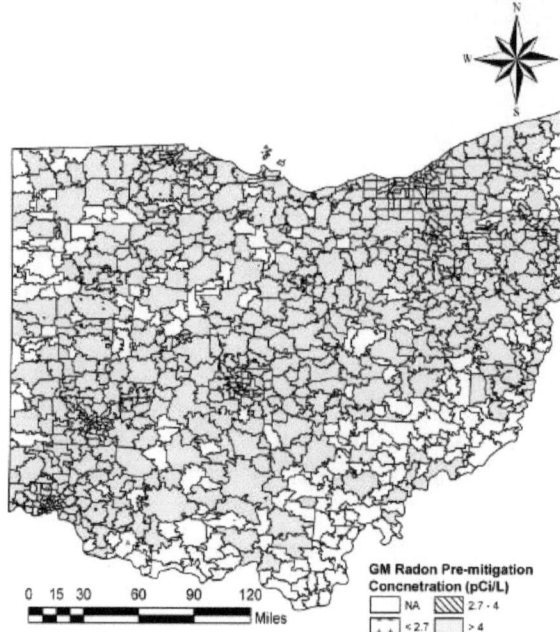

Figure 4a: GM [222]Rn Pre-mitigation concentrations in Ohio zip codes based on the WHO-USEPA classification.

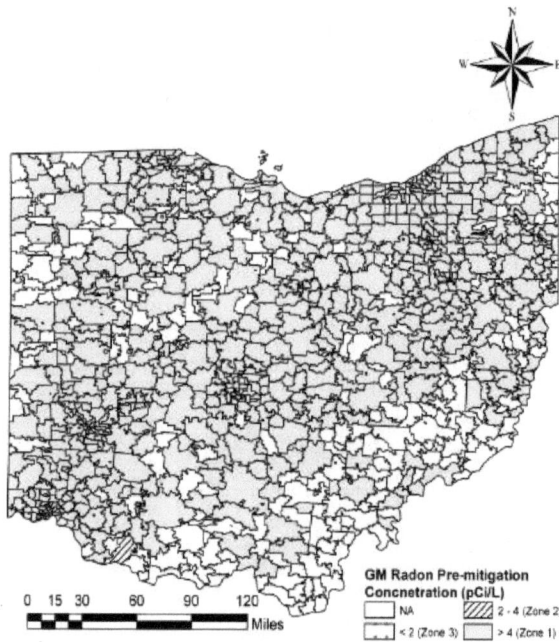

Figure 4b: GM [222]Rn Pre-mitigation concentrations in Ohio zip codes based on the USEPA classification.

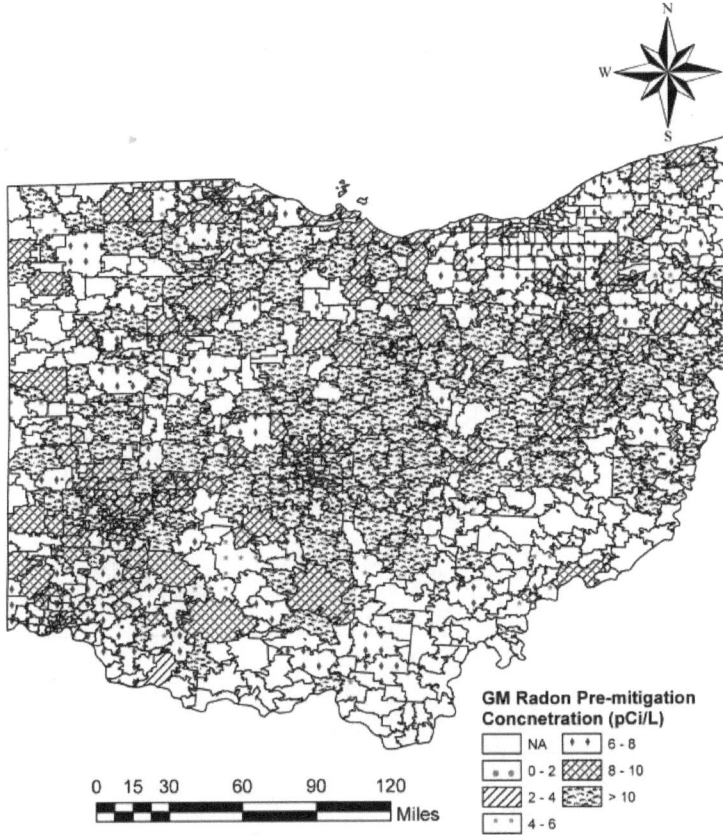

Figure 4c: GM [222]Rn Pre-mitigation concentrations in Ohio zip codes based on the 2 pCi/L breakdown classifications.

Table 4: Post-mitigation [222]Rn statistics for Ohio zip codes

Zip Code	No.	Min.	Max.	AM	GM	SD	CV	Md	Q1	Q3
43001	8	0.40	3.50	1.69	1.43	0.95	0.56	1.70	1.00	1.90
43003	4	1.10	3.40	2.33	2.15	0.94	0.41	2.40	1.10	2.40
43004	205	0.00	8.70	1.26	0.92	1.05	0.83	0.90	0.60	1.60
43005	1	2.40	2.40	2.40	2.40	0.00	0.00	2.40	2.40	2.40
43009	6	0.20	2.30	1.07	0.86	0.68	0.64	0.95	0.90	1.10
43011	13	0.30	3.30	1.44	1.14	0.99	0.69	1.30	0.60	1.90
43012	3	1.90	2.70	2.23	2.21	0.42	0.19	2.10	1.90	2.70
43013	3	0.30	1.40	0.67	0.50	0.64	0.95	0.30	0.30	1.40
43014	3	0.70	2.90	1.63	1.38	1.14	0.70	1.30	0.70	2.90
43015	379	0.06	4.70	1.50	1.19	0.97	0.65	1.20	0.70	2.20
43016	422	0.10	10.90	1.46	1.13	1.26	0.86	1.10	0.70	1.80

Zip Code	No.	Min.	Max.	AM	GM	SD	CV	Md	Q1	Q3
43017	1155	0.00	35.00	1.56	1.21	1.52	0.98	1.20	0.70	2.20
43018	1	1.10	1.10	1.10	1.10	0.00	0.00	1.10	1.10	1.10
43019	9	0.30	3.20	1.50	1.21	0.97	0.65	1.20	0.80	2.30
43020	1	2.50	2.50	2.50	2.50	0.00	0.00	2.50	2.50	2.50
43021	173	0.20	3.70	1.44	1.20	0.85	0.59	1.20	0.80	2.00
43022	10	0.50	3.30	1.85	1.62	0.90	0.49	1.75	1.20	2.40
43023	190	0.10	29.30	2.14	1.48	2.82	1.31	1.45	0.90	2.60
43024	3	0.40	2.00	1.40	1.13	0.87	0.62	1.80	0.40	2.00
43025	11	0.40	3.80	1.51	1.24	1.01	0.67	1.20	0.70	2.00
43026	707	0.10	5.60	1.20	0.91	0.90	0.75	0.90	0.60	1.60
43027	1	1.10	1.10	1.10	1.10	0.00	0.00	1.10	1.10	1.10
43028	114	0.30	12.90	1.71	1.29	1.66	0.97	1.30	0.80	2.00
43029	2	1.50	3.10	2.30	2.16	1.13	0.49	2.30	1.50	3.10
43030	1	1.50	1.50	1.50	1.50	0.00	0.00	1.50	1.50	1.50
43031	38	0.40	3.20	1.52	1.28	0.91	0.60	1.20	0.80	2.10
43033	3	0.50	3.10	1.97	1.53	1.33	0.68	2.30	0.50	3.10
43035	413	0.10	3.70	1.15	0.94	0.76	0.66	0.90	0.60	1.50
43036	1	0.70	0.70	0.70	0.70	0.00	0.00	0.70	0.70	0.70
43039	1	0.50	0.50	0.50	0.50	0.00	0.00	0.50	0.50	0.50
43040	80	0.10	4.05	1.38	1.06	0.96	0.70	1.05	0.60	2.00
43041	1	3.50	3.50	3.50	3.50	0.00	0.00	3.50	3.50	3.50
43042	1	1.60	1.60	1.60	1.60	0.00	0.00	1.60	1.60	1.60
43043	1	3.60	3.60	3.60	3.60	0.00	0.00	3.60	3.60	3.60
43044	4	0.40	3.90	2.33	1.63	1.79	0.77	2.50	0.40	3.80
43045	12	0.40	3.10	1.95	1.68	0.91	0.46	2.15	1.00	2.50
43046	3	0.70	2.60	1.33	1.08	1.10	0.82	0.70	0.70	2.60
43047	1	2.30	2.30	2.30	2.30	0.00	0.00	2.30	2.30	2.30
43050	44	0.30	50.00	2.76	1.52	7.34	2.66	1.50	1.00	2.10
43052	2	0.80	3.70	2.25	1.72	2.05	0.91	2.25	0.80	3.70
43053	2	2.50	14.00	8.25	5.92	8.13	0.99	8.25	2.50	14.00
43054	583	0.10	6.30	1.34	1.08	0.86	0.64	1.10	0.70	1.90
43055	235	0.01	57.50	2.04	1.39	3.97	1.95	1.40	0.80	2.50
43056	34	0.40	4.90	1.52	1.22	1.09	0.71	1.20	0.80	2.10
43058	2	1.00	3.20	2.10	1.79	1.56	0.74	2.10	1.00	3.20
43060	3	0.30	2.20	1.17	0.87	0.96	0.82	1.00	0.30	2.20
43061	20	0.20	3.90	1.59	1.31	1.00	0.63	1.25	0.90	2.00

Zip Code	No.	Min.	Max.	AM	GM	SD	CV	Md	Q1	Q3
43062	124	0.20	3.90	1.25	1.02	0.82	0.66	1.00	0.60	1.60
43064	52	0.30	3.30	1.41	1.17	0.81	0.57	1.15	0.80	2.00
43065	886	0.10	7.70	1.38	1.12	0.90	0.65	1.20	0.70	1.80
43066	4	0.30	2.60	1.43	0.97	1.19	0.84	1.40	0.30	2.30
43067	1	1.30	1.30	1.30	1.30	0.00	0.00	1.30	1.30	1.30
43068	170	0.10	3.90	1.15	0.88	0.85	0.74	0.90	0.50	1.60
43071	1	0.90	0.90	0.90	0.90	0.00	0.00	0.90	0.90	0.90
43072	1	1.30	1.30	1.30	1.30	0.00	0.00	1.30	1.30	1.30
43074	54	0.30	3.70	1.52	1.29	0.88	0.57	1.30	0.80	2.00
43075	1	0.50	0.50	0.50	0.50	0.00	0.00	0.50	0.50	0.50
43076	3	1.50	20.00	8.17	4.48	10.28	1.26	3.00	1.50	20.00
43078	22	0.90	4.20	2.06	1.86	0.96	0.46	1.75	1.40	2.50
43080	4	0.80	3.50	2.70	2.33	1.29	0.48	3.25	0.80	3.50
43081	499	0.10	23.00	1.29	1.00	1.30	1.00	1.00	0.60	1.60
43082	696	0.10	9.40	1.28	1.04	0.85	0.67	1.10	0.60	1.70
43084	1	0.50	0.50	0.50	0.50	0.00	0.00	0.50	0.50	0.50
43085	474	0.20	61.00	1.81	1.36	2.94	1.62	1.50	0.80	2.30
43086	1	1.20	1.20	1.20	1.20	0.00	0.00	1.20	1.20	1.20
43089	1	1.30	1.30	1.30	1.30	0.00	0.00	1.30	1.30	1.30
43103	12	0.40	3.90	1.59	1.32	0.99	0.62	1.40	1.00	1.90
43105	18	0.40	3.80	1.57	1.34	0.95	0.60	1.20	1.00	1.90
43106	1	0.30	0.30	0.30	0.30	0.00	0.00	0.30	0.30	0.30
43107	3	2.40	2.90	2.70	2.69	0.26	0.10	2.80	2.40	2.90
43110	116	0.10	6.10	1.53	1.24	0.96	0.62	1.35	0.70	2.20
43112	8	0.30	3.10	1.86	1.54	0.94	0.51	1.90	0.90	2.30
43113	22	0.40	3.80	1.67	1.36	0.99	0.60	1.70	0.70	2.10
43115	1	1.00	1.00	1.00	1.00	0.00	0.00	1.00	1.00	1.00
43116	2	1.90	2.50	2.20	2.18	0.42	0.19	2.20	1.90	2.50
43117	1	1.10	1.10	1.10	1.10	0.00	0.00	1.10	1.10	1.10
43119	148	0.00	3.80	1.03	0.79	0.76	0.74	0.80	0.50	1.20
43123	141	0.20	11.00	1.39	1.05	1.28	0.92	1.00	0.60	1.80
43124	1	1.80	1.80	1.80	1.80	0.00	0.00	1.80	1.80	1.80
43125	22	0.40	3.00	1.40	1.17	0.84	0.60	1.05	0.70	1.90
43130	56	0.20	8.20	1.63	1.19	1.38	0.85	1.30	0.70	2.20
43132	1	0.30	0.30	0.30	0.30	0.00	0.00	0.30	0.30	0.30
43135	2	0.40	2.20	1.30	0.94	1.27	0.98	1.30	0.40	2.20

Zip Code	No.	Min.	Max.	AM	GM	SD	CV	Md	Q1	Q3
43136	3	0.80	3.80	1.90	1.50	1.65	0.87	1.10	0.80	3.80
43138	3	1.10	2.90	1.83	1.69	0.95	0.52	1.50	1.10	2.90
43140	44	0.30	9.00	1.70	1.30	1.46	0.86	1.30	0.80	2.40
43142	2	1.70	3.40	2.55	2.40	1.20	0.47	2.55	1.70	3.40
43143	4	0.80	2.00	1.45	1.36	0.55	0.38	1.50	0.80	1.80
43146	7	0.90	2.30	1.60	1.52	0.54	0.33	1.60	1.00	2.10
43147	381	0.10	7.30	1.16	0.95	0.78	0.67	0.90	0.60	1.50
43148	4	1.10	3.20	2.23	2.07	0.90	0.40	2.30	1.10	2.60
43149	2	1.00	3.40	2.20	1.84	1.70	0.77	2.20	1.00	3.40
43150	1	3.50	3.50	3.50	3.50	0.00	0.00	3.50	3.50	3.50
43154	2	1.70	2.10	1.90	1.89	0.28	0.15	1.90	1.70	2.10
43155	2	1.10	3.20	2.15	1.88	1.48	0.69	2.15	1.10	3.20
43160	1	3.10	3.10	3.10	3.10	0.00	0.00	3.10	3.10	3.10
43162	9	0.40	2.30	1.07	0.95	0.55	0.52	1.10	0.70	1.20
43164	1	3.60	3.60	3.60	3.60	0.00	0.00	3.60	3.60	3.60
43174	1	1.80	1.80	1.80	1.80	0.00	0.00	1.80	1.80	1.80
43201	76	0.10	4.10	1.31	1.01	0.88	0.67	1.15	0.50	1.90
43202	60	0.40	3.80	1.75	1.45	1.04	0.59	1.45	0.80	2.60
43203	12	1.00	3.80	1.75	1.61	0.83	0.47	1.55	1.10	1.70
43204	54	0.10	3.50	1.35	1.00	0.96	0.71	1.00	0.60	1.80
43205	2	1.70	1.80	1.75	1.75	0.07	0.04	1.75	1.70	1.80
43206	72	0.20	3.90	1.76	1.43	0.99	0.56	1.80	0.90	2.40
43207	18	0.30	5.10	2.18	1.78	1.27	0.58	2.15	1.10	2.70
43208	1	3.20	3.20	3.20	3.20	0.00	0.00	3.20	3.20	3.20
43209	266	0.20	5.90	1.98	1.69	1.02	0.51	1.90	1.10	2.70
43210	1	1.70	1.70	1.70	1.70	0.00	0.00	1.70	1.70	1.70
43211	3	1.50	2.60	2.03	1.98	0.55	0.27	2.00	1.50	2.60
43212	93	0.30	3.70	1.85	1.54	1.03	0.56	1.60	0.90	2.70
43213	55	0.30	3.90	1.55	1.28	0.89	0.58	1.30	0.80	2.10
43214	229	0.20	9.90	1.85	1.46	1.24	0.67	1.50	0.90	2.80
43215	25	0.20	10.00	2.18	1.54	1.98	0.91	2.35	0.70	2.90
43216	1	3.20	3.20	3.20	3.20	0.00	0.00	3.20	3.20	3.20
43217	2	1.70	1.70	1.70	1.70	0.00	0.00	1.70	1.70	1.70
43219	19	0.20	3.90	1.52	1.04	1.25	0.82	1.00	0.50	2.40
43220	304	0.30	13.30	2.05	1.69	1.39	0.68	1.90	1.10	2.80
43221	615	0.10	13.00	1.88	1.53	1.16	0.62	1.70	1.00	2.70

Zip Code	No.	Min.	Max.	AM	GM	SD	CV	Md	Q1	Q3
43222	1	1.50	1.50	1.50	1.50	0.00	0.00	1.50	1.50	1.50
43223	22	0.20	2.20	0.92	0.72	0.62	0.67	0.75	0.40	1.40
43224	36	0.20	3.70	1.78	1.39	1.09	0.62	1.45	0.70	2.70
43225	1	0.50	0.50	0.50	0.50	0.00	0.00	0.50	0.50	0.50
43226	4	1.00	2.60	1.50	1.38	0.76	0.50	1.20	1.00	1.40
43227	7	0.60	2.60	1.10	0.96	0.71	0.65	0.80	0.60	1.30
43228	152	0.10	3.60	0.96	0.75	0.73	0.76	0.80	0.50	1.10
43229	73	0.30	3.90	1.35	1.11	0.85	0.63	1.10	0.70	1.90
43230	451	0.10	12.00	1.24	0.97	0.99	0.80	1.00	0.60	1.60
43231	60	0.20	3.70	1.26	0.98	0.93	0.74	0.90	0.60	1.70
43232	33	0.30	3.90	1.70	1.39	1.05	0.62	1.30	0.90	2.50
43235	446	0.10	5.10	1.50	1.21	0.95	0.64	1.20	0.70	2.10
43240	16	0.10	2.20	0.73	0.59	0.54	0.74	0.60	0.40	0.80
43255	1	0.90	0.90	0.90	0.90	0.00	0.00	0.90	0.90	0.90
43258	1	0.60	0.60	0.60	0.60	0.00	0.00	0.60	0.60	0.60
43290	1	1.40	1.40	1.40	1.40	0.00	0.00	1.40	1.40	1.40
43302	53	0.20	3.80	1.65	1.39	0.89	0.54	1.60	1.00	2.30
43303	2	0.20	1.00	0.60	0.45	0.57	0.94	0.60	0.20	1.00
43311	37	0.30	8.20	2.13	1.74	1.42	0.66	2.10	1.20	2.90
43313	1	2.70	2.70	2.70	2.70	0.00	0.00	2.70	2.70	2.70
43315	3	0.40	2.30	1.23	0.97	0.97	0.79	1.00	0.40	2.30
43316	4	0.30	0.80	0.50	0.46	0.24	0.49	0.45	0.30	0.60
43319	2	0.90	1.70	1.30	1.24	0.57	0.44	1.30	0.90	1.70
43320	2	2.00	3.80	2.90	2.76	1.27	0.44	2.90	2.00	3.80
43324	3	0.30	1.30	0.90	0.75	0.53	0.59	1.10	0.30	1.30
43326	2	0.30	1.70	1.00	0.71	0.99	0.99	1.00	0.30	1.70
43331	1	3.40	3.40	3.40	3.40	0.00	0.00	3.40	3.40	3.40
43334	7	0.20	2.40	1.36	1.08	0.79	0.58	1.00	0.90	2.00
43338	13	0.30	2.80	1.35	1.18	0.68	0.50	1.30	0.80	1.80
43342	3	0.80	1.50	1.10	1.06	0.36	0.33	1.00	0.80	1.50
43344	2	0.70	1.40	1.05	0.99	0.49	0.47	1.05	0.70	1.40
43345	1	1.40	1.40	1.40	1.40	0.00	0.00	1.40	1.40	1.40
43351	4	0.70	2.00	1.20	1.10	0.59	0.50	1.05	0.70	1.30
43356	2	0.50	1.20	0.85	0.77	0.49	0.58	0.85	0.50	1.20
43357	25	0.30	6.20	2.26	1.91	1.27	0.56	2.00	1.50	3.10
43360	5	1.00	3.40	1.94	1.78	0.92	0.47	1.60	1.50	2.20

Zip Code	No.	Min.	Max.	AM	GM	SD	CV	Md	Q1	Q3
43402	105	0.30	16.00	1.46	1.07	1.72	1.18	0.90	0.70	1.80
43403	2	0.50	0.85	0.68	0.65	0.25	0.37	0.68	0.50	0.85
43406	1	0.50	0.50	0.50	0.50	0.00	0.00	0.50	0.50	0.50
43407	1	1.60	1.60	1.60	1.60	0.00	0.00	1.60	1.60	1.60
43410	15	0.40	3.00	1.10	0.92	0.73	0.67	0.90	0.60	1.50
43412	7	0.30	3.10	1.37	1.14	0.86	0.62	1.20	1.00	1.50
43413	1	0.50	0.50	0.50	0.50	0.00	0.00	0.50	0.50	0.50
43416	2	1.00	1.10	1.05	1.05	0.07	0.07	1.05	1.00	1.10
43420	48	0.30	3.80	1.35	1.04	0.96	0.71	1.10	0.60	1.80
43421	1	1.96	1.96	1.96	1.96	0.00	0.00	1.96	1.96	1.96
43430	7	0.30	2.90	1.24	0.98	0.90	0.72	1.10	0.50	1.90
43431	3	0.60	1.30	0.90	0.85	0.36	0.40	0.80	0.60	1.30
43432	2	0.80	1.00	0.90	0.89	0.14	0.16	0.90	0.80	1.00
43435	1	0.30	0.30	0.30	0.30	0.00	0.00	0.30	0.30	0.30
43440	13	0.30	3.40	1.40	1.08	1.02	0.73	1.20	0.70	1.70
43443	3	0.90	2.10	1.70	1.58	0.69	0.41	2.10	0.90	2.10
43445	1	0.60	0.60	0.60	0.60	0.00	0.00	0.60	0.60	0.60
43446	1	1.60	1.60	1.60	1.60	0.00	0.00	1.60	1.60	1.60
43447	6	0.10	2.40	1.13	0.79	0.81	0.71	1.10	0.60	1.50
43449	9	0.30	2.70	1.43	1.21	0.74	0.52	1.60	0.90	1.80
43450	4	0.70	1.40	1.00	0.97	0.29	0.29	0.95	0.70	1.00
43452	36	0.30	3.90	1.76	1.32	1.20	0.68	1.50	0.60	3.10
43460	17	0.10	2.50	0.98	0.75	0.66	0.67	1.00	0.50	1.30
43462	1	1.10	1.10	1.10	1.10	0.00	0.00	1.10	1.10	1.10
43465	10	0.30	3.70	1.38	0.99	1.24	0.90	0.90	0.50	1.50
43466	1	0.60	0.60	0.60	0.60	0.00	0.00	0.60	0.60	0.60
43501	1	0.40	0.40	0.40	0.40	0.00	0.00	0.40	0.40	0.40
43502	6	0.20	3.10	1.36	0.87	1.32	0.97	0.68	0.50	3.00
43504	1	0.70	0.70	0.70	0.70	0.00	0.00	0.70	0.70	0.70
43506	4	0.30	0.80	0.58	0.53	0.26	0.46	0.60	0.30	0.80
43511	3	1.10	1.30	1.23	1.23	0.12	0.09	1.30	1.10	1.30
43512	6	0.30	2.90	0.93	0.68	0.98	1.06	0.53	0.50	0.80
43515	9	0.40	1.20	0.79	0.75	0.26	0.33	0.80	0.60	1.00
43520	1	0.50	0.50	0.50	0.50	0.00	0.00	0.50	0.50	0.50
43522	6	0.30	2.80	1.42	1.05	1.11	0.78	1.00	0.60	2.80
43523	1	1.20	1.20	1.20	1.20	0.00	0.00	1.20	1.20	1.20

Zip Code	No.	Min.	Max.	AM	GM	SD	CV	Md	Q1	Q3
43525	21	0.10	2.70	0.77	0.57	0.62	0.80	0.80	0.30	1.00
43526	2	0.80	1.20	1.00	0.98	0.28	0.28	1.00	0.80	1.20
43528	52	0.30	2.60	1.04	0.91	0.54	0.52	1.00	0.60	1.30
43530	1	0.60	0.60	0.60	0.60	0.00	0.00	0.60	0.60	0.60
43531	1	3.60	3.60	3.60	3.60	0.00	0.00	3.60	3.60	3.60
43532	3	1.00	2.30	1.73	1.63	0.67	0.38	1.90	1.00	2.30
43533	1	2.00	2.00	2.00	2.00	0.00	0.00	2.00	2.00	2.00
43534	1	0.30	0.30	0.30	0.30	0.00	0.00	0.30	0.30	0.30
43535	1	1.20	1.20	1.20	1.20	0.00	0.00	1.20	1.20	1.20
43537	291	0.10	3.90	1.03	0.82	0.76	0.73	0.80	0.50	1.20
43540	4	0.30	0.90	0.70	0.64	0.27	0.39	0.80	0.30	0.80
43542	52	0.30	2.90	0.96	0.80	0.60	0.62	0.90	0.50	1.20
43543	1	3.80	3.80	3.80	3.80	0.00	0.00	3.80	3.80	3.80
43545	2	0.30	1.00	0.65	0.55	0.49	0.76	0.65	0.30	1.00
43551	529	0.10	9.00	1.21	0.95	0.94	0.77	1.00	0.60	1.60
43552	1	0.80	0.80	0.80	0.80	0.00	0.00	0.80	0.80	0.80
43556	2	0.30	1.00	0.65	0.55	0.49	0.76	0.65	0.30	1.00
43557	2	0.50	0.80	0.65	0.63	0.21	0.33	0.65	0.50	0.80
43558	11	0.30	2.90	1.22	0.98	0.85	0.70	1.10	0.50	1.60
43560	185	0.10	3.90	0.96	0.80	0.64	0.67	0.80	0.60	1.10
43566	87	0.20	3.80	1.24	0.96	0.91	0.74	0.90	0.60	1.50
43567	2	1.80	1.90	1.85	1.85	0.07	0.04	1.85	1.80	1.90
43569	2	0.60	1.20	0.90	0.85	0.42	0.47	0.90	0.60	1.20
43571	28	0.20	3.60	0.89	0.73	0.67	0.75	0.75	0.50	0.90
43577	1	1.90	1.90	1.90	1.90	0.00	0.00	1.90	1.90	1.90
43601	2	0.30	1.90	1.10	0.75	1.13	1.03	1.10	0.30	1.90
43604	2	0.70	1.90	1.30	1.15	0.85	0.65	1.30	0.70	1.90
43605	5	0.30	2.20	0.94	0.75	0.75	0.80	0.70	0.50	1.00
43606	31	0.10	2.70	1.07	0.84	0.71	0.66	0.90	0.50	1.40
43607	4	0.40	3.40	1.55	1.17	1.30	0.84	1.20	0.40	1.40
43608	2	0.50	0.90	0.70	0.67	0.28	0.40	0.70	0.50	0.90
43609	3	0.10	1.50	0.80	0.49	0.70	0.88	0.80	0.10	1.50
43610	1	0.40	0.40	0.40	0.40	0.00	0.00	0.40	0.40	0.40
43611	17	0.10	1.80	0.77	0.63	0.45	0.58	0.70	0.40	1.10
43612	16	0.20	2.00	0.74	0.61	0.51	0.68	0.55	0.30	0.90
43613	8	0.10	1.40	0.55	0.43	0.40	0.73	0.45	0.30	0.70

Zip Code	No.	Min.	Max.	AM	GM	SD	CV	Md	Q1	Q3
43614	95	0.10	3.80	1.15	0.84	0.98	0.85	0.80	0.40	1.40
43615	48	0.20	3.30	1.42	1.06	0.99	0.70	1.35	0.60	2.10
43616	39	0.10	3.50	1.17	0.89	0.81	0.69	0.90	0.60	1.70
43617	9	0.10	1.60	0.73	0.58	0.45	0.61	0.60	0.50	1.00
43618	3	0.60	1.80	1.13	1.03	0.61	0.54	1.00	0.60	1.80
43619	17	0.30	3.30	1.24	1.07	0.70	0.56	1.10	0.90	1.40
43623	6	0.30	1.50	0.93	0.79	0.52	0.56	0.95	0.50	1.40
43635	1	0.30	0.30	0.30	0.30	0.00	0.00	0.30	0.30	0.30
43661	1	0.30	0.30	0.30	0.30	0.00	0.00	0.30	0.30	0.30
43701	71	0.20	14.50	2.04	1.52	2.25	1.11	1.50	1.00	2.60
43713	4	0.50	1.20	0.90	0.86	0.29	0.33	0.95	0.50	1.00
43716	5	1.60	3.90	2.56	2.45	0.86	0.33	2.60	2.10	2.60
43718	8	0.80	3.40	2.38	2.10	1.04	0.44	2.55	0.90	3.30
43719	2	0.90	2.30	1.60	1.44	0.99	0.62	1.60	0.90	2.30
43720	1	0.30	0.30	0.30	0.30	0.00	0.00	0.30	0.30	0.30
43723	3	1.50	1.60	1.57	1.57	0.06	0.04	1.60	1.50	1.60
43725	8	0.80	3.80	1.94	1.73	0.96	0.50	2.00	0.90	2.10
43739	3	0.40	2.30	1.50	1.18	0.98	0.66	1.80	0.40	2.30
43749	2	1.30	2.70	2.00	1.87	0.99	0.49	2.00	1.30	2.70
43760	1	1.80	1.80	1.80	1.80	0.00	0.00	1.80	1.80	1.80
43762	7	0.40	3.50	2.06	1.65	1.23	0.60	1.90	0.90	3.20
43767	1	1.40	1.40	1.40	1.40	0.00	0.00	1.40	1.40	1.40
43777	1	3.90	3.90	3.90	3.90	0.00	0.00	3.90	3.90	3.90
43783	1	2.20	2.20	2.20	2.20	0.00	0.00	2.20	2.20	2.20
43812	9	1.00	3.40	2.23	2.04	0.94	0.42	2.30	1.60	3.20
43821	2	0.30	2.90	1.60	0.93	1.84	1.15	1.60	0.30	2.90
43822	3	1.20	1.60	1.43	1.42	0.21	0.15	1.50	1.20	1.60
43824	1	0.90	0.90	0.90	0.90	0.00	0.00	0.90	0.90	0.90
43830	12	0.30	1.70	0.97	0.84	0.48	0.50	0.90	0.60	1.20
43832	7	0.30	3.80	1.84	1.31	1.34	0.73	1.90	0.40	3.20
43837	4	1.40	2.80	2.18	2.10	0.64	0.30	2.25	1.40	2.60
43840	2	1.80	2.60	2.20	2.16	0.57	0.26	2.20	1.80	2.60
43842	1	2.70	2.70	2.70	2.70	0.00	0.00	2.70	2.70	2.70
43845	1	2.80	2.80	2.80	2.80	0.00	0.00	2.80	2.80	2.80
43867	1	0.80	0.80	0.80	0.80	0.00	0.00	0.80	0.80	0.80
43901	2	0.30	1.00	0.65	0.55	0.49	0.76	0.65	0.30	1.00

Zip Code	No.	Min.	Max.	AM	GM	SD	CV	Md	Q1	Q3
43903	1	1.50	1.50	1.50	1.50	0.00	0.00	1.50	1.50	1.50
43906	2	0.70	1.00	0.85	0.84	0.21	0.25	0.85	0.70	1.00
43907	2	0.80	3.40	2.10	1.65	1.84	0.88	2.10	0.80	3.40
43912	4	0.80	6.40	2.90	2.08	2.59	0.89	2.20	0.80	3.30
43913	1	0.80	0.80	0.80	0.80	0.00	0.00	0.80	0.80	0.80
43916	2	0.20	1.50	0.85	0.55	0.92	1.08	0.85	0.20	1.50
43920	27	0.40	3.20	1.80	1.51	0.92	0.51	2.10	0.90	2.60
43930	1	0.70	0.70	0.70	0.70	0.00	0.00	0.70	0.70	0.70
43935	3	0.30	0.70	0.50	0.47	0.20	0.40	0.50	0.30	0.70
43942	5	0.50	1.90	1.20	1.10	0.49	0.41	1.20	1.20	1.20
43943	2	1.00	1.00	1.00	1.00	0.00	0.00	1.00	1.00	1.00
43945	2	0.60	1.30	0.95	0.88	0.49	0.52	0.95	0.60	1.30
43947	4	0.50	1.80	0.90	0.77	0.62	0.68	0.65	0.50	0.80
43950	41	0.30	3.70	1.80	1.48	0.99	0.55	1.80	0.90	2.50
43952	14	0.05	3.80	1.28	0.87	1.04	0.81	0.95	0.70	1.60
43953	8	0.30	3.20	1.50	1.11	1.07	0.71	1.25	0.30	2.40
43964	2	1.50	1.60	1.55	1.55	0.07	0.05	1.55	1.50	1.60
43968	1	3.50	3.50	3.50	3.50	0.00	0.00	3.50	3.50	3.50
43977	1	0.50	0.50	0.50	0.50	0.00	0.00	0.50	0.50	0.50
43988	11	0.90	3.90	2.31	2.06	1.07	0.46	2.20	1.50	3.50
44001	49	0.10	3.30	1.11	0.83	0.78	0.70	0.90	0.50	1.60
44004	10	0.30	3.60	1.18	0.92	0.96	0.82	0.90	0.60	1.30
44010	4	0.40	2.10	1.28	1.05	0.81	0.63	1.30	0.40	1.80
44011	315	0.10	3.90	0.93	0.76	0.62	0.67	0.80	0.50	1.20
44012	396	0.00	3.80	1.18	0.88	0.86	0.73	0.90	0.50	1.50
44015	1	1.20	1.20	1.20	1.20	0.00	0.00	1.20	1.20	1.20
44017	49	0.00	3.90	1.04	0.65	0.95	0.91	0.60	0.40	1.70
44020	3	1.50	3.40	2.27	2.13	1.00	0.44	1.90	1.50	3.40
44021	7	1.50	3.10	2.07	2.00	0.63	0.30	1.90	1.60	2.80
44022	79	0.20	4.40	1.62	1.36	0.94	0.58	1.40	0.90	2.10
44023	111	0.30	3.10	1.32	1.10	0.76	0.58	1.20	0.70	1.90
44024	41	0.10	2.90	0.98	0.74	0.72	0.73	0.80	0.50	1.30
44026	24	0.60	3.60	1.48	1.27	0.87	0.59	1.10	0.70	2.10
44028	19	0.20	2.20	1.19	1.05	0.54	0.45	1.20	0.80	1.60
44030	8	0.30	2.20	0.91	0.76	0.62	0.68	0.75	0.50	0.90
44033	1	2.50	2.50	2.50	2.50	0.00	0.00	2.50	2.50	2.50

Zip Code	No.	Min.	Max.	AM	GM	SD	CV	Md	Q1	Q3
44034	2	1.10	1.60	1.35	1.33	0.35	0.26	1.35	1.10	1.60
44035	61	0.10	9.40	1.21	0.78	1.45	1.20	0.80	0.40	1.10
44037	1	0.70	0.70	0.70	0.70	0.00	0.00	0.70	0.70	0.70
44039	97	0.20	2.50	0.72	0.59	0.47	0.66	0.60	0.40	0.90
44040	23	0.40	3.50	1.91	1.67	0.86	0.45	2.00	1.15	2.50
44041	2	0.60	0.70	0.65	0.65	0.07	0.11	0.65	0.60	0.70
44044	24	0.30	9.70	1.53	1.03	1.90	1.24	0.95	0.60	1.70
44045	1	1.60	1.60	1.60	1.60	0.00	0.00	1.60	1.60	1.60
44046	2	0.80	0.80	0.80	0.80	0.00	0.00	0.80	0.80	0.80
44047	4	0.50	2.10	1.30	1.15	0.65	0.50	1.30	0.50	1.30
44048	2	1.30	1.55	1.43	1.42	0.18	0.12	1.43	1.30	1.55
44049	2	1.00	1.50	1.25	1.22	0.35	0.28	1.25	1.00	1.50
44050	14	0.30	2.00	1.01	0.92	0.43	0.42	1.03	0.70	1.30
44052	10	0.30	3.10	1.02	0.81	0.81	0.79	0.85	0.50	1.30
44053	20	0.20	2.80	0.90	0.71	0.68	0.75	0.60	0.40	1.00
44054	37	0.00	2.50	1.05	0.83	0.59	0.56	0.90	0.70	1.10
44055	4	0.40	1.10	0.80	0.75	0.29	0.37	0.85	0.40	0.90
44056	56	0.10	3.10	1.08	0.87	0.72	0.67	0.80	0.50	1.40
44057	24	0.20	3.70	1.37	1.03	1.01	0.74	1.05	0.50	1.80
44060	144	0.10	3.40	1.12	0.88	0.78	0.70	0.90	0.50	1.60
44062	6	0.60	2.90	1.45	1.20	1.02	0.70	0.90	0.80	2.60
44064	2	2.80	3.00	2.90	2.90	0.14	0.05	2.90	2.80	3.00
44065	7	0.60	2.20	1.13	1.02	0.58	0.51	0.80	0.70	1.50
44067	117	0.10	3.50	1.16	0.94	0.78	0.68	0.90	0.60	1.40
44068	2	1.30	1.30	1.30	1.30	0.00	0.00	1.30	1.30	1.30
44070	103	0.10	3.80	1.22	0.92	0.94	0.77	0.90	0.50	1.60
44072	6	0.70	3.50	1.58	1.38	1.00	0.63	1.30	1.00	1.70
44074	24	0.20	3.30	1.07	0.89	0.68	0.64	0.90	0.60	1.40
44076	1	0.80	0.80	0.80	0.80	0.00	0.00	0.80	0.80	0.80
44077	151	0.20	3.70	1.03	0.85	0.68	0.66	0.85	0.50	1.30
44080	1	1.70	1.70	1.70	1.70	0.00	0.00	1.70	1.70	1.70
44081	20	0.30	1.80	0.78	0.69	0.41	0.53	0.60	0.50	0.90
44084	8	0.70	2.70	1.29	1.18	0.63	0.49	1.15	0.90	1.20
44085	1	0.30	0.30	0.30	0.30	0.00	0.00	0.30	0.30	0.30
44086	2	1.00	1.40	1.20	1.18	0.28	0.24	1.20	1.00	1.40
44087	89	0.30	3.40	1.18	1.02	0.67	0.57	0.90	0.70	1.50

Zip Code	No.	Min.	Max.	AM	GM	SD	CV	Md	Q1	Q3
44088	1	2.85	2.85	2.85	2.85	0.00	0.00	2.85	2.85	2.85
44089	37	0.30	3.70	1.19	0.99	0.79	0.66	0.80	0.70	1.60
44090	12	0.50	1.80	1.15	1.06	0.45	0.39	1.05	0.80	1.60
44092	8	0.20	3.70	1.59	1.23	1.07	0.68	1.25	1.00	2.10
44094	57	0.30	3.60	1.27	1.06	0.76	0.60	1.10	0.60	1.80
44095	8	0.30	2.50	1.08	0.85	0.79	0.73	0.80	0.50	1.20
44101	4	0.50	2.10	1.23	1.06	0.72	0.59	1.15	0.50	1.50
44102	11	0.40	2.80	1.71	1.42	0.93	0.55	1.70	0.80	2.60
44103	12	0.20	3.10	1.18	0.81	1.00	0.85	0.90	0.30	1.10
44105	7	0.60	3.70	1.87	1.61	1.05	0.56	2.10	0.90	2.30
44106	76	0.10	3.30	0.88	0.65	0.73	0.84	0.50	0.40	1.00
44107	15	0.00	3.30	1.37	0.76	1.11	0.81	1.00	0.30	2.60
44109	10	0.20	2.30	1.24	1.00	0.72	0.58	1.25	0.60	1.80
44110	1	2.70	2.70	2.70	2.70	0.00	0.00	2.70	2.70	2.70
44111	20	0.10	2.90	1.31	0.89	1.04	0.80	0.70	0.40	2.30
44112	2	1.30	1.30	1.30	1.30	0.00	0.00	1.30	1.30	1.30
44113	12	0.20	2.30	1.11	0.87	0.71	0.64	0.95	0.40	1.60
44114	9	0.40	2.20	1.00	0.84	0.64	0.64	0.90	0.50	1.20
44116	50	0.20	3.70	1.47	1.18	0.91	0.62	1.15	0.70	2.10
44117	5	0.40	3.30	1.52	1.22	1.10	0.72	1.40	0.90	1.60
44118	19	0.50	3.00	1.56	1.40	0.74	0.47	1.40	1.00	2.10
44119	1	1.40	1.40	1.40	1.40	0.00	0.00	1.40	1.40	1.40
44120	6	0.30	2.20	1.40	1.18	0.68	0.49	1.60	0.90	1.80
44121	14	0.30	2.55	1.09	0.93	0.62	0.57	1.05	0.60	1.40
44122	45	0.30	3.60	1.82	1.51	0.96	0.53	1.80	0.80	2.60
44123	3	0.60	2.45	1.42	1.21	0.94	0.67	1.20	0.60	2.45
44124	69	0.30	3.30	1.48	1.26	0.80	0.54	1.30	0.80	2.00
44125	9	0.50	8.80	2.78	1.67	3.15	1.14	1.20	0.80	2.50
44126	32	0.20	3.60	1.46	1.09	1.02	0.70	1.20	0.50	2.10
44128	2	1.10	1.10	1.10	1.10	0.00	0.00	1.10	1.10	1.10
44129	7	0.10	1.70	0.93	0.71	0.53	0.58	1.00	0.60	1.30
44130	18	0.60	3.40	1.87	1.60	0.96	0.51	2.05	1.00	2.80
44131	31	0.30	3.80	1.72	1.48	0.91	0.53	1.50	1.00	2.20
44132	1	1.10	1.10	1.10	1.10	0.00	0.00	1.10	1.10	1.10
44133	62	0.30	3.70	1.30	1.10	0.73	0.56	1.20	0.70	1.60
44134	10	0.50	3.80	2.30	1.97	1.12	0.49	2.50	1.60	3.00

Zip Code	No.	Min.	Max.	AM	GM	SD	CV	Md	Q1	Q3
44135	6	0.20	3.30	1.88	1.42	1.10	0.58	1.95	1.40	2.50
44136	57	0.40	3.80	1.73	1.50	0.89	0.52	1.50	1.00	2.40
44137	3	0.80	2.10	1.57	1.45	0.68	0.43	1.80	0.80	2.10
44138	77	0.20	2.70	0.90	0.72	0.62	0.68	0.70	0.40	1.30
44139	98	0.20	3.40	1.38	1.13	0.83	0.60	1.10	0.70	1.90
44140	227	0.10	3.90	1.78	1.41	1.06	0.59	1.60	0.80	2.60
44141	68	0.20	3.90	1.60	1.32	0.88	0.55	1.55	0.80	2.20
44142	2	1.20	2.50	1.85	1.73	0.92	0.50	1.85	1.20	2.50
44143	62	0.20	3.30	1.42	1.15	0.90	0.64	1.05	0.70	1.90
44144	1	1.70	1.70	1.70	1.70	0.00	0.00	1.70	1.70	1.70
44145	334	0.10	3.70	1.25	1.01	0.79	0.63	1.00	0.70	1.80
44146	10	0.40	2.40	1.13	0.92	0.72	0.64	1.10	0.50	1.50
44147	83	0.30	3.50	1.50	1.27	0.81	0.54	1.40	0.80	2.00
44149	34	0.60	3.30	1.59	1.42	0.74	0.47	1.60	0.90	2.00
44154	2	0.80	1.60	1.20	1.13	0.57	0.47	1.20	0.80	1.60
44184	1	1.20	1.20	1.20	1.20	0.00	0.00	1.20	1.20	1.20
44188	1	3.40	3.40	3.40	3.40	0.00	0.00	3.40	3.40	3.40
44201	5	0.70	1.50	1.00	0.96	0.32	0.32	0.90	0.80	1.10
44202	103	0.10	3.50	1.10	0.95	0.61	0.56	0.90	0.70	1.40
44203	90	0.20	3.40	1.39	1.09	0.92	0.67	1.10	0.60	1.90
44210	3	0.60	1.70	1.07	0.97	0.57	0.53	0.90	0.60	1.70
44212	78	0.10	3.30	0.96	0.79	0.65	0.67	0.75	0.50	1.30
44214	5	0.30	1.60	0.96	0.83	0.49	0.51	1.00	0.70	1.20
44215	7	0.30	1.10	0.61	0.57	0.25	0.41	0.60	0.40	0.70
44216	33	0.30	2.80	1.08	0.94	0.60	0.56	1.00	0.70	1.20
44217	17	0.50	3.30	1.42	1.16	0.93	0.66	1.00	0.70	2.50
44219	1	1.80	1.80	1.80	1.80	0.00	0.00	1.80	1.80	1.80
44220	1	0.80	0.80	0.80	0.80	0.00	0.00	0.80	0.80	0.80
44221	60	0.20	3.70	1.08	0.83	0.82	0.75	0.75	0.40	1.40
44223	76	0.20	3.50	1.23	1.01	0.79	0.64	0.90	0.70	1.90
44224	250	0.03	4.50	0.96	0.76	0.72	0.75	0.70	0.50	1.20
44230	27	0.30	2.30	0.86	0.75	0.49	0.56	0.70	0.60	1.10
44231	6	0.60	1.90	1.12	1.04	0.47	0.42	1.10	0.70	1.30
44232	8	0.50	1.80	1.06	0.95	0.52	0.49	0.95	0.50	1.40
44233	23	0.30	3.10	1.59	1.40	0.75	0.47	1.60	0.90	2.10
44234	6	0.40	3.30	1.37	1.05	1.12	0.82	0.90	0.60	2.10

Zip Code	No.	Min.	Max.	AM	GM	SD	CV	Md	Q1	Q3
44235	7	0.50	2.30	1.17	1.03	0.64	0.54	1.00	0.70	1.60
44236	320	0.20	4.00	1.05	0.87	0.71	0.67	0.83	0.50	1.30
44237	2	1.10	1.80	1.45	1.41	0.49	0.34	1.45	1.10	1.80
44238	1	0.80	0.80	0.80	0.80	0.00	0.00	0.80	0.80	0.80
44239	1	0.80	0.80	0.80	0.80	0.00	0.00	0.80	0.80	0.80
44240	102	0.30	3.50	1.17	0.93	0.80	0.68	1.00	0.50	1.60
44241	62	0.30	3.10	1.07	0.89	0.68	0.64	0.85	0.60	1.30
44242	4	0.40	3.40	1.73	1.28	1.35	0.78	1.55	0.40	2.20
44243	1	0.50	0.50	0.50	0.50	0.00	0.00	0.50	0.50	0.50
44244	2	0.80	1.30	1.05	1.02	0.35	0.34	1.05	0.80	1.30
44251	6	1.20	2.70	1.87	1.81	0.51	0.27	1.90	1.50	2.00
44253	12	0.40	3.20	0.92	0.74	0.78	0.85	0.65	0.40	0.80
44254	12	0.30	3.70	1.64	1.33	1.00	0.61	1.53	0.60	2.00
44255	10	0.50	3.50	1.43	1.12	1.10	0.77	0.95	0.60	2.40
44256	472	0.20	6.00	1.32	1.07	0.89	0.67	1.10	0.70	1.70
44260	46	0.30	2.90	1.08	0.93	0.60	0.56	0.80	0.70	1.50
44261	2	1.10	2.90	2.00	1.79	1.27	0.64	2.00	1.10	2.90
44262	20	0.30	2.90	0.86	0.69	0.72	0.84	0.60	0.50	0.90
44264	10	0.40	2.20	1.30	1.14	0.64	0.49	1.20	0.80	1.90
44265	1	2.50	2.50	2.50	2.50	0.00	0.00	2.50	2.50	2.50
44266	46	0.20	3.20	0.87	0.71	0.60	0.69	0.70	0.50	1.10
44270	14	0.30	3.30	1.09	0.85	0.85	0.78	0.65	0.50	1.60
44272	12	0.20	2.60	0.82	0.59	0.75	0.92	0.40	0.30	0.90
44273	30	0.20	3.60	1.36	1.09	0.84	0.62	1.15	0.80	1.90
44274	1	1.10	1.10	1.10	1.10	0.00	0.00	1.10	1.10	1.10
44275	3	0.55	1.60	0.92	0.81	0.59	0.65	0.60	0.55	1.60
44278	48	0.20	3.30	0.93	0.70	0.78	0.84	0.70	0.40	1.00
44280	23	0.30	3.80	1.43	1.18	0.91	0.64	1.10	0.80	2.10
44281	177	0.20	3.80	1.10	0.88	0.79	0.72	0.80	0.60	1.40
44284	1	0.30	0.30	0.30	0.30	0.00	0.00	0.30	0.30	0.30
44286	37	0.30	3.30	1.37	1.15	0.80	0.58	1.10	0.80	2.10
44287	16	0.30	2.50	1.13	0.96	0.66	0.58	0.90	0.60	1.70
44288	1	0.40	0.40	0.40	0.40	0.00	0.00	0.40	0.40	0.40
44291	1	0.30	0.30	0.30	0.30	0.00	0.00	0.30	0.30	0.30
44296	1	3.10	3.10	3.10	3.10	0.00	0.00	3.10	3.10	3.10
44301	6	0.20	2.20	1.00	0.75	0.72	0.72	1.05	0.30	1.20

Zip Code	No.	Min.	Max.	AM	GM	SD	CV	Md	Q1	Q3
44302	3	1.00	2.80	1.93	1.78	0.90	0.47	2.00	1.00	2.80
44303	20	0.30	3.10	1.37	0.99	1.06	0.77	0.85	0.40	2.10
44304	1	0.30	0.30	0.30	0.30	0.00	0.00	0.30	0.30	0.30
44305	16	0.30	3.00	0.78	0.60	0.71	0.91	0.45	0.30	0.90
44306	7	0.50	2.50	0.89	0.73	0.73	0.83	0.50	0.50	0.90
44307	2	1.20	1.40	1.30	1.30	0.14	0.11	1.30	1.20	1.40
44309	2	0.30	0.80	0.55	0.49	0.35	0.64	0.55	0.30	0.80
44310	23	0.20	15.00	1.80	1.07	2.98	1.66	1.10	0.60	1.80
44312	86	0.30	3.50	1.05	0.88	0.68	0.65	0.85	0.60	1.30
44313	145	0.10	3.80	1.31	1.03	0.92	0.71	0.90	0.60	1.70
44314	9	0.20	3.70	1.78	1.45	0.96	0.54	1.70	1.30	1.80
44317	1	0.40	0.40	0.40	0.40	0.00	0.00	0.40	0.40	0.40
44319	67	0.10	3.80	1.12	0.84	0.87	0.77	0.80	0.50	1.50
44320	18	0.30	3.70	1.43	1.15	0.99	0.69	0.95	0.70	2.00
44321	196	0.10	3.80	0.94	0.77	0.64	0.68	0.80	0.50	1.10
44324	1	0.30	0.30	0.30	0.30	0.00	0.00	0.30	0.30	0.30
44333	211	0.00	3.90	1.31	1.06	0.85	0.65	1.10	0.70	1.70
44336	1	0.50	0.50	0.50	0.50	0.00	0.00	0.50	0.50	0.50
44401	1	0.50	0.50	0.50	0.50	0.00	0.00	0.50	0.50	0.50
44405	1	1.80	1.80	1.80	1.80	0.00	0.00	1.80	1.80	1.80
44406	44	0.20	3.20	1.09	0.90	0.70	0.65	0.90	0.60	1.40
44408	19	0.30	3.40	1.24	1.07	0.75	0.60	1.00	0.80	1.80
44410	25	0.40	2.20	0.95	0.88	0.41	0.43	0.90	0.70	1.10
44411	1	1.00	1.00	1.00	1.00	0.00	0.00	1.00	1.00	1.00
44412	1	1.20	1.20	1.20	1.20	0.00	0.00	1.20	1.20	1.20
44413	8	0.40	2.80	1.48	1.25	0.84	0.57	1.35	0.70	1.90
44420	11	0.20	5.50	1.41	1.02	1.43	1.02	1.30	0.60	1.50
44423	1	1.40	1.40	1.40	1.40	0.00	0.00	1.40	1.40	1.40
44425	6	0.30	1.90	0.82	0.62	0.65	0.79	0.65	0.30	1.10
44427	2	2.80	3.80	3.30	3.26	0.71	0.21	3.30	2.80	3.80
44429	4	0.70	1.30	0.85	0.82	0.30	0.35	0.70	0.70	0.70
44431	2	0.60	2.40	1.50	1.20	1.27	0.85	1.50	0.60	2.40
44432	11	0.30	4.40	1.99	1.57	1.30	0.65	1.30	1.10	3.00
44436	1	1.80	1.80	1.80	1.80	0.00	0.00	1.80	1.80	1.80
44437	2	0.40	3.70	2.05	1.22	2.33	1.14	2.05	0.40	3.70
44438	1	1.30	1.30	1.30	1.30	0.00	0.00	1.30	1.30	1.30

Zip Code	No.	Min.	Max.	AM	GM	SD	CV	Md	Q1	Q3
44440	3	0.60	0.80	0.73	0.73	0.12	0.16	0.80	0.60	0.80
44441	3	0.30	0.90	0.57	0.51	0.31	0.54	0.50	0.30	0.90
44442	4	0.30	1.00	0.68	0.62	0.30	0.44	0.70	0.30	0.80
44443	1	2.20	2.20	2.20	2.20	0.00	0.00	2.20	2.20	2.20
44444	1	0.20	0.20	0.20	0.20	0.00	0.00	0.20	0.20	0.20
44445	3	0.30	1.80	0.93	0.72	0.78	0.83	0.70	0.30	1.80
44451	1	0.30	0.30	0.30	0.30	0.00	0.00	0.30	0.30	0.30
44452	3	0.90	1.60	1.27	1.23	0.35	0.28	1.30	0.90	1.60
44460	30	0.40	3.20	1.31	1.10	0.81	0.62	1.10	0.70	1.60
44466	1	0.70	0.70	0.70	0.70	0.00	0.00	0.70	0.70	0.70
44471	1	0.80	0.80	0.80	0.80	0.00	0.00	0.80	0.80	0.80
44477	1	2.40	2.40	2.40	2.40	0.00	0.00	2.40	2.40	2.40
44481	5	0.60	1.30	0.92	0.89	0.26	0.28	0.90	0.80	1.00
44483	2	1.20	1.80	1.50	1.47	0.42	0.28	1.50	1.20	1.80
44484	13	0.30	2.50	1.22	1.04	0.63	0.52	1.20	0.60	1.60
44485	1	3.00	3.00	3.00	3.00	0.00	0.00	3.00	3.00	3.00
44491	1	0.40	0.40	0.40	0.40	0.00	0.00	0.40	0.40	0.40
44505	5	0.30	1.90	0.86	0.70	0.63	0.74	0.80	0.40	0.90
44510	9	0.30	2.80	1.16	0.90	0.82	0.71	1.00	0.40	1.50
44511	6	0.40	1.60	0.72	0.62	0.48	0.66	0.50	0.40	0.90
44512	25	0.40	1.30	0.77	0.73	0.25	0.32	0.70	0.60	0.90
44514	34	0.40	3.30	1.10	0.91	0.77	0.70	0.85	0.60	1.20
44515	14	0.30	1.20	0.71	0.66	0.28	0.40	0.65	0.50	1.00
44517	1	0.50	0.50	0.50	0.50	0.00	0.00	0.50	0.50	0.50
44535	1	2.60	2.60	2.60	2.60	0.00	0.00	2.60	2.60	2.60
44555	1	0.30	0.30	0.30	0.30	0.00	0.00	0.30	0.30	0.30
44572	2	0.80	0.80	0.80	0.80	0.00	0.00	0.80	0.80	0.80
44601	41	0.30	3.80	1.44	1.17	0.97	0.67	1.10	0.80	1.70
44606	9	0.40	1.50	0.94	0.85	0.45	0.48	0.70	0.60	1.40
44607	2	0.60	0.70	0.65	0.65	0.07	0.11	0.65	0.60	0.70
44608	6	0.60	3.60	1.85	1.45	1.29	0.70	1.60	0.70	3.00
44609	1	0.80	0.80	0.80	0.80	0.00	0.00	0.80	0.80	0.80
44610	2	2.70	2.90	2.80	2.80	0.14	0.05	2.80	2.70	2.90
44611	3	0.90	2.40	1.47	1.33	0.81	0.56	1.10	0.90	2.40
44612	51	0.10	3.60	1.66	1.32	0.99	0.59	1.40	0.90	2.40
44614	59	0.10	3.40	0.82	0.69	0.52	0.64	0.70	0.50	1.00

Zip Code	No.	Min.	Max.	AM	GM	SD	CV	Md	Q1	Q3
44615	14	0.40	20.70	3.21	1.54	5.34	1.67	1.25	0.60	3.20
44616	1	1.70	1.70	1.70	1.70	0.00	0.00	1.70	1.70	1.70
44618	11	0.40	2.60	1.53	1.29	0.85	0.56	1.40	0.90	2.50
44619	2	3.70	5.10	4.40	4.34	0.99	0.22	4.40	3.70	5.10
44620	5	1.10	3.60	2.58	2.40	0.92	0.36	2.80	2.50	2.90
44621	11	0.60	3.60	1.81	1.55	1.00	0.55	1.80	0.90	2.30
44622	198	0.30	5.30	1.59	1.33	0.91	0.58	1.35	0.90	2.10
44624	4	0.70	3.40	1.73	1.37	1.30	0.75	1.40	0.70	2.10
44626	11	0.40	3.90	1.41	1.08	1.18	0.84	0.90	0.60	2.10
44628	1	1.20	1.20	1.20	1.20	0.00	0.00	1.20	1.20	1.20
44629	11	0.50	3.60	1.60	1.41	0.84	0.53	1.50	1.00	2.10
44631	1	1.80	1.80	1.80	1.80	0.00	0.00	1.80	1.80	1.80
44632	42	0.30	3.90	1.23	1.02	0.81	0.66	1.00	0.60	1.60
44633	2	1.90	2.50	2.20	2.18	0.42	0.19	2.20	1.90	2.50
44634	1	1.20	1.20	1.20	1.20	0.00	0.00	1.20	1.20	1.20
44636	1	1.00	1.00	1.00	1.00	0.00	0.00	1.00	1.00	1.00
44637	2	1.10	2.80	1.95	1.75	1.20	0.62	1.95	1.10	2.80
44638	2	0.90	1.20	1.05	1.04	0.21	0.20	1.05	0.90	1.20
44641	89	0.10	3.70	0.86	0.70	0.58	0.68	0.70	0.40	1.20
44643	12	0.20	3.20	1.36	0.97	1.04	0.76	1.00	0.40	2.20
44644	44	0.20	4.80	1.72	1.40	1.07	0.62	1.50	0.90	2.40
44645	8	0.50	1.60	1.14	1.07	0.40	0.35	1.05	0.80	1.50
44646	408	0.00	3.70	1.13	0.92	0.76	0.67	0.90	0.60	1.40
44647	34	0.30	2.80	1.11	0.95	0.64	0.58	0.95	0.60	1.50
44648	1	1.00	1.00	1.00	1.00	0.00	0.00	1.00	1.00	1.00
44651	2	0.70	1.10	0.90	0.88	0.28	0.31	0.90	0.70	1.10
44653	3	1.30	3.00	2.17	2.05	0.85	0.39	2.20	1.30	3.00
44654	25	0.60	3.80	1.97	1.69	1.04	0.53	1.90	1.10	2.90
44656	7	0.50	2.80	1.73	1.46	0.88	0.51	1.90	0.60	2.50
44657	9	0.80	2.80	1.64	1.51	0.70	0.42	1.50	1.20	2.30
44661	1	2.80	2.80	2.80	2.80	0.00	0.00	2.80	2.80	2.80
44662	18	0.30	3.60	1.78	1.37	1.17	0.66	1.40	1.00	3.00
44663	236	0.10	48.90	1.94	1.46	3.24	1.67	1.50	0.90	2.50
44666	2	0.50	0.50	0.50	0.50	0.00	0.00	0.50	0.50	0.50
44667	19	0.20	3.60	1.38	1.10	0.96	0.69	0.90	0.80	1.90
44671	2	1.70	3.10	2.40	2.30	0.99	0.41	2.40	1.70	3.10

Zip Code	No.	Min.	Max.	AM	GM	SD	CV	Md	Q1	Q3
44672	1	1.60	1.60	1.60	1.60	0.00	0.00	1.60	1.60	1.60
44675	7	0.50	3.40	1.81	1.54	1.03	0.57	1.50	1.00	2.90
44676	6	0.60	2.10	1.17	1.08	0.53	0.46	1.05	0.80	1.40
44677	3	0.90	3.50	2.17	1.88	1.30	0.60	2.10	0.90	3.50
44680	40	0.20	4.90	1.79	1.43	1.09	0.61	1.75	0.90	2.30
44681	6	0.70	3.50	1.88	1.60	1.14	0.61	1.55	1.00	3.00
44682	6	0.70	3.70	1.95	1.67	1.15	0.59	1.60	1.20	2.90
44683	16	0.90	3.50	1.98	1.80	0.85	0.43	1.90	1.10	2.60
44685	332	0.20	7.00	0.92	0.72	0.77	0.84	0.70	0.40	1.00
44688	3	0.70	2.70	1.73	1.50	1.00	0.58	1.80	0.70	2.70
44689	1	2.60	2.60	2.60	2.60	0.00	0.00	2.60	2.60	2.60
44691	240	0.10	3.80	1.59	1.30	0.94	0.59	1.30	0.80	2.30
44695	6	0.30	3.80	1.98	1.48	1.38	0.70	1.75	0.90	3.40
44696	1	0.80	0.80	0.80	0.80	0.00	0.00	0.80	0.80	0.80
44697	3	2.90	3.80	3.33	3.31	0.45	0.14	3.30	2.90	3.80
44698	1	1.40	1.40	1.40	1.40	0.00	0.00	1.40	1.40	1.40
44699	1	1.40	1.40	1.40	1.40	0.00	0.00	1.40	1.40	1.40
44701	3	1.00	1.70	1.30	1.27	0.36	0.28	1.20	1.00	1.70
44703	19	0.30	2.90	0.81	0.65	0.65	0.80	0.60	0.40	0.90
44704	3	0.80	3.90	1.97	1.55	1.69	0.86	1.20	0.80	3.90
44705	37	0.20	3.80	1.25	1.01	0.82	0.65	1.10	0.60	1.70
44706	37	0.20	3.50	1.12	0.89	0.87	0.77	0.80	0.50	1.10
44707	20	0.20	3.00	0.87	0.62	0.81	0.94	0.55	0.30	0.90
44708	211	0.07	4.40	1.24	0.99	0.87	0.70	1.00	0.60	1.60
44709	110	0.20	3.70	1.17	0.92	0.82	0.70	0.90	0.60	1.50
44710	18	0.20	1.90	0.98	0.85	0.51	0.51	0.95	0.60	1.10
44711	1	1.20	1.20	1.20	1.20	0.00	0.00	1.20	1.20	1.20
44714	69	0.20	3.60	1.08	0.89	0.69	0.64	0.90	0.60	1.20
44718	180	0.20	3.90	1.19	0.97	0.78	0.66	1.00	0.60	1.60
44719	1	0.20	0.20	0.20	0.20	0.00	0.00	0.20	0.20	0.20
44720	660	0.10	7.20	1.14	0.93	0.80	0.70	0.90	0.60	1.40
44721	195	0.10	3.70	1.08	0.89	0.71	0.66	0.90	0.60	1.30
44730	15	0.10	3.30	0.83	0.58	0.82	0.99	0.50	0.30	1.20
44746	2	0.70	2.00	1.35	1.18	0.92	0.68	1.35	0.70	2.00
44770	1	0.40	0.40	0.40	0.40	0.00	0.00	0.40	0.40	0.40
44774	1	1.60	1.60	1.60	1.60	0.00	0.00	1.60	1.60	1.60

Zip Code	No.	Min.	Max.	AM	GM	SD	CV	Md	Q1	Q3
44802	1	1.30	1.30	1.30	1.30	0.00	0.00	1.30	1.30	1.30
44803	2	1.50	3.60	2.55	2.32	1.48	0.58	2.55	1.50	3.60
44804	1	0.90	0.90	0.90	0.90	0.00	0.00	0.90	0.90	0.90
44805	114	0.20	3.90	1.43	1.17	0.89	0.62	1.10	0.80	1.90
44811	21	0.20	3.70	1.56	1.20	1.03	0.66	1.40	0.60	2.10
44813	20	0.30	3.00	1.47	1.28	0.74	0.51	1.30	0.90	1.80
44814	9	0.50	21.10	5.46	2.46	7.58	1.39	2.00	1.30	4.10
44817	1	0.30	0.30	0.30	0.30	0.00	0.00	0.30	0.30	0.30
44818	1	0.35	0.35	0.35	0.35	0.00	0.00	0.35	0.35	0.35
44820	7	0.30	3.80	1.86	1.51	1.09	0.58	1.60	1.20	2.30
44822	5	0.30	3.50	1.40	1.03	1.24	0.88	1.00	0.80	1.40
44824	4	0.30	7.20	2.43	1.27	3.21	1.32	1.10	0.30	1.20
44825	1	3.00	3.00	3.00	3.00	0.00	0.00	3.00	3.00	3.00
44827	3	1.10	1.60	1.43	1.41	0.29	0.20	1.60	1.10	1.60
44830	19	0.30	3.80	1.62	1.15	1.20	0.74	1.30	0.55	2.50
44833	12	0.60	2.10	1.26	1.17	0.49	0.39	1.15	0.80	1.60
44836	3	0.30	1.00	0.70	0.62	0.36	0.52	0.80	0.30	1.00
44837	1	1.90	1.90	1.90	1.90	0.00	0.00	1.90	1.90	1.90
44839	56	0.30	4.10	1.39	1.09	0.98	0.71	1.10	0.70	1.80
44840	7	0.60	1.50	1.06	1.00	0.39	0.37	0.80	0.80	1.50
44841	1	1.70	1.70	1.70	1.70	0.00	0.00	1.70	1.70	1.70
44842	20	0.60	3.50	1.45	1.25	0.88	0.61	1.05	0.80	1.70
44843	8	0.70	2.20	1.23	1.15	0.51	0.41	1.05	0.90	1.10
44846	16	0.60	3.50	1.74	1.58	0.77	0.45	1.55	1.20	2.20
44847	6	0.50	3.20	1.60	1.36	0.95	0.59	1.60	0.80	1.90
44851	6	0.60	2.50	1.10	0.96	0.72	0.65	0.80	0.70	1.20
44853	1	1.40	1.40	1.40	1.40	0.00	0.00	1.40	1.40	1.40
44856	1	0.30	0.30	0.30	0.30	0.00	0.00	0.30	0.30	0.30
44857	31	0.30	3.70	1.33	1.15	0.73	0.55	1.10	1.00	1.80
44859	3	1.30	3.60	2.07	1.83	1.33	0.64	1.30	1.30	3.60
44862	5	0.80	1.80	1.24	1.20	0.38	0.30	1.10	1.10	1.40
44864	10	0.30	3.50	1.33	1.05	0.99	0.74	1.00	0.70	1.50
44866	1	0.50	0.50	0.50	0.50	0.00	0.00	0.50	0.50	0.50
44870	48	0.30	5.30	1.22	1.00	0.93	0.77	0.90	0.60	1.40
44875	13	0.40	3.50	1.27	1.04	0.89	0.70	1.00	0.60	1.90
44878	2	0.30	1.00	0.65	0.55	0.49	0.76	0.65	0.30	1.00

Zip Code	No.	Min.	Max.	AM	GM	SD	CV	Md	Q1	Q3
44882	2	0.30	3.50	1.90	1.02	2.26	1.19	1.90	0.30	3.50
44883	65	0.16	6.10	1.72	1.34	1.17	0.68	1.55	0.80	2.40
44886	1	2.70	2.70	2.70	2.70	0.00	0.00	2.70	2.70	2.70
44888	1	1.30	1.30	1.30	1.30	0.00	0.00	1.30	1.30	1.30
44889	4	0.60	2.70	1.43	1.18	0.98	0.69	1.20	0.60	1.70
44890	4	0.90	3.40	1.65	1.42	1.17	0.71	1.15	0.90	1.20
44901	2	0.50	2.90	1.70	1.20	1.70	1.00	1.70	0.50	2.90
44903	75	0.40	3.40	1.49	1.32	0.73	0.49	1.30	1.00	1.80
44904	102	0.30	5.20	1.48	1.24	0.91	0.61	1.20	0.80	1.90
44905	13	0.50	3.00	1.42	1.18	0.91	0.64	1.00	0.80	2.50
44906	35	0.20	3.10	1.51	1.30	0.77	0.51	1.30	0.90	2.10
44907	32	0.20	3.40	1.73	1.49	0.84	0.49	1.60	0.90	2.40
44920	1	1.00	1.00	1.00	1.00	0.00	0.00	1.00	1.00	1.00
44961	3	0.80	2.20	1.37	1.25	0.74	0.54	1.10	0.80	2.20
45001	1	2.60	2.60	2.60	2.60	0.00	0.00	2.60	2.60	2.60
45002	23	0.50	3.10	1.33	1.16	0.71	0.54	1.10	0.70	1.70
45005	66	0.30	3.50	1.38	1.14	0.83	0.61	1.10	0.80	1.90
45006	2	0.80	3.90	2.35	1.77	2.19	0.93	2.35	0.80	3.90
45010	1	1.20	1.20	1.20	1.20	0.00	0.00	1.20	1.20	1.20
45011	478	0.20	3.80	1.11	0.97	0.61	0.55	1.00	0.70	1.40
45012	1	0.70	0.70	0.70	0.70	0.00	0.00	0.70	0.70	0.70
45013	79	0.20	4.80	1.42	1.19	0.87	0.61	1.20	0.80	1.80
45014	165	0.10	4.50	1.35	1.12	0.86	0.64	1.10	0.80	1.70
45015	11	0.30	1.90	1.08	0.99	0.42	0.39	1.00	0.80	1.40
45018	1	0.70	0.70	0.70	0.70	0.00	0.00	0.70	0.70	0.70
45019	1	1.40	1.40	1.40	1.40	0.00	0.00	1.40	1.40	1.40
45028	1	1.20	1.20	1.20	1.20	0.00	0.00	1.20	1.20	1.20
45029	1	0.40	0.40	0.40	0.40	0.00	0.00	0.40	0.40	0.40
45030	50	0.30	3.40	1.33	1.15	0.75	0.56	1.10	0.80	1.60
45032	1	0.80	0.80	0.80	0.80	0.00	0.00	0.80	0.80	0.80
45034	15	0.60	3.40	1.32	1.13	0.90	0.68	1.10	0.80	1.20
45036	250	0.30	7.40	1.28	1.07	0.90	0.70	1.10	0.70	1.60
45038	1	0.90	0.90	0.90	0.90	0.00	0.00	0.90	0.90	0.90
45039	363	0.20	3.80	1.13	0.97	0.67	0.59	1.00	0.70	1.40
45040	1034	0.20	4.50	1.18	1.03	0.66	0.56	1.10	0.70	1.50
45041	1	0.70	0.70	0.70	0.70	0.00	0.00	0.70	0.70	0.70

Zip Code	No.	Min.	Max.	AM	GM	SD	CV	Md	Q1	Q3
45042	39	0.40	3.60	1.55	1.34	0.89	0.58	1.30	0.90	2.00
45044	260	0.10	4.40	1.17	0.97	0.77	0.66	1.00	0.60	1.40
45049	1	1.20	1.20	1.20	1.20	0.00	0.00	1.20	1.20	1.20
45050	57	0.20	3.00	0.96	0.81	0.57	0.60	0.90	0.60	1.10
45051	1	0.80	0.80	0.80	0.80	0.00	0.00	0.80	0.80	0.80
45052	3	0.60	6.10	2.50	1.43	3.12	1.25	0.80	0.60	6.10
45053	6	0.50	2.90	1.63	1.45	0.78	0.48	1.60	1.40	1.80
45054	13	0.60	3.90	1.50	1.32	0.87	0.58	1.30	1.00	1.90
45056	90	0.30	7.50	1.37	1.15	0.97	0.71	1.10	0.80	1.70
45058	1	1.10	1.10	1.10	1.10	0.00	0.00	1.10	1.10	1.10
45061	1	0.10	0.10	0.10	0.10	0.00	0.00	0.10	0.10	0.10
45062	1	0.40	0.40	0.40	0.40	0.00	0.00	0.40	0.40	0.40
45064	5	0.40	1.10	0.90	0.85	0.29	0.32	1.00	0.90	1.10
45065	59	0.30	3.80	1.47	1.29	0.72	0.49	1.40	0.90	1.90
45066	405	0.20	6.00	1.12	0.94	0.73	0.65	0.90	0.70	1.30
45067	41	0.30	5.90	1.23	1.00	0.97	0.79	0.90	0.70	1.40
45068	87	0.20	3.90	1.45	1.26	0.79	0.54	1.30	0.90	1.80
45069	659	0.10	3.70	1.17	1.01	0.65	0.56	1.00	0.70	1.50
45102	74	0.30	6.10	1.07	0.91	0.81	0.75	0.80	0.60	1.20
45103	148	0.20	3.90	1.08	0.94	0.63	0.58	1.00	0.70	1.40
45104	2	0.60	1.70	1.15	1.01	0.78	0.68	1.15	0.60	1.70
45106	7	0.80	3.10	1.89	1.70	0.85	0.45	2.30	0.90	2.30
45107	6	0.30	2.60	1.13	0.93	0.78	0.69	1.05	0.70	1.10
45111	1	0.50	0.50	0.50	0.50	0.00	0.00	0.50	0.50	0.50
45113	8	0.70	3.30	1.53	1.30	0.97	0.64	1.20	0.70	1.50
45118	1	1.00	1.00	1.00	1.00	0.00	0.00	1.00	1.00	1.00
45120	2	1.70	1.90	1.80	1.80	0.14	0.08	1.80	1.70	1.90
45121	1	0.50	0.50	0.50	0.50	0.00	0.00	0.50	0.50	0.50
45122	13	0.70	3.30	1.55	1.35	0.85	0.55	1.50	0.80	2.10
45133	4	0.80	2.94	2.19	1.96	0.95	0.43	2.50	0.80	2.50
45135	1	2.90	2.90	2.90	2.90	0.00	0.00	2.90	2.90	2.90
45140	764	0.20	8.10	1.24	1.06	0.80	0.65	1.10	0.70	1.60
45146	2	1.30	1.30	1.30	1.30	0.00	0.00	1.30	1.30	1.30
45147	1	1.90	1.90	1.90	1.90	0.00	0.00	1.90	1.90	1.90
45148	2	1.70	1.90	1.80	1.80	0.14	0.08	1.80	1.70	1.90
45150	194	0.20	5.70	1.23	1.08	0.71	0.58	1.10	0.70	1.50

Zip Code	No.	Min.	Max.	AM	GM	SD	CV	Md	Q1	Q3
45152	101	0.40	67.00	2.15	1.29	6.67	3.10	1.20	0.90	1.60
45154	5	1.10	3.80	2.08	1.84	1.18	0.57	1.50	1.20	2.80
45157	13	0.40	3.10	1.32	1.08	0.88	0.67	1.00	0.70	2.20
45159	2	1.50	1.50	1.50	1.50	0.00	0.00	1.50	1.50	1.50
45171	3	0.80	1.90	1.40	1.32	0.56	0.40	1.50	0.80	1.90
45174	77	0.10	7.00	1.53	1.23	1.10	0.72	1.30	0.80	1.70
45176	4	0.60	2.20	1.05	0.90	0.77	0.73	0.70	0.60	0.70
45177	22	0.30	3.60	1.40	1.21	0.75	0.53	1.35	0.80	1.70
45202	11	0.50	3.80	1.43	1.10	1.18	0.83	1.20	0.60	1.90
45203	3	0.60	0.80	0.73	0.73	0.12	0.16	0.80	0.60	0.80
45205	3	0.60	3.20	1.57	1.20	1.42	0.91	0.90	0.60	3.20
45206	5	0.50	3.10	1.76	1.51	0.93	0.53	1.60	1.60	2.00
45208	120	0.10	4.10	1.50	1.23	0.88	0.59	1.30	0.80	2.10
45209	24	0.20	1.60	1.03	0.93	0.38	0.38	1.05	0.80	1.30
45211	20	0.30	3.20	1.41	1.22	0.76	0.54	1.30	0.90	1.70
45212	14	0.30	2.20	1.05	0.89	0.59	0.56	1.00	0.60	1.60
45213	22	0.10	3.10	1.37	1.05	0.88	0.64	1.10	0.60	2.10
45214	4	0.40	0.80	0.58	0.56	0.17	0.30	0.55	0.40	0.60
45215	78	0.40	5.70	1.64	1.37	1.00	0.61	1.35	1.00	2.00
45216	1	2.80	2.80	2.80	2.80	0.00	0.00	2.80	2.80	2.80
45217	10	0.40	1.90	1.12	0.94	0.63	0.56	1.10	0.50	1.80
45218	5	1.00	2.00	1.38	1.34	0.38	0.27	1.30	1.20	1.40
45219	1	3.90	3.90	3.90	3.90	0.00	0.00	3.90	3.90	3.90
45220	12	0.70	2.90	1.30	1.20	0.61	0.47	1.05	0.90	1.30
45221	1	0.90	0.90	0.90	0.90	0.00	0.00	0.90	0.90	0.90
45223	17	0.40	2.90	1.18	1.03	0.65	0.55	1.00	0.70	1.60
45224	21	0.30	3.80	1.45	1.24	0.85	0.59	1.40	0.90	1.70
45225	3	0.50	0.70	0.57	0.56	0.12	0.20	0.50	0.50	0.70
45226	22	0.30	9.10	1.43	1.01	1.81	1.27	0.95	0.60	1.50
45227	86	0.30	3.60	1.40	1.16	0.87	0.62	1.10	0.70	1.80
45229	6	1.00	3.00	1.92	1.70	0.98	0.51	1.80	1.00	2.90
45230	140	0.20	4.80	1.18	0.98	0.75	0.64	1.00	0.60	1.60
45231	42	0.30	2.90	1.29	1.08	0.74	0.58	1.15	0.70	1.80
45232	2	0.80	3.10	1.95	1.57	1.63	0.83	1.95	0.80	3.10
45233	16	0.50	2.70	1.14	1.02	0.61	0.54	1.00	0.60	1.30
45234	1	0.70	0.70	0.70	0.70	0.00	0.00	0.70	0.70	0.70

Zip Code	No.	Min.	Max.	AM	GM	SD	CV	Md	Q1	Q3
45235	3	0.80	0.90	0.83	0.83	0.06	0.07	0.80	0.80	0.90
45236	66	0.20	3.90	1.31	1.07	0.88	0.67	1.00	0.70	1.60
45237	20	0.40	3.70	1.82	1.51	1.02	0.56	1.55	0.90	2.60
45238	11	0.10	2.30	1.05	0.77	0.70	0.67	0.90	0.70	1.40
45239	21	0.30	2.00	1.03	0.92	0.49	0.47	1.00	0.70	1.30
45240	19	0.30	3.00	1.45	1.19	0.82	0.57	1.40	0.70	1.90
45241	173	0.20	3.90	1.28	1.09	0.78	0.61	1.10	0.70	1.60
45242	241	0.00	5.60	1.38	1.12	0.90	0.65	1.20	0.70	1.80
45243	120	0.20	3.50	1.58	1.35	0.86	0.54	1.35	0.90	2.10
45244	254	0.20	7.00	1.22	1.03	0.78	0.64	1.00	0.70	1.50
45245	90	0.00	3.30	1.04	0.89	0.52	0.50	1.00	0.70	1.40
45246	29	0.20	3.40	1.48	1.19	0.84	0.56	1.60	0.80	1.80
45247	29	0.30	4.50	1.72	1.42	1.04	0.60	1.40	0.90	2.40
45248	32	0.60	3.10	1.53	1.34	0.78	0.51	1.20	0.90	2.20
45249	157	0.30	3.50	1.30	1.13	0.70	0.53	1.20	0.80	1.70
45250	1	1.50	1.50	1.50	1.50	0.00	0.00	1.50	1.50	1.50
45251	14	0.30	2.20	1.13	0.99	0.55	0.49	1.10	0.70	1.50
45252	6	0.50	1.40	0.92	0.87	0.31	0.33	0.85	0.80	1.10
45255	161	0.30	3.60	1.16	1.02	0.60	0.52	1.00	0.70	1.50
45273	1	2.20	2.20	2.20	2.20	0.00	0.00	2.20	2.20	2.20
45277	2	1.00	2.20	1.60	1.48	0.85	0.53	1.60	1.00	2.20
45302	8	0.70	2.40	1.44	1.29	0.68	0.47	1.35	0.70	1.80
45304	6	0.30	3.30	1.13	0.80	1.14	1.01	0.60	0.50	1.50
45305	186	0.10	3.80	1.31	1.07	0.84	0.64	1.05	0.70	1.70
45306	1	0.68	0.68	0.68	0.68	0.00	0.00	0.68	0.68	0.68
45307	1	0.60	0.60	0.60	0.60	0.00	0.00	0.60	0.60	0.60
45308	4	0.70	2.90	1.73	1.52	0.93	0.54	1.65	0.70	1.90
45309	19	0.70	3.00	1.71	1.54	0.80	0.47	1.40	1.10	2.50
45312	4	0.90	2.30	1.38	1.27	0.66	0.48	1.15	0.90	1.40
45314	9	0.30	1.70	1.04	0.89	0.53	0.51	1.10	0.50	1.50
45315	33	0.20	3.60	1.08	0.89	0.73	0.68	0.90	0.70	1.20
45316	1	1.90	1.90	1.90	1.90	0.00	0.00	1.90	1.90	1.90
45317	1	1.20	1.20	1.20	1.20	0.00	0.00	1.20	1.20	1.20
45318	5	0.30	2.70	1.10	0.78	0.98	0.89	1.10	0.30	1.10
45319	2	1.00	1.40	1.20	1.18	0.28	0.24	1.20	1.00	1.40
45320	16	0.70	2.40	1.38	1.30	0.46	0.33	1.30	1.00	1.70

Zip Code	No.	Min.	Max.	AM	GM	SD	CV	Md	Q1	Q3
45321	1	0.70	0.70	0.70	0.70	0.00	0.00	0.70	0.70	0.70
45322	72	0.10	3.80	1.28	1.01	0.87	0.68	1.00	0.60	1.80
45323	27	0.30	3.80	1.67	1.32	1.12	0.67	1.30	0.90	2.00
45324	203	0.10	3.70	1.16	0.93	0.79	0.68	0.90	0.60	1.60
45325	4	1.20	3.30	1.88	1.73	0.96	0.51	1.50	1.20	1.60
45326	2	1.20	2.00	1.60	1.55	0.57	0.35	1.60	1.20	2.00
45327	26	0.40	3.60	1.47	1.25	0.91	0.62	1.10	0.80	1.80
45331	16	0.40	3.40	1.90	1.70	0.82	0.43	1.80	1.20	2.40
45334	1	1.20	1.20	1.20	1.20	0.00	0.00	1.20	1.20	1.20
45335	12	0.10	3.60	1.14	0.71	1.00	0.88	1.05	0.20	1.30
45337	2	1.80	1.90	1.85	1.85	0.07	0.04	1.85	1.80	1.90
45338	6	0.20	3.30	1.55	1.04	1.25	0.81	1.40	0.60	2.40
45339	4	1.10	3.60	2.63	2.40	1.07	0.41	2.90	1.10	2.90
45340	2	0.90	2.20	1.55	1.41	0.92	0.59	1.55	0.90	2.20
45341	11	0.70	8.00	2.35	1.77	2.12	0.90	1.70	0.80	3.10
45342	126	0.00	3.70	0.95	0.74	0.76	0.80	0.70	0.50	1.10
45343	3	0.80	1.50	1.07	1.03	0.38	0.35	0.90	0.80	1.50
45344	38	0.30	3.80	1.68	1.39	1.00	0.60	1.35	1.00	2.30
45345	4	1.20	3.20	2.13	2.00	0.83	0.39	2.05	1.20	2.20
45349	2	0.90	2.60	1.75	1.53	1.20	0.69	1.75	0.90	2.60
45354	1	2.60	2.60	2.60	2.60	0.00	0.00	2.60	2.60	2.60
45356	33	0.20	3.90	1.39	1.04	1.08	0.78	0.90	0.60	1.90
45357	1	0.50	0.50	0.50	0.50	0.00	0.00	0.50	0.50	0.50
45358	2	2.10	3.60	2.85	2.75	1.06	0.37	2.85	2.10	3.60
45359	2	0.80	3.60	2.20	1.70	1.98	0.90	2.20	0.80	3.60
45360	1	1.20	1.20	1.20	1.20	0.00	0.00	1.20	1.20	1.20
45363	1	2.70	2.70	2.70	2.70	0.00	0.00	2.70	2.70	2.70
45365	52	0.10	3.80	1.35	1.13	0.76	0.56	1.20	0.80	1.60
45368	4	0.60	2.20	1.35	1.21	0.70	0.52	1.30	0.60	1.60
45369	5	0.90	3.30	1.72	1.48	1.08	0.63	1.10	0.90	2.40
45370	24	0.40	2.60	1.20	1.08	0.56	0.47	1.05	0.80	1.40
45371	144	0.30	6.00	1.43	1.17	0.95	0.66	1.15	0.72	1.90
45372	2	0.70	2.10	1.40	1.21	0.99	0.71	1.40	0.70	2.10
45373	237	0.20	6.00	1.43	1.16	0.94	0.66	1.20	0.80	1.90
45375	2	0.50	1.30	0.90	0.81	0.57	0.63	0.90	0.50	1.30
45377	49	0.20	3.90	1.19	0.96	0.80	0.67	1.00	0.60	1.50

Zip Code	No.	Min.	Max.	AM	GM	SD	CV	Md	Q1	Q3
45380	9	0.50	2.50	1.40	1.23	0.70	0.50	1.10	1.00	2.00
45381	5	0.50	3.90	1.68	1.31	1.34	0.80	1.40	0.80	1.80
45382	1	0.90	0.90	0.90	0.90	0.00	0.00	0.90	0.90	0.90
45383	10	0.60	3.20	1.39	1.21	0.83	0.59	1.05	0.80	1.70
45385	235	0.30	11.50	1.37	1.12	1.04	0.76	1.10	0.70	1.80
45387	73	0.10	5.30	1.83	1.41	1.18	0.65	1.60	0.80	2.80
45389	2	0.90	1.00	0.95	0.95	0.07	0.07	0.95	0.90	1.00
45402	14	0.40	3.20	1.36	1.09	0.98	0.72	1.05	0.60	1.90
45403	12	0.30	3.50	1.10	0.88	0.88	0.80	0.80	0.50	1.20
45404	8	0.50	3.00	1.33	1.14	0.82	0.62	1.15	0.70	1.30
45405	15	0.30	2.70	1.29	1.02	0.83	0.65	1.40	0.50	1.90
45406	10	0.30	1.30	0.71	0.64	0.31	0.44	0.75	0.40	0.90
45407	30	0.10	2.30	0.71	0.54	0.57	0.80	0.50	0.30	1.00
45408	5	0.10	2.50	0.70	0.37	1.01	1.44	0.30	0.30	0.30
45409	47	0.30	3.70	1.19	0.96	0.86	0.72	0.90	0.60	1.40
45410	27	0.30	9.00	1.84	1.33	1.74	0.95	1.30	0.70	2.80
45413	1	0.60	0.60	0.60	0.60	0.00	0.00	0.60	0.60	0.60
45414	56	0.40	3.80	1.53	1.32	0.84	0.55	1.40	0.90	2.00
45415	50	0.20	6.89	1.51	1.15	1.24	0.82	1.10	0.70	2.00
45416	2	0.10	2.00	1.05	0.45	1.34	1.28	1.05	0.10	2.00
45417	20	0.30	3.00	0.56	0.41	0.67	1.21	0.30	0.30	0.30
45418	6	0.30	3.30	1.57	1.15	1.21	0.77	1.25	0.60	2.70
45419	186	0.20	4.50	1.34	1.03	0.96	0.72	1.00	0.60	1.80
45420	75	0.20	3.60	0.96	0.79	0.68	0.71	0.80	0.50	1.10
45423	1	1.20	1.20	1.20	1.20	0.00	0.00	1.20	1.20	1.20
45424	160	0.10	3.90	1.07	0.87	0.75	0.71	0.80	0.60	1.20
45426	10	0.30	2.80	1.31	1.09	0.78	0.60	1.15	0.70	1.60
45428	1	0.40	0.40	0.40	0.40	0.00	0.00	0.40	0.40	0.40
45429	226	0.10	6.00	1.30	0.99	1.00	0.77	1.00	0.60	1.70
45430	93	0.30	3.90	1.16	0.95	0.84	0.73	0.90	0.60	1.20
45431	135	0.20	3.50	1.13	0.91	0.79	0.70	0.90	0.60	1.40
45432	107	0.20	7.00	1.18	0.93	0.94	0.80	0.90	0.60	1.50
45433	2	0.60	2.10	1.35	1.12	1.06	0.79	1.35	0.60	2.10
45434	219	0.10	5.60	1.33	1.09	0.87	0.65	1.00	0.70	1.60
45439	22	0.30	3.60	1.27	0.98	1.01	0.79	0.85	0.60	1.40
45440	182	0.20	18.40	1.44	1.13	1.53	1.06	1.05	0.80	1.70

Zip Code	No.	Min.	Max.	AM	GM	SD	CV	Md	Q1	Q3
45449	40	0.10	4.50	1.21	0.98	0.81	0.67	1.00	0.60	1.60
45453	1	0.70	0.70	0.70	0.70	0.00	0.00	0.70	0.70	0.70
45458	292	0.20	4.10	1.27	1.04	0.84	0.66	1.00	0.70	1.50
45459	276	0.10	3.90	1.44	1.19	0.90	0.63	1.20	0.80	1.90
45480	1	0.30	0.30	0.30	0.30	0.00	0.00	0.30	0.30	0.30
45485	1	0.50	0.50	0.50	0.50	0.00	0.00	0.50	0.50	0.50
45499	1	0.70	0.70	0.70	0.70	0.00	0.00	0.70	0.70	0.70
45501	1	0.80	0.80	0.80	0.80	0.00	0.00	0.80	0.80	0.80
45502	79	0.30	4.00	1.61	1.31	1.03	0.64	1.20	0.80	2.40
45503	79	0.20	3.50	1.47	1.17	0.94	0.64	1.20	0.70	2.30
45504	37	0.10	3.90	1.88	1.58	0.94	0.50	1.80	1.20	2.50
45505	9	0.40	3.30	1.72	1.38	1.03	0.60	1.80	1.10	2.20
45506	12	0.20	2.40	1.32	1.08	0.72	0.55	1.20	0.70	1.80
45560	1	0.45	0.45	0.45	0.45	0.00	0.00	0.45	0.45	0.45
45585	1	0.60	0.60	0.60	0.60	0.00	0.00	0.60	0.60	0.60
45601	57	0.40	3.60	1.71	1.49	0.88	0.52	1.60	1.00	2.30
45612	4	0.70	1.40	1.08	1.03	0.33	0.31	1.10	0.70	1.30
45628	1	1.10	1.10	1.10	1.10	0.00	0.00	1.10	1.10	1.10
45634	1	0.50	0.50	0.50	0.50	0.00	0.00	0.50	0.50	0.50
45640	10	0.70	3.40	2.08	1.87	0.90	0.43	2.05	1.30	3.00
45644	1	0.60	0.60	0.60	0.60	0.00	0.00	0.60	0.60	0.60
45647	1	3.50	3.50	3.50	3.50	0.00	0.00	3.50	3.50	3.50
45648	1	3.60	3.60	3.60	3.60	0.00	0.00	3.60	3.60	3.60
45656	1	0.90	0.90	0.90	0.90	0.00	0.00	0.90	0.90	0.90
45662	2	0.70	3.30	2.00	1.52	1.84	0.92	2.00	0.70	3.30
45690	4	0.30	1.20	0.88	0.77	0.39	0.45	1.00	0.30	1.00
45694	2	0.90	1.00	0.95	0.95	0.07	0.07	0.95	0.90	1.00
45697	1	2.40	2.40	2.40	2.40	0.00	0.00	2.40	2.40	2.40
45701	7	0.80	7.60	2.06	1.40	2.49	1.21	1.00	0.80	2.20
45714	12	0.70	3.30	1.73	1.55	0.81	0.47	1.65	0.90	2.30
45716	1	1.30	1.30	1.30	1.30	0.00	0.00	1.30	1.30	1.30
45723	1	1.60	1.60	1.60	1.60	0.00	0.00	1.60	1.60	1.60
45732	1	2.10	2.10	2.10	2.10	0.00	0.00	2.10	2.10	2.10
45742	3	1.80	3.00	2.50	2.44	0.62	0.25	2.70	1.80	3.00
45743	1	3.80	3.80	3.80	3.80	0.00	0.00	3.80	3.80	3.80
45750	31	0.30	4.10	1.84	1.50	1.08	0.59	1.50	1.10	2.70

Zip Code	No.	Min.	Max.	AM	GM	SD	CV	Md	Q1	Q3
45766	1	3.90	3.90	3.90	3.90	0.00	0.00	3.90	3.90	3.90
45780	4	0.90	3.60	1.88	1.63	1.21	0.64	1.50	0.90	1.80
45784	4	1.80	3.40	2.55	2.48	0.70	0.27	2.50	1.80	2.80
45786	1	1.40	1.40	1.40	1.40	0.00	0.00	1.40	1.40	1.40
45801	3	0.30	1.10	0.70	0.61	0.40	0.57	0.70	0.30	1.10
45802	1	0.50	0.50	0.50	0.50	0.00	0.00	0.50	0.50	0.50
45804	2	0.40	1.05	0.73	0.65	0.46	0.63	0.73	0.40	1.05
45805	34	0.30	3.85	1.59	1.30	0.95	0.60	1.35	0.80	2.30
45806	9	0.30	3.10	1.45	1.14	1.00	0.69	1.00	0.90	1.90
45807	15	0.30	1.60	0.86	0.78	0.38	0.44	0.80	0.70	1.10
45809	1	0.30	0.30	0.30	0.30	0.00	0.00	0.30	0.30	0.30
45810	2	0.30	1.70	1.00	0.71	0.99	0.99	1.00	0.30	1.70
45813	1	1.50	1.50	1.50	1.50	0.00	0.00	1.50	1.50	1.50
45814	6	0.60	1.90	1.30	1.23	0.42	0.32	1.35	1.20	1.40
45816	1	0.30	0.30	0.30	0.30	0.00	0.00	0.30	0.30	0.30
45817	17	0.30	3.80	1.26	0.92	1.09	0.87	0.80	0.60	1.40
45822	2	1.60	1.60	1.60	1.60	0.00	0.00	1.60	1.60	1.60
45826	1	1.00	1.00	1.00	1.00	0.00	0.00	1.00	1.00	1.00
45828	3	0.80	2.80	2.13	1.84	1.15	0.54	2.80	0.80	2.80
45830	5	0.30	3.90	1.87	1.22	1.69	0.91	0.95	0.70	3.50
45833	5	0.20	2.00	0.84	0.65	0.68	0.81	0.60	0.60	0.80
45840	472	0.10	4.20	1.20	0.94	0.83	0.69	0.90	0.60	1.60
45841	1	0.60	0.60	0.60	0.60	0.00	0.00	0.60	0.60	0.60
45843	2	1.10	5.40	3.25	2.44	3.04	0.94	3.25	1.10	5.40
45844	1	1.10	1.10	1.10	1.10	0.00	0.00	1.10	1.10	1.10
45845	2	0.70	2.00	1.35	1.18	0.92	0.68	1.35	0.70	2.00
45846	1	1.70	1.70	1.70	1.70	0.00	0.00	1.70	1.70	1.70
45850	1	0.30	0.30	0.30	0.30	0.00	0.00	0.30	0.30	0.30
45856	3	0.70	2.50	1.45	1.26	0.94	0.65	1.15	0.70	2.50
45858	10	0.30	3.00	1.57	1.32	0.81	0.52	1.60	0.90	2.00
45865	4	0.10	3.15	1.49	0.87	1.26	0.84	1.35	0.10	1.40
45868	1	1.10	1.10	1.10	1.10	0.00	0.00	1.10	1.10	1.10
45869	12	0.55	2.20	1.18	1.08	0.51	0.43	1.05	0.80	1.60
45872	3	0.80	1.20	0.97	0.95	0.21	0.22	0.90	0.80	1.20
45873	1	1.10	1.10	1.10	1.10	0.00	0.00	1.10	1.10	1.10
45875	15	0.30	2.55	0.98	0.81	0.64	0.66	0.90	0.50	1.15

Zip Code	No.	Min.	Max.	AM	GM	SD	CV	Md	Q1	Q3
45877	2	0.30	0.45	0.38	0.37	0.11	0.28	0.38	0.30	0.45
45879	1	0.70	0.70	0.70	0.70	0.00	0.00	0.70	0.70	0.70
45881	3	0.30	0.80	0.57	0.52	0.25	0.44	0.60	0.30	0.80
45883	2	0.30	1.30	0.80	0.62	0.71	0.88	0.80	0.30	1.30
45884	1	3.70	3.70	3.70	3.70	0.00	0.00	3.70	3.70	3.70
45885	1	0.70	0.70	0.70	0.70	0.00	0.00	0.70	0.70	0.70
45889	4	0.30	0.90	0.65	0.60	0.28	0.43	0.70	0.30	0.85
45890	1	0.70	0.70	0.70	0.70	0.00	0.00	0.70	0.70	0.70
45895	9	0.40	3.30	1.13	0.87	1.01	0.90	0.65	0.60	0.80
Unknown	488	0.10	7.40	1.31	1.05	0.90	0.69	1.00	0.70	1.80

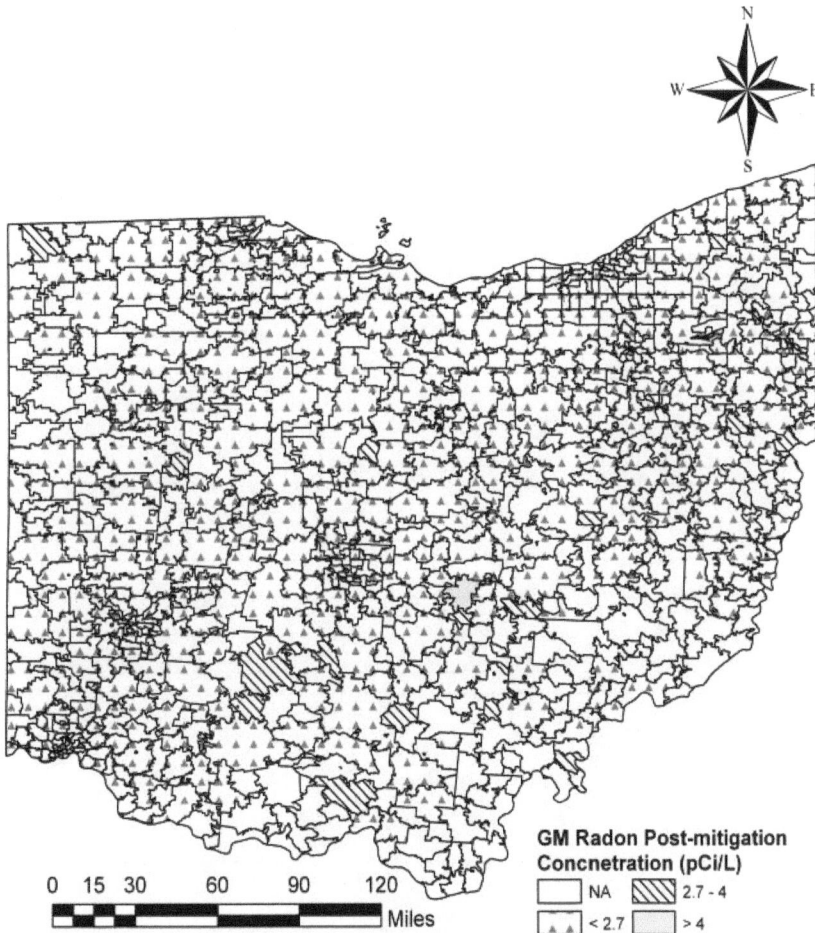

Figure 5a: GM ^{222}Rn Post-mitigation concentrations in Ohio zip codes based on the WHO-USEPA classification.

Figure 5b: GM ^{222}Rn Post-mitigation concentrations in Ohio zip codes based on the USEPA classification.

Figure 5c: GM ^{222}Rn Post-mitigation concentrations in Ohio zip codes based on the 2 pCi/L breakdown classifications.

4.4. PERFORMANCE OF MITIGATION SYSTEMS

In view of the available ^{222}Rn control techniques for existing constructions (as summarized in Table **1** of Chapter 1), one may broadly classify the ^{222}Rn mitigation techniques for existing constructions under the following five categories:

- Sub-slab suction, also known as sub-slab depressurization: SSD

- Drain tile suction, also known as drain tile depressurization: DTD

- Block-wall suction, also known as block-wall depressurization: BWD

- Sump-hole suction, also known as sump-pit perimeter depressurization: SUMP

- Submembrane depressurization in a crawlspace: SMD

The Ohio mitigation database largely comprised of all the above five mentioned mitigation systems either individually or in combinations. A brief summary of the working mechanism for each of the above mentioned mitigation system is discussed below.

The SSD system is a ^{222}Rn control technique that is designed to achieve a lower sub-slab air pressure in comparison with the indoor air pressure. The working mechanism of the SSD systems involves the collection of ^{222}Rn gas from the soil underneath the concrete slab in a basement, slab-below-grade or slab-on-grade house, by drawing out air with a fan-powered vent and discharging the air into the environment. The SSD systems are predominately categorized as the 'low pressure-high flow' or the 'high pressure-low flow' systems depending on the type of soil in the foundation of the building. The SSD systems work best, if air can move easily in the material underneath the slab (ASTM Publication 1993; EPA 2001). The DTD system works similar to the SSD system, with the exemption that the suction point piping attaches to a drain tile or the gas permeable located near the drain tile. The drain tile may be outside or inside the footings of the building (ASTM Publication 1993). The BWD is used when the foundation wall consists of hollow block construction on a poured concrete footer. The poly vinyl chloride (PVC) piping is penetrated in the foundation wall and suction is applied to the wall. The BWD system can be applied from the interior or the exterior of a building. It can be used only in homes with hollow block-walls and requires proper sealing of major openings (ASTM Publication 1993). The SUMP system is adopted in houses with a sump pit. The pit usually has drain tile from the perimeter of the house, which all drains into the sump. This technique involves covering the sump pit opening with an airtight cover and a removable section for ease of access in the cases of pit being ever repaired. A vent pipe is then installed in the sump pit and the ^{222}Rn gas is exhausted from around the perimeter of the home to the outside (ASTM Publication 1993). The SUMP system is vital when the water table in the ground is elevated where moisture is discharged out through the system. The SMD system achieves lower sub membrane air pressure relative to the crawl space air pressure using a vent and drawing the air beneath the soil gas retarder membrane (plastic sheets). The SMD system has less heat loss than natural ventilation in winter climates (ASTM Publication 1993).

Sections '4.1' and '4.2' of this chapter have shown that the installed mitigation systems in Ohio homes were largely successful in controlling the in-house ^{222}Rn concentrations. This section summarizes the development of a novel statistical procedure that may be used in comparing the performance of installed mitigation systems in Ohio.

4.4.1. Three-Pronged Approach to Rank the Performance of Mitigation Systems

The novel method developed for determining the performance of mitigation systems involves a three-pronged approach that incorporates:

- Development of a comprehensive mitigation database with 'complete details' on an yearly basis,

- Examination of the statistical significance of the performance of installed mitigation systems using the two sample t-test, and

- Allocation of points on the basis of statistical significance of the mean percentage removal for each mitigation system against the numerically better performing mitigation system for the categorized yearly mitigation system data points. The allotted points are subsequently used to develop a unique cumulative performance index for each mitigation system for the entire duration of the study.

Of the total 39,858 mitigation data points representing Ohio homes installed mitigation systems, about 37,470 records (94.01% of the total records) were designated as 'complete records' indicating that there are there are no missing values for any of the details in relation to test house location (county, zip code), and mitigation (type of the mitigation system installed, pre-mitigation ^{222}Rn level, post-mitigation ^{222}Rn level, year). Table **5** enlists a summary of the number of complete records per mitigation system type (and their combinations) that were noted in the Ohio mitigation database. To ensure adequate representativeness for each mitigation system when comparing the performance of different control techniques, the authors have chosen to evaluate and compare the performance of only the control techniques that had a minimum of 10 complete records in any given year. Table **6** presents a summary of the breakdown of mitigation systems installed in Ohio that were considered to be analyzed further (on the basis of a minimum representative sample of 10 records per control technique in any given year) using statistical analyses. Resultantly, a total of 21 mitigation systems are analyzed using the mitigation database with 37,236 data points between 2002 and 2012.

Table 5: Total Mitigation Systems Installed in Ohio from 2002-2012

Control Technique	No. of Records
BWD	18
BWD, DTD	1
BWD, DTD, SUMP	5
BWD, SSD	79
CRAWL SPACE BARRIER	5
DTD	1320
DTD, SMD	85
OTHER	1044
SMD	243

Control Technique	No. of Records
SSD	22091
SSD, ATTIC	54
SSD, CRAWL	12
SSD, DTD	490
SSD, DTD, SUMP	555
SSD, EXT	303
SSD, GARAGE	6
SSD, OTHER	4
SSD, SMD	4072
SSD, SMD, BWD	2
SSD, SMD, DTD	13
SSD, SMD, SUMP	11
SSD, SMD, SUMP, DTD	112
SSD, SMD, SUMP, DTD, BWD	1
SSD, SUMP	464
SSD, SUMP, ATTIC	13
SSD, SUMP, EXT	110
SSD, SUMP, VENTILATION	1
SUMP	425
SUMP, ATTIC	6
SUMP, CRAWL	4
SUMP, DTD	4841
SUMP, EXT	29
SUMP, SMD	82
SUMP, SMD, DTD	751
SUMP, VENTILATION	218
Total Number of Records	*37470*

Table 6: Mitigation Systems Installed in Ohio from 2002-2012 (with N_{min}=10 per control technique in any selected year)

Control Technique	No. of Records per Control Technique	No. of Records per Control Technique per Year										
		2002	2003	2004	2005	2006	2007	2008	2009	2010	2011	2012
BWD	15	---	---	---	---	---	---	---	15	---	---	---
BWD, SSD	36	---	---	---	---	---	---	---	14	22	---	---
DTD	1320	---	---	---	221	226	166	164	124	269	103	47

Control Technique	No. of Records per Control Technique	No. of Records per Control Technique per Year										
		2002	2003	2004	2005	2006	2007	2008	2009	2010	2011	2012
DTD, SMD	58	---	---	---	---	---	---	---	---	---	23	35
OTHER	1037	168	429	75	16	65	20	38	37	52	137	---
SMD	236	---	19	---	17	14	18	19	21	36	58	34
SSD	22091	599	1335	1147	1669	1763	1849	2663	2579	3462	3325	1700
SSD, ATTIC	53	---	---	---	12	30	---	---	---	---	11	---
SSD, DTD	490	---	---	---	63	24	26	71	100	71	114	21
SSD, DTD, SUMP	555	80	178	75	22	19	19	24	31	18	69	20
SSD, EXT	303	---	---	---	35	160	---	---	---	32	76	
SSD, SMD	4072	60	192	167	219	450	349	579	353	605	722	376
SSD, SMD, SUMP, DTD	89	---	12	---	36	---	---	---	---	11	30	---
SSD, SUMP	464	16	58	59	19	1	11	37	95	30	122	16
SSD, SUMP, EXT	107	---	---	---	28	79	---	---	---	---	---	---
SUMP	416	12	59	20	28	20	---	---	20	96	52	109
SUMP, DTD	4841	213	525	361	190	339	294	620	629	521	715	434
SUMP, EXT	29	---	---	---	---	---	---	---	---	15	14	---
SUMP, SMD	63	---	---	---	---	---	---	---	---	46	17	---
SUMP, SMD, DTD	751	32	36	25	151	14	29	45	88	93	132	106
SUMP, VENTILATION	210	---	---	---	---	---	32	43	39	65	31	---

Table **7** presents a summary of the mean percentage removal efficiencies of 21 mitigation systems installed in Ohio homes from 2002 through 2012. From Table **7**, one can note that the numerically better performing mitigation systems are: BWD in 2009; DTD in 2008; SSD, DTD, SUMP in 2006 and 2012; SSD, SMD in 2005 and 2007; SSD, SMD, SUMP, DTD in 2003; and SUMP, SMD, DTD in 2002, 2004, 2010, and 2011. However, the performance of the mitigation systems need to be statistically evaluated to determine if the performance of one type of mitigation system in reducing in-house ^{222}Rn concentrations is significantly different (higher/lower) from another type of mitigation system.

The authors have chosen to use the two sample t-test to determine if there is statistically significant difference in the mean percentage removal efficiencies between two mitigation systems. In the two sample t-test, the null hypothesis (H_0) is that the mean removal efficiencies of the two considered mitigation systems are statistically the same and the alternative hypothesis (H_a) is that the mean removal efficiencies of the two considered mitigation systems are statistically different. One needs to reject the null hypothesis and accept the alternate hypothesis that the mean removal efficiencies of the two considered mitigation systems is not zero and are statistically different if the p-value < 0.05 (α). One needs to accept the null hypothesis and reject the alternate hypothesis that the mean removal efficiencies of the two considered mitigation systems is zero and the differences in mean percentage removals is not statistically significant if p-value > 0.05 (α). MINITAB $15^{®}$ was used to perform this test.

The numerically better performing mitigation system (NBPMS), *i.e.*, the mitigation system with highest numerical mean removal percentage, is identified for each year on the basis of Table **7**. In stage 1, the identified NBPMS is designated as NBPMS1 (suffix '1' indicative of the stage number) and the performance of each other monitored mitigation system is analyzed statistically using the two sample t-test. If a mitigation system yielded a p-value > 0.05 when analyzed using the two sample t-test with respect to NBPMS1, then it may be concluded there are statistically no significant differences in the mean percentage removals between the two mitigation systems. The differences in mean percentage removals are considered to have been occurred by chance. All such mitigation systems which are not statistically different to NBPMS1 are not considered to proceed for subsequent stages. If a mitigation system had a p-value < 0.05 when analyzed using the two sample t-test with respect to NBPMS1, when analyzed with the NBPMS1, then it may be concluded there are statistically significant differences in the mean percentage removals between the two mitigation systems. Only those mitigation systems which are statistically significantly different to NBPMS1 are considered to proceed to the subsequent stages. The complete procedure adopted in stage 1 (identification of NBPMS and determination of mitigation systems with statistical significance) is repeated in subsequent stages until all the mitigation systems in a given year have been evaluated. Tables **8** through **18** summarize the two sample t-test results in determining the statistical significance of each mitigation system against the NBPMS in each stage.

On the basis of the two sample t-test results of significance, each mitigation system is ranked and points are allocated to develop a unique performance index, referred as the 'Mitigation System Rank Index'. The NBPMS1 in each year is ranked '1' and assigned the highest score of '21' (considering that there are a total of 21 mitigation systems from 2002-2012). All the mitigation systems which are not statistically significantly different to NBPMS1 are assigned the same rank and number of points. All the mitigation systems which are statistically significantly different to NBPMS1 are assigned a rank and number of points only in the subsequent stages. The NBPMS2 in each year is ranked 2 and assigned the second-highest score of '20'. All the mitigation systems which are not statistically significantly different to NBPMS2 are assigned the same rank '2' and the number of points '20', while those statistically significantly different are ranked and points allotted only in subsequent stages. This process is continued until all the mitigation systems have been ranked and points allotted in any given year. Tables 4.8 through 4.18 summarize the ranks and points allotted to each mitigation system in accordance with their performance based on the two sample t-test results. Table **19** presents a summary of the points allotted to each mitigation system from 2002-2012. The points allotted for each mitigation system across different years are accumulated to obtain the 'Cumulative Score' as can be noted from Table **19**. The Mitigation System Rank Index (MSRI) is then computed using equation 4.2. For determining the maximum achievable score in equation 4.2, multiply the number of years for which the mitigation system has been installed with the highest achievable score of '21'. This provides a corresponding maximum achievable score for each mitigation system from 2002 through 2012. The computed MSRI is then used to rank the performance of mitigation system in ascending order as can be seen from Table **19**. It may be noted that while the two sample t-test accounted for determining the initial variations amongst mitigation system for a particular year, the use of MSRI had accounted for the variations across different years in providing a new methodology to rank the performance of mitigation systems that have been installed across different years.

$$\text{Mitigation System Rank Index } (MSRI) = \frac{Cumulative\ Score\ Obtained\ for\ a\ Mitigation\ System}{Maximum\ Achievable\ Score} \tag{4.2}$$

where,

Cumulative Score Obtained for a Mitigation System is computed by the summation of all allocated points for a given mitigation system from 2002 to

2012; and *Maximum Achievable Score* is the maximum number of points that a mitigation system may attain.

Based on the computed MSRI from Table **19**, one can note that 'BWD' performed the best, followed by SUMP, EXT; SSD, DTD, SUMP; SUMP, SMD, DTD; SSD, SMD, SUMP, DTD; SSD, ATTIC; SSD, SUMP, EXT; SUMP; SMD; BWD, SSD; SSD, SMD; DTD; SSD, SUMP; SUMP, DTD; DTD, SMD; SUMP, VENTILATION; SSD, DTD; SSD; OTHER; SUMP, SMD; and SSD, EXT.

Table 7: Mean Percentage Removal Efficiency of Mitigation Systems Installed in Ohio from 2002-2012

Control Technique	Mean Percentage Removal										
	2002	2003	2004	2005	2006	2007	2008	2009	2010	2011	2012
BWD	---	---	---	---	---	---	---	87.79	---	---	---
BWD, SSD	---	---	---	---	---	---	---	70.78	76.82	---	---
DTD	---	---	---	85.5	85.1	86.68	87.66	86.06	86.04	85.72	84.58
DTD, SMD	---	---	---	---	---	---	---	---	---	87.67	84.67
OTHER	84.08	83.49	80.71	72.28	77.39	78.64	75.33	78.35	77.18	79.77	---
SMD	---	80.67	---	77.64	76.11	71.09	78.93	75.58	79.27	83.94	84.99
SSD	81.6	84.43	82.61	82.93	84.08	83.83	82.83	83.32	82.18	81.57	81.41
SSD, ATTIC	---	---	---	77.12	77.25	---	---	---	---	80.56	---
SSD, DTD	---	---	---	82.26	82.82	83.12	80.99	84.92	85.76	88.11	85.54
SSD, DTD, SUMP	83.06	87.61	79.12	84.47	90.01	85.16	86.33	87.72	83.37	86.46	88.19
SSD, EXT	---	---	---	75.69	80.39	---	---	---	73.49	78.12	---
SSD, SMD	78.51	89.67	86.94	87.42	87.29	87.82	86.49	85.16	84.67	86.34	85.57
SSD, SMD, SUMP, DTD	---	94.2	---	86.69	---	---	---	---	87.24	87.05	---
SSD, SUMP	85.21	85.81	84.86	83.11	---	84.68	79.53	85.74	78.59	83.41	84.53
SSD, SUMP, EXT	---	---	---	85.76	86.27	---	---	---	---	---	---
SUMP	86.33	86.23	88.88	86.72	86.21	---	---	84.17	84.58	84.14	85.86
SUMP, DTD	85.05	85.89	86.67	84.12	84.25	84.44	85.42	85.06	84.64	85.43	85.53
SUMP, EXT	---	---	---	---	---	---	---	---	74.81	78.22	---
SUMP, SMD	---	---	---	---	---	---	---	---	83.95	88.25	---
SUMP, SMD, DTD	88.44	91.98	89.04	87.21	89.18	82	83.27	85.76	87.48	88.56	87.72
SUMP, VENTILATION	---	---	---	---	---	86.6	87.21	87.7	82.24	85.95	---

Table 8: Statistical Significance and Ranking of the Mitigation Systems Based on their Performances against NBPMS for 2002

Control Technique	Mean % Removal	Is there Significant Difference in Mean Percentage Removal?				Rank	Points
		Stage 1	Stage 2	Stage 3	Stage 4		
OTHER	84.08	Yes (0.018)	NBPMS2	---	---	2	20
SSD	81.6	No	---	---	---	1	21
SSD, DTD, SUMP	83.06	Yes (0.013)	Yes	NBPMS3	---	3	19
SSD, SMD	78.51	Yes (0.042)	Yes	Yes	NBPMS4	4	18
SSD, SUMP	85.21	No	---	---	---	1	21
SUMP	86.33	No	---	---	---	1	21
SUMP, DTD	85.05	No	---	---	---	1	21
SUMP, SMD, DTD	88.44	NBPMS1	---	---	---	1	21

Table 9: Statistical Significance and Ranking of the Mitigation Systems Based on their Performances against NBPMS for 2003

Control Technique	Mean % Removal	Is there Significant Difference in Mean Percentage Removal?		Rank	Points
		Stage 1	Stage 2		
OTHER	83.49	Yes	No	2	20
SMD	80.67	Yes	No	2	20
SSD	84.43	No	---	1	21
SSD, DTD, SUMP	87.61	No	---	1	21
SSD, SMD	89.67	No	---	1	21
SSD, SMD, SUMP, DTD	94.2	NBPMS1	---	1	21
SSD, SUMP	85.81	No	---	1	21
SUMP	86.23	No	---	1	21
SUMP, DTD	85.89	No	---	1	21
SUMP, SMD, DTD	91.98	Yes	NBPMS2	2	20

Table 10: Statistical Significance and Ranking of the Mitigation Systems Based on their Performances against NBPMS for 2004

Control Technique	Mean % Removal	Is there Significant Difference in Mean Percentage Removal?				Rank	Points
		Stage 1	Stage 2	Stage 3	Stage 4		
OTHER	80.71	Yes	Yes	Yes	Yes	5	17
SSD	82.61	No	---	---	---	1	21
SSD, DTD, SUMP	79.12	No	---	---	---	1	21
SSD, SMD	86.94	Yes	No	---	---	2	20
SSD, SUMP	84.86	Yes	Yes	Yes	NBPMS4	4	18
SUMP	88.88	Yes	NBPMS2	---	---	2	20
SUMP, DTD	86.67	Yes	Yes	NBPMS3	---	3	19
SUMP, SMD, DTD	89.04	NBPMS1	---	---	---	1	21

Table 11: Statistical Significance and Ranking of the Mitigation Systems Based on their Performances against NBPMS for 2005

Control Technique	Mean % Removal	Is there Significant Difference in Mean Percentage Removal?							Rank	Points
		Stage 1	Stage 2	Stage 3	Stage 4	Stage 5	Stage 6	Stage 7		
DTD	85.5	Yes	Yes	Yes	NBPMS4	---	---	---	4	18
OTHER	72.28	No	---	---	---	---	---	---	1	21
SMD	77.64	No	---	---	---	---	---	---	1	21
SSD	82.93	Yes	Yes	Yes	Yes	Yes	NBPMS6	---	6	16
SSD, ATTIC	77.12	No	---	---	---	---	---	---	1	21
SSD, DTD	82.26	Yes	Yes	Yes	Yes	Yes	Yes	NBPMS7	7	15
SSD, DTD, SUMP	84.47	No	---	---	---	---	---	---	1	21
SSD, EXT	75.69	Yes	Yes	Yes	Yes	Yes	Yes	Yes	8	14
SSD, SMD	87.42	NBPMS1	---	---	---	---	---	---	1	21
SSD, SMD, SUMP, DTD	86.69	Yes	Yes	NBPMS3	---	---	---	---	3	19
SSD,	83.11	No	---	---	---	---	---	---	1	21

Control Technique	Mean % Removal	Is there Significant Difference in Mean Percentage Removal?							Rank	Points
		Stage 1	Stage 2	Stage 3	Stage 4	Stage 5	Stage 6	Stage 7		
SUMP										
SSD, SUMP, EXT	85.76	No	---	---	---	---	---	---	1	21
SUMP	86.72	No	---	---	---	---	---	---	1	21
SUMP, DTD	84.12	Yes	Yes	Yes	Yes	NBPMS5	---	---	5	17
SUMP, SMD, DTD	87.21	Yes	NBPMS2	---	---	---	---	---	2	20

Table 12: Statistical Significance and Ranking of the Mitigation Systems Based on their Performances against NBPMS for 2006

Control Technique	Mean % Removal	Is there Significant Difference in Mean Percentage Removal?						Rank	Points
		Stage 1	Stage 2	Stage 3	Stage 4	Stage 5	Stage 6		
DTD	85.1	Yes	Yes	Yes	NBPMS4	---	---	4	18
OTHER	77.39	Yes	Yes	Yes	Yes	Yes	Yes	7	15
SMD	76.11	No	---	---	---	---	---	1	21
SSD	84.08	No	---	---	---	---	---	1	21
SSD, ATTIC	77.25	Yes	Yes	Yes	No	---	---	4	18
SSD, DTD	82.82	Yes	No	---	---	---	---	2	20
SSD, DTD, SUMP	90.01	NBPMS1	---	---	---	---	---	1	21
SSD, EXT	80.39	Yes	Yes	Yes	Yes	Yes	NBPMS6	6	16
SSD, SMD	87.29	Yes	NBPMS2	---	---	---	---	2	20
SSD, SUMP, EXT	86.27	Yes	Yes	NBPMS3	---	---	---	3	19
SUMP	86.21	No	---	---	---	---	---	1	21
SUMP, DTD	84.25	Yes	Yes	Yes	Yes	NBPMS5	---	5	17
SUMP, SMD, DTD	89.18	No	---	---	---	---	---	1	21

Table 13: Statistical Significance and Ranking of the Mitigation Systems Based on their Performances against NBPMS for 2007

Control Technique	Mean % Removal	Is there Significant Difference in Mean Percentage Removal?			Rank	Points
		Stage 1	Stage 2	Stage 3		
DTD	86.68	Yes	NBPMS2	---	2	20
OTHER	78.64	Yes	No	---	2	20
SMD	71.09	Yes	No	---	2	20
SSD	83.83	Yes	Yes	No	3	19
SSD, DTD	83.12	Yes	No	---	2	20
SSD, DTD, SUMP	85.16	Yes	No	---	2	20
SSD, SMD	87.82	NBPMS1	---	---	1	21
SSD, SUMP	84.68	No	---	---	1	21
SUMP, DTD	84.44	Yes	Yes	No	3	19
SUMP, SMD, DTD	82	Yes	Yes	No	3	19
SUMP, VENTILATION	86.6	Yes	Yes	NBPMS3	3	19

Table 14: Statistical Significance and Ranking of the Mitigation Systems Based on their Performances against NBPMS for 2008

Control Technique	Mean % Removal	Is there Significant Difference in Mean Percentage Removal?					Rank	Points
		Stage 1	Stage 2	Stage 3	Stage 4	Stage 5		
DTD	87.66	NBPMS1	---	---	---	---	1	21
OTHER	75.33	Yes	No	---	---	---	2	20
SMD	78.93	No	---	---	---	---	1	21
SSD	82.83	Yes	Yes	Yes	NBPMS4	---	4	18
SSD, DTD	80.99	Yes	Yes	Yes	Yes	NBPMS5	5	17
SSD, DTD, SUMP	86.33	No	---	---	---	---	1	21
SSD, SMD	86.49	Yes	Yes	NBPMS3	---	---	3	19
SSD, SUMP	79.53	Yes	Yes	Yes	Yes	No	5	17
SUMP, DTD	85.42	Yes	No	---	---	---	2	20
SUMP, SMD, DTD	83.27	Yes	No	---	---	---	2	20
SUMP, VENTILATION	87.21	Yes	NBPMS2	---	---	---	2	20

Table 15: Statistical Significance and Ranking of the Mitigation Systems Based on their Performances against NBPMS for 2009

Control Technique	Mean % Removal	Is there Significant Difference in Mean Percentage Removal?								Rank	Points
		Stage 1	Stage 2	Stage 3	Stage 4	Stage 5	Stage 6	Stage 7	Stage 8		
BWD	87.79	NBPMS1	---	---	---	---	---	---	---	1	21
BWD, SSD	70.78	No	---	---	---	---	---	---	---	1	21
DTD	86.06	Yes	Yes	NBPMS3	---	---	---	---	---	3	19
OTHER	78.35	Yes	Yes	Yes	No	---	---	---	---	4	18
SMD	75.58	No	---	---	---	---	---	---	---	1	21
SSD	83.32	Yes	Yes	Yes	Yes	Yes	Yes	Yes	Yes	9	13
SSD, DTD	84.92	Yes	Yes	Yes	Yes	Yes	Yes	Yes	NBPMS8	8	14
SSD, DTD, SUMP	87.72	Yes	NBPMS2	---	---	---	---	---	---	2	20
SSD, SMD	85.16	Yes	Yes	Yes	Yes	Yes	Yes	NBPMS7	---	7	15
SSD, SUMP	85.74	Yes	Yes	Yes	Yes	NBPMS5	---	---	---	5	17
SUMP	84.17	No	---	---	---	---	---	---	---	1	21
SUMP, DTD	85.06	Yes	Yes	Yes	Yes	Yes	NBPMS6	---	---	6	16
SUMP, SMD, DTD	85.76	Yes	Yes	Yes	NBPMS4	---	---	---	---	4	18
SUMP, VENTILATION	87.7	No	---	---	---	---	---	---	---	1	21

Table 16: Statistical Significance and Ranking of the Mitigation Systems Based on their Performances against NBPMS for 2010

Control Technique	Mean % Removal	Is there Significant Difference in Mean Percentage Removal?										Rank	Points
		Stage 1	Stage 2	Stage 3	Stage 4	Stage 5	Stage 6	Stage 7	Stage 8	Stage 9	Stage 10		
BWD, SSD	76.82	Yes	Yes	Yes	No	---	---	---	---	---	---	4	18
DTD	86.04	Yes	NBPMS2	---	---	---	---	---	---	---	---	2	20
OTHER	77.18	Yes	Yes	Yes	Yes	Yes	Yes	Yes	Yes	Yes	NBPMS10	10	12
SMD	79.27	Yes	Yes	Yes	Yes	Yes	Yes	Yes	No	---	---	8	14
SSD	82.18	Yes	Yes	Yes	Yes	Yes	Yes	Yes	Yes	NBPMS9	---	9	13
SSD, DTD	85.76	Yes	Yes	NBPMS3	---	---	---	---	---	---	---	3	19
SSD, DTD, SUMP	83.37	Yes	No	---	---	---	---	---	---	---	---	2	20
SSD, EXT	73.49	Yes	Yes	Yes	Yes	Yes	Yes	Yes	Yes	Yes	Yes	11	11
SSD, SMD	84.67	Yes	Yes	Yes	NBPMS4	---	---	---	---	---	---	4	18
SSD, SMD, SUMP, DTD	87.24	No	---	---	---	---	---	---	---	---	---	1	21
SSD, SUMP	78.59	Yes	Yes	Yes	Yes	Yes	Yes	Yes	No	---	---	8	14
SUMP	84.58	Yes	Yes	Yes	Yes	Yes	NBPMS6	---	---	---	---	6	16
SUMP, DTD	84.64	Yes	Yes	Yes	Yes	NBPMS5	---	---	---	---	---	5	17
SUMP, EXT	74.81	Yes	No	---	---	---	---	---	---	---	---	2	20
SUMP, SMD	83.95	Yes	Yes	Yes	Yes	Yes	Yes	NBPMS7	---	---	---	7	15
SUMP, SMD, DTD	87.48	NBPMS1	---	---	---	---	---	---	---	---	---	1	21
SUMP, VENTILATION	82.24	Yes	Yes	Yes	Yes	Yes	Yes	Yes	NBPMS8	---	---	8	14

Table 17: Statistical Significance and Ranking of the Mitigation Systems Based on their Performances against NBPMS for 2011

Control Technique	Mean % Removal	Is there Significant Difference in Mean Percentage Removal?					Rank	Points
		Stage 1	Stage 2	Stage 3	Stage 4	Stage 5		
DTD	85.72	Yes	No	---	---	---	2	20
DTD, SMD	87.67	Yes	No	---	---	---	2	20
OTHER	79.77	Yes	Yes	Yes	Yes	No	5	17
SMD	83.94	Yes	Yes	Yes	No	---	4	18
SSD	81.57	Yes	No	---	---	---	2	20
SSD, ATTIC	80.56	No	---	---	---	---	1	21
SSD, DTD	88.11	Yes	No	---	---	---	2	20
SSD, DTD, SUMP	86.46	Yes	Yes	NBPMS3	---	---	3	19
SSD, EXT	78.12	Yes	No	---	---	---	2	20
SSD, SMD	86.34	Yes	Yes	Yes	NBPMS4	---	4	18
SSD, SMD, SUMP, DTD	87.05	Yes	No	---	---	---	2	20
SSD, SUMP	83.41	Yes	Yes	Yes	Yes	No	5	17
SUMP	84.14	Yes	Yes	Yes	Yes	NBPMS5	5	17
SUMP, DTD	85.43	Yes	No	---	---	---	2	20
SUMP, EXT	78.22	No	---	---	---	---	1	21
SUMP, SMD	88.25	Yes	NBPMS2	---	---	---	2	20
SUMP, SMD, DTD	88.56	NBPMS1	---	---	---	---	1	21
SUMP, VENTILATION	85.95	Yes	Yes	Yes	No	---	4	18

Table 18: Statistical Significance and Ranking of the Mitigation Systems Based on their Performances against NBPMS for 2012

Control Technique	Mean % Removal	Is there Significant Difference in Mean Percentage Removal?						Rank	Points
		Stage 1	Stage 2	Stage 3	Stage 4	Stage 5	Stage 6		
DTD	84.58	Yes	Yes	Yes	Yes	Yes	NBPMS6	6	16
DTD, SMD	84.67	Yes	Yes	Yes	Yes	NBPMS5	---	5	17
SMD	84.99	No	---	---	---	---	---	1	21
SSD	81.41	Yes	Yes	Yes	Yes	Yes	No	6	16
SSD, DTD	85.54	No	---	---	---	---	---	1	21
SSD, DTD,	88.19	NBPMS1	---	---	---	---	---	1	21

Control Technique	Mean % Removal	Is there Significant Difference in Mean Percentage Removal?						Rank	Points
		Stage 1	Stage 2	Stage 3	Stage 4	Stage 5	Stage 6		
SUMP									
SSD, SMD	85.57	Yes	Yes	NBPMS3	---	---	---	3	19
SSD, SUMP	84.53	No	---	---	---	---	---	1	21
SUMP	85.86	Yes	NBPMS2	---	---	---	---	2	20
SUMP, DTD	85.53	Yes	Yes	Yes	NBPMS4	---	---	4	18
SUMP, SMD, DTD	87.72	No	---	---	---	---	---	1	21

Table 19: Mean Percentage Removal Efficiency of Mitigation Systems Installed in Ohio from 2002-2012

Control Technique	Allocated Points per Year											Cum. Score	MSRI	Rank
	2002	2003	2004	2005	2006	2007	2008	2009	2010	2011	2012			
BWD	---	---	---	---	---	---	---	21	---	---	---	21	1.00	1
BWD, SSD	---	---	---	---	---	---	---	21	18	---	---	39	0.93	10
DTD	---	---	---	18	18	20	21	19	20	20	16	152	0.90	12
DTD, SMD	---	---	---	---	---	---	---	---	---	20	17	37	0.88	15
OTHER	20	20	17	21	15	20	20	18	12	17	---	180	0.86	19
SMD	---	20	---	21	21	20	21	21	14	18	21	177	0.94	9
SSD	21	21	21	16	21	19	18	13	13	20	16	199	0.86	18
SSD, ATTIC	---	---	---	21	18	---	---	---	---	21	---	60	0.95	6
SSD, DTD	---	---	---	15	20	20	17	14	19	20	21	146	0.87	17
SSD, DTD, SUMP	19	21	21	21	21	20	21	20	20	19	21	224	0.97	3
SSD, EXT	---	---	---	14	16	---	---	---	11	20	---	61	0.73	21
SSD, SMD	18	21	20	21	20	21	19	15	18	18	19	210	0.91	11
SSD, SMD, SUMP, DTD	---	21	---	19	---	---	---	---	21	20	---	81	0.96	5
SSD, SUMP	21	21	18	21	---	21	17	17	14	17	21	188	0.90	13
SSD, SUMP, EXT	---	---	---	21	19	---	---	---	---	---	---	40	0.95	7
SUMP	21	21	20	21	21	---	---	21	16	17	20	178	0.94	8
SUMP, DTD	21	21	19	17	17	19	20	16	17	20	18	205	0.89	14
SUMP, EXT	---	---	---	---	---	---	---	---	20	21	---	41	0.98	2
SUMP, SMD	---	---	---	---	---	---	---	---	15	20	---	35	0.83	20
SUMP, SMD, DTD	21	20	21	20	21	19	20	18	21	21	21	223	0.97	4
SUMP, VENTILATION	---	---	---	---	---	19	20	21	14	18	---	92	0.88	16

CHAPTER 5

Conclusions

This book provided a comprehensive review of the ^{222}Rn problem in Ohio homes complimented with an analysis on the performance of installed mitigation systems by documenting different statistical metrics and GIS-based maps on a county level basis and a zip code level basis for the state of Ohio.

5.1. KEY FINDINGS FROM THE ANALYSIS OF ^{222}RN PROBLEM IN OHIO HOMES

The homes database from the ORIS (maintained by the APRG of the Civil Engineering Department at The University of Toledo, Toledo, Ohio, USA), which comprised of 219,114 ^{222}Rn data points, collected from 1988 to 2012, was used to examine the extent of ^{222}Rn problem in Ohio homes. This book proved to be successful in providing a comprehensive insight into the distribution of ^{222}Rn concentrations across Ohio using GIS maps on the basis of the WHO and the USEPA action levels of 2.7 pCi/L and 4 pCi/L, respectively. GIS maps were also developed on a much finer ^{222}Rn distribution scale, *i.e.*, the 2 pCi/L breakdown to determine the specific areas of concern. Some of the key finding associated with the research on ^{222}Rn in Ohio homes are:

- Ohio had a GM ^{222}Rn concentration of 3.49 pCi/L.

- 67% of the Ohio counties had the GM ^{222}Rn concentrations exceeding the WHO action level.

- 33% of the Ohio counties had the GM ^{222}Rn concentrations exceeding the USEPA action level.

- Licking county had the highest GM ^{222}Rn concentration (8.13 pCi/L), followed by Pickaway (7.57 pCi/L), Knox (6.95 pCi/L), and Harrison (6.55 pCi/L).

- 61% of the Ohio homes tested for ^{222}Rn had the GM ^{222}Rn concentrations exceeding the WHO action level.

- 47% of the Ohio homes tested for ^{222}Rn had the GM ^{222}Rn concentrations exceeding the USEPA action level.

- 56% of the monitored Ohio zip codes had the GM ^{222}Rn concentrations exceeding the WHO action level.

- 32% of the monitored Ohio zip codes had the GM ^{222}Rn concentrations exceeding the USEPA action level.

- Of the 457 zip codes that had the GM ^{222}Rn concentrations greater than 4 pCi/L, 96 zip codes had the GM ^{222}Rn concentrations exceeding 8 pCi/L and 13 zip codes had the GM ^{222}Rn concentrations exceeding 20 pCi/L.

- Of the monitored zip codes with at least five observations, zip code 43916 had the highest GM ^{222}Rn concentration (25.91 pCi/L), followed by the zip codes 43988 (21.87 pCi/L), 43216 (19.5 pCi/L), and 43033 (13.89 pCi/L).

5.2. KEY FINDINGS FROM THE ANALYSIS OF MITIGATION SYSTEMS PERFORMANCE IN OHIO HOMES

The mitigation database from the ORIS, which comprised of 39,858 data points, gathered from 2002 to 2012, was used to analyze the performance of mitigation systems in reducing in-house ^{222}Rn levels in Ohio apart from developing a systematic procedure to rank the performance of 21 installed mitigation systems in Ohio (with a minimum representative sample of 10 points). Some of the key finding associated with the research on performance of ^{222}Rn mitigation systems in Ohio homes are:

- 98.78% of the Ohio counties that had installed mitigation systems had the post-mitigation GM ^{222}Rn concentrations that did not exceed the WHO action level.

- 100% of the Ohio counties that had installed mitigation systems had the post-mitigation GM ^{222}Rn concentrations that did not exceed the USEPA action level.

- Meigs county had the highest post-mitigation GM ^{222}Rn concentration (3.8 pCi/L), followed by Coshocton (2.37 pCi/L), Fayette (2.3 pCi/L), and Highland (2.22 pCi/L).

- 95.54% of Ohio zip codes that had installed mitigation systems had post-mitigation GM ^{222}Rn concentrations that did not exceed the WHO action level.

- 99.57% of Ohio zip codes that had installed mitigation systems had post-mitigation GM ^{222}Rn concentrations that did not exceed the USEPA action level.

- The post-mitigation GM ^{222}Rn concentrations were the highest in zip code 43053 (5.92 pCi/L), followed by the zip codes 43076 (4.48 pCi/L), and 44619 (4.34 pCi/L).

- Based on the computed MSRI values, the chorological order of the mitigation systems in ascending order of the rank are BWD; SUMP, EXT; SSD, DTD, SUMP; SUMP, SMD, DTD; SSD, SMD, SUMP, DTD; SSD, ATTIC; SSD, SUMP, EXT; SUMP; SMD; BWD, SSD; SSD, SMD; DTD; SSD, SUMP; SUMP, DTD; DTD, SMD; SUMP, VENTILATION; SSD, DTD; SSD; OTHER; SUMP, SMD; and SSD, EXT.

It is hoped that health professionals around the globe will be able to manage the radon problem in their area by following the approach discussed in this book.

REFERENCES

Akerblom, G. 1986. Investigation and mapping of radon risk areas. Swedish Geological Co., Luela, Sweden, Report IRAP 86036.

Alexander, W.G., Devocelle, L.L. Mapping indoor radon potential using geology and soil permeability. In Proceedings of AARST international research symposium, Cincinnati, Ohio, 16 pp., November 1997.

Andersen, C.E. Numerical modelling of radon-222 entry into houses: an outline of techniques and results. *The Science of the Total Environment*, 2001, 272, 33-42.

Arvela, H. Radon mitigation in blocks of flats. *The science of the Total Environment*. 2001, 272, 137.

ASTM (American Society of Testing and Materials) Publication. 1993. Modeling of indoor air quality and exposure. ASTM Publication STP1205-EB, Philadelphia, PA.

Bates, T.F., Strahl, E.O. Mineralogy and chemistry of uranium-bearing black shales. In *Proceedings of the second United Nations international conference on the peaceful uses of atomic energy*, vol. 2, 407-411, 1958.

BEIR (Biological Effects of Ionizing Radiation) VI Report. 1999. Health effects of exposure to indoor radon. BEIR Report, National Academy Press, Washington D.C. ISBN: 0-309-52374-5, 516 pp.

Coskeran, T., Denman, A., Phillips, P. The costs of radon mitigation in domestic properties. *Health Policy*, 2001, 57, 97-109.

Damkjaer, A., Korsbech, U. 1985. Measurement of the emanation of radon-222 from Danish soils. *The Science of the Total Environment*, 1985, 45, 343-350.

Denman, A., Harris, E.P., Hermann, M.R., Phillips, P. Auditing the cost effectiveness of radon mitigation in the workplace. *Managerial Auditing Journal*, 2000, 15,153-160.

Deyuan, T. Indoor and outdoor air radon concentration level in China. In *Proceedings of the 6th international conference on indoor air quality and climate*, Helsinki, Finland, vol. 4,459-463, July 1993.

Dixon, D.W. Radon exposures from the use of natural gas in buildings. Radiation Protection Dosimetry, 2001, 97, 259–264.

Dyck, W. The mobility and concentration of uranium and its decay products in temperature surficial environments. In Kimberly, M.M., (Ed.), Uranium deposits, their mineralogy, and origin, Mineral Association of Canada, Short Course Handbook, 3, pp. 57-100, 1978.

Gadd, M.S., Borak, T.B. *In situ* determination of the diffusion coefficient of Rn-222 in concrete. *Health physics*, 1995, 68, 817-822.

Gadgil, A.J. Models of radon entry. *Radiation Protection Dosimetry*, 1992, 45, 373-380.

Garbesi, K., Sextro, R.G. Modeling and field evidence of pressure-driven entry of soil gas into a house through permeable below-grade walls. *Environmental Science and Technology*, 1989, 23, 1481-1487.

Gessell, T.F. Background atmospheric Rn-222 concentrations outdoors and indoors: a review. *Health physics*, 1983, 45, 289-302.

Gessell, T.F., Prichard, H.M. The contribution of radon in tap water to indoor radon concentrations. In Gessell, T.F., Lowder, W.M., (Eds.), Natural radiation environment III, U.S. Department of Commerce, National Technical Information Service, Springfield, VA, p. 1347, 1980.

Grasty, R.L. Summer outdoor radon variations in Canada and their relation to soil moisture. *Health Physics*, 1994, 66, 185-193.

Gundersen, L.C.S., Schumann, R.R., Otton, J.K., Owen, D.E., Dubiel, R.F., Dickinson, K.A. Geology of radon in the United States. In Gates, A.E., and Gundersen, L.C.S., (Eds.), Geologic controls on radon, Geological Society of America Special Paper 271, pp. 1-16, 1992.

Harrell, J.A., Belsito, M.E., Kumar, A. Radon hazards associated with outcrops of Ohio Shale in Ohio. *Environmental Geology and Water Sciences*, 1991, 18, 17-26.

Harrell, J.A., Kumar, A. 1988. Radon hazards associated with outcrops of the Devonian Ohio shale. A final report for a grant awarded by the Ohio Air Quality Development Authority, Columbus, Ohio, August 1988, 163 p.

Harrell, J.A., Kumar, A. Multivariate stepwise regression analysis of indoor radon data from Ohio, U.S.A. *Journal of Official Statistics*, 5, 409-420.

Hess, C.T., Weiffenbach, C.V., Norton, S.A. Variations of airborne and waterborne Rn-222 in houses in Maine. *Environment International*, 1982, 8, 59-66.

Heydinger, A.; Kumar, A.; Harrell, J. A. Development of an indoor radon information system. *Environmental Software*, 1991, 6, 194-201.

Hoffmann, R.L., May, M.J. Radon contamination of residential structures III: an overview of factors influencing infiltration rates. In *Proceedings of AARST international research symposium*, Cincinnati, Ohio, 15 pp., November 1997.

Hopper, R.D., Levy, R.A., Rankin, R.C., Boyd, M.A. National ambient radon study. In *Proceedings of the international symposium on radon and radon reduction technology*, Philadelphia, Pennsylvania, 1991.

Jiranek, M., Neznal, M. Mitigation of ineffective measures against radon. *Radiation Protection Dosimetry*, 2008, 130, 68-71.

Korhonen, P., Kokotti, H., Kalliokoski, P. Survey and mitigation of occupational exposure of radon in work places. *Building and Environment*. 2000, 35, 555-562.

Kraemer, T.F. Radon in unconventional natural gas from gulf coast geopressured-geothermal reservoirs. *Environmental Science and Technology*, 1986, 20, 939-942.

Kranrod, C., Tokonami, S., Ishikawa, T., Sorimachi, A., Janik, M., Shingaki, R., Furukawa, M., Chanyota, S., Chankow,N. Mitigation of the effective dose of radon decay products through the use of an air cleaner in a dwelling in Okinawa, Japan. *Applied Radiation and Isotopes*, 2009, 67, 1127-1132.

Kumar, A., Heydinger, A., Harrell, J. A. Development of an indoor radon information system for Ohio and its application in the study of the geology of radon in Ohio. A final report submitted to the Ohio Air Quality Development Authority, Columbus, Ohio, November 1990, 192 p.

Kumar, A., Varadarajan, C. Development of Ohio radon information system. In *Proceedings of environmental data analysis - assessing health and environmental impacts, developing policy, and achieving regulatory compliance conference by the Air & Waste Management Association*, Oak Brook, Illinois, October 2005.

Kumar, A., Varadarajan, C., Kadiyala, A. Management of radon data in the state of Ohio. *The Open Environmental & Biological Monitoring Journal*, 2011, 4, 57-71.

Lindmark, A., Rosen, B. Radon in soil gas - exhalation tests and *in situ* measurements. *The Science of the Total Environment*, 1985, 45, 397-404.

Loureiro, C.O., Abriola, L.M., Martin, J.E., Sextro, R.G. Three-dimensional simulation of radon transport into houses with basements under constant negative pressure. *Environmental Science and Technology*, 1990, 24, 1338-1348.

Manthena, D. V., Kadiyala, A., Kumar, A. Interpolation of radon concentrations using GIS based kriging and cokriging techniques. *Environmental Progress and Sustainable Energy*, 2009, 28, 487-492.

Maringer, F.J., Kaineder, H., Nadschlager, E., Sperker, S. Standards and experience in radon measurement and regulation of radon mitigation in Austria. *Applied Radiation and Isotopes*, 2008, 66, 1644-1649.

NCRP (National Council on Radiation Protection and Measurements) Report. 1988. Measurement of radon and radon daughters in air. NCRP, Bethesda, Maryland, Report No. 97, 174 p.

Nazaroff, W.W. Predicting the rate of ^{222}Rn entry from soil into the basement of a dwelling due to pressure-driven flow. *Radiation Protection Dosimetry*, 1988, 24, 199-202.

Nazaroff, W.W., Doyle, S.M., Nero, Jr., A.V., Sextro, R.G. Radon entry *via* potable water. In Nazaroff, W.W., and Nero Jr., A.V., (Eds.) Radon and its decay products in indoor air, John Wiley & Sons, Inc.: New York, pp.57-112, 1988b.

Nazaroff, W.W., Moed, B.A., Sextro, R.G. Soil as a source of indoor radon: generation, migration, and entry. In Nazaroff, W.W., and Nero Jr., A.V., (Eds.) Radon and its decay products in indoor air, John Wiley & Sons, Inc.: New York, pp.57-112, 1988a.

Nero, Jr., A.V. Radon and its decay products in indoor air: an overview. In Nazaroff, W.W., and Nero Jr., A.V., (Eds.) Radon and its decay products in indoor air, John Wiley & Sons, Inc.: New York, pp.1-56, 1988.

OECD (Organization for Economic Cooperation and Development) Report. 1985. Metrology and monitoring of radon, thoron and their daughter products. OECD Publications, Paris, 148 p.

Ohio Department of Geological Survey. 2006. Geologic map and cross section of Ohio. Ohio Department of Natural Resources, Division of Geological Survey, page-size map, 1p., 1:2,000,000.

Ohio Department of Health. 2001. Radon licensing program rules and regulations. Ohio Administrative Code, Chapter 3701-69, Ohio Department of Health.

Osborne, M., Harrison, J. An overview of indoor radon risk reduction in the United States. *Journal of Radioanalytical and Nuclear Chemistry*, 1992, 161, 265-272.

Paridaens, J., Saint-Georges, L., Vanmarcke H. Mitigation of a radon-rich Belgian dwelling using active subslab depressurization. *Journal of Environmental Radioactivity*, 2005, 79, 25-37.

Partridge, J.E., Horton, T.R., Sensintaffar, E.L. 1979. A study of radon-222 released from water during typical household activities. USEPA Report No. ORP-EERF-79-1, Eastern Environmental Radiation Facility, Montgomery, AL.

Pennsylvania Bureau of Radiation Protection. 2012. Available at http://www.dep.state.pa.us/brp/radon_division/PA_Radon_Story1.htm. Accessed on December 14 2012.

Price, J.G., Rigby, J.G., Christensen, L., Hess, R., LaPointe, D.D., Ramelli, A.R., Desilets, M., Hopper, R.D., Kluesner, T., Marshall, S. Radon in outdoor air in Nevada. *Health Physics*, 1994, 66, 433-438.

Prill, R.J., Fisk, W.J., Turk, B.H. Evaluation of radon mitigation systems in 14 houses over a two-year period. *Journal of the Air & Waste Management Association*, 1990, 40, 740-746.

Renken, K.J., Rosenberg, T. Laboratory measurements of the transport of radon through concrete samples. *Health physics*, 1995, 68, 800-808.

Revzan, K.L., Fisk, W.J., Gadgil, A.J. Modeling radon entry into houses with basements: model description and verification. *Indoor Air*, 1991, 1, 173-189.

Ringer, W., Simader, M., Bernreiter, M., Kainder, H. Mitigation of three water supplies with high radon exposure to the employees. *Radiation Protection Dosimetry*, 2008, 130, 26-29.

Robe, M.C., Rannou, A., Le Bronec, J. Radon measurement in the environment in France. Radiation Protection Dosimetry, 45, 455-457.

Roelofs, L.M.M., Scholten, L.C. The effect of aging, humidity, and fly-ash additive on the radon exhalation from concrete. *Health physics*, 1994, 67, 266-271.

Rogers, V.C., Nielson, K.K. Data and models for radon transport through concrete. In *Proceedings of the international symposium on radon and radon reduction technology*, Minneapolis, Minnesota, 1992.

Rogers, V.C., Nielson, K.K., Holt, R.B. Radon diffusion coefficient for aged residential concretes. *Health Physics*, 1995, 68, 832-834.

Rogers, V.C., Nielson, K.K., Holt, R.B., Snoddy, R. Radon diffusion coefficient for residential concretes. *Health Physics*, 1994, 67, 261-265.

Rutherford, P.M., Dudas, M.J., Arocena, J.M. Radon emanation coefficients for phosphogypsum. *Health Physics*, 1995, 69, 513-520.

Stranden, E. Building materials as a source of indoor radon. In Nazaroff, W.W., and Nero Jr., A.V., (Eds.) Radon and its decay products in indoor air, John Wiley & Sons, Inc.: New York, pp.113-130, 1988.

Sud, A. Update and analysis of a residential radon database for the state of Ohio. M.S. Thesis, The University of Toledo, Toledo, 1998.

Swanson, V.E. Oil yield and uranium content of black shales. *U.S. Geological Survey*, Professional paper 356-A, 1960.

Tanner, A.B. Geological factors that influence radon availability. In *Proceedings of APCA international specialty conference on indoor radon*, Pittsburgh, Pennsylvania, APCA special publication SP-54, 1-12, February 1986.

Tanner, A.B. Methods of characterization of ground for assessment of indoor radon potential at a site. In Gundersen, L.C.S., and Wanty, R.B., (Eds.), Field studies of radon in rocks, soils, and water, CRC Press Inc.: Boca Raton, Florida, pp. 1-13, 1993.

Tracy, B.M. 1983. The characterization if the chemical and trace element content of selected oil shales of Ohio. M.S. thesis, The University of Toledo, Toledo, Ohio.

Tso, M.W., Chor-yi, N.G., Leung, J.K. Radon release from building materials in Hong Kong. *Health physics*, 1994, 67, 378-384.

Tyson, J.L., Fairey, P.W., Withers, C.R. Elevated radon levels in ambient air. In Proce*edings of the 6th international conference on indoor air quality and climate*, Helsinki, Finland, vo1.4, 443-448, July 1993.

UNSCEAR (United Nations Scientific Committee on the Effects of Atomic Radiation) Report. 1977. Sources and effects of ionizing radiation. UNSCEAR, New York, NY.

UNSCEAR (United Nations Scientific Committee on the Effects of Atomic Radiation) Report. 1988. Sources, effects and risks of ionizing radiation. UNSCEAR, New York, NY.

UNSCEAR (United Nations Scientific Committee on the Effects of Atomic Radiation) Report. 1993. Sources and effects of ionizing radiation. UNSCEAR, New York, NY.

USEPA (United States Environmental Protection Agency) Report. 1986. Final rule for radon – 222 emissions from licensed uranium mill tailings. USEPA Publication 520/1-86-009, Washington, D.C.

USEPA (United States Environmental Protection Agency) Report. 1992. Indoor radon and radon decay product measurement device protocols. USEPA Publication 402-R-92-004, Washington, D.C.

USEPA (United States Environmental Protection Agency) Report. 1993a. Protocols for radon and radon decay product measurement in homes. USEPA Publication 402-R-92-003. Washington, D.C.

USEPA (United States Environmental Protection Agency) Report. 1993b. Radon reduction techniques for existing detached houses: technical guidance (third edition) for active soil depressurization. USEPA Publication 625-R-93-011, Washington, D.C.

USEPA (United States Environmental Protection Agency) Report. 1997. National radon proficiency program: guidance on quality assurance. USEPA Publication 402-R-95-012, Montgomery, Alabama.

USEPA (United States Environmental Protection Agency) Report. 2001. Building radon out: a step by step guide on how to build radon-resistant homes. USEPA Publication 402-K-01-002, Washington, D.C.

USEPA (United States Environmental Protection Agency) Report. 2003. Consumer's guide to radon reduction. USEPA Publication 402-K-03-002, Washington D.C.

USEPA (United States Environmental Protection Agency) Report. 2010. Consumer's guide to radon reduction: how to fix your home. USEPA Publication 402-K-10-005, Washington D.C.

USEPA (United States Environmental Protection Agency) Report. 2012. A citizen's guide to radon: the guide to protecting yourself and your family from radon. USEPA Publication 402-K-12-002, Washington D.C.

Wang, F., Ward, I.C. The development of a radon entry model for a house with a cellar. *Building and Environment*, 2000, 35, 615-631.

Wang, F., Ward, I.C. Radon entry, migration and reduction in houses with cellars. *Building and Environment*, 2002, 37, 1153-1165.

Ward, D.C., Borak, T.B., Gadd, M.S. Characterization of Rn entry into a basement structure surrounded by low-permeability soil. *Health physics*, 1993, 65, 1-11.

WHO (World health Organization) Report. 2007. International radon project: survey on radon guidelines, programmes and activities. WHO Press, Geneva, Switzerland.

WHO (World health Organization) Report. 2009. WHO handbook on indoor radon: a public heath perspective. WHO Press, Geneva, Switzerland.

Author Index

A

Abriola, L.M., 368
Alexander, W.G., 7-8,367
Akerblom, G., 27, 367
Andersen, C.E., 13, 367
Arocena, J.M., 370
Arvela, H., 22, 367

B

Bates, T.F., 25, 367
Belsito, M.E., 368
Bernreiter, M., 370
Borak, T.B, 14-15, 367, 371
Boyd, M.A., 368

C

Chankow, N., 368
Chanyota, S., 368
Chor-yi, N.G., 370
Christensen, L., 369
Coskeran, T., 22, 367

D

Damkjaer, A., 9, 367
Denman, A., 22, 367
Desilets, M., 369
Devocelle, L.L., 7-8, 367
Deyuan, T., 18, 367
Dickinson, K.A., 368
Dixon, D.W., 18, 367
Doyle, S.M., 369

Dubiel, R.F., 368
Dudas, M.J., 370
Dyck, W., 3, 367

F

Fairey, P.W., 370
Fisk, W.J., 369
Furukawa, M., 368

G

Gadd, M.S., 14-15, 367, 371
Gadgil, A.J., 13, 367, 369
Garbesi, K., 13, 367
Gessell, T.F., 17-18, 367
Grasty, R.L., 18, 368
Gundersen, L.C.S., 7, 368

H

Harrell, J.A., 26-28, 368
Harris, E.P., 367
Harrison, J., 22, 369
Hermann, M.R., 367
Hess, C.T., 17, 368
Hess, R., 369
Heydinger, A., 28, 368
Hoffmann, R.L., 10-11, 368
Holt, R.B., 370
Hopper, R.D., 18, 368-369
Horton, T.R., 369

I

Ishikawa, T., 368

Subject Index

A

accumulation, 6, 11, 278
activated charcoal detectors (ACDs), 19
aeration, 24
air
- ambient, 18, 370
- crawl space, 348
- lower sub-slab, 347
- pore, 11
- supply, 11

air-conditioning ducts, 9
air exchange, 11
air exchange factor, 27
air exchange rates, 10, 14, 16-17
air movement, 9
air pumps, 19
alpha particles, 3-4, 8, 20
alpha track, 19
alpha track detectors (ATDs), 19
ATTIC, 350-351, 354, 356-357, 360-361, 365

B

barometric pressure, 10-11
basement, 9, 11, 23-24, 27, 347, 368-369, 371
blowers, 9-10
building materials, 3, 6, 9, 13-15, 18, 22, 370
building occupants, 6-7, 9
building ventilation rate, 22
building walls, 11
block-wall depressurization/suction (BWD), 23, 347-351, 354, 359, 361, 365

C

cancer, 3, 21

coefficient of variation, 25, 28, 87, 277
composition, 7
concentration statistics
- arithmetic mean radon, 25, 87, 277
- geometric mean radon, 25, 87, 277
- maximum radon, 25, 87, 277
- median radon, 25, 87, 277
- minimum radon, 25, 87, 277
- quartile radon, 25, 87, 277

constructions, 3, 19, 22-23, 347
control technique, 22-23, 347, 349-351, 354-361
convection, 10, 12-13
correlations, 27
cost efficiency, 22
county based analysis, 29, 278
county name, 30-32, 37-39, 280-282, 286-288
cracks, 6, 9
crawlspace, 23, 347
continuous radon monitors (CRMs), 19

D

decay, 3-5, 7-8, 12-14, 16, 20-21, 367-371
density, 12-14
diffusion, 12-13
diffusion coefficient, 12, 14-15, 367, 370
diluting factor, 18
disintegrations, 5
drain tile depressurization/suction (DTD), 23, 347, 349-351, 354-361, 365
dwelling, 17, 19-20, 368-369

E

electret ion chambers (EICs), 19
electronic integrating devices (EIDs), 19